Composite Materials - Science and Engineering

Edited by Petrica Vizureanu

Published in London, United Kingdom

Composite Materials - Science and Engineering
http://dx.doi.org/10.5772/intechopen.1004693
Edited by Petrica Vizureanu

Contributors
Ahmet Hakan Yilmaz, Alin Dinita, Andrei Victor Sandu, Bengisu Arı, Bilge Ersöz, Brecken Beyer, Bulend Ortaç, Caleb Huffmeyer, Changfang Zhao, Clayton Ness, Dragos Cristian Achitei, Dumitru Doru Burduhos Nergis, Elsayed G. Khater, Jiacheng Li, Manar E. Elashry, Manuela Cristina Perju, Marcin Nabiałek, Maria Tanase, Marina Bunea, Mădălina Simona Bălțatu, Owen Nifong, Petrică Vizureanu, Robert M. Stwalley III, Saliha Mutlu, Samir A. Ali, Sevil Savaskan Yilmaz, Thomas Barnes, Tyler J. McPheron, Wyatt Griffey

First published in London, United Kingdom, 2025 by IntechOpen
IntechOpen is the global imprint of INTECHOPEN LIMITED, registered in England and Wales, registration number: 11086078, 167-169 Great Portland Street, London, W1W 5PF, United Kingdom

For EU product safety concerns: IN TECH d.o.o., Prolaz Marije Krucifikse Kozulić 3, 51000 Rijeka, Croatia, info@intechopen.com or visit our website at intechopen.com.

British Library Cataloguing-in-Publication Data
A catalogue record for this book is available from the British Library

Composite Materials - Science and Engineering
Edited by Petrica Vizureanu
p. cm.

This title is part of the Materials Science Book Series, Volume 10
Topic: Composite Materials
Series Editor: Chonghe Li
Topic Editor: Anna Boczkowska
Associate Topic Editors: Kamil Dydek and Bartlomiej Przybyszewski

Print ISBN 978-1-83634-365-3
Online ISBN 978-1-83634-364-6
eBook (PDF) ISBN 978-1-83634-366-0
ISSN 3049-8856

If disposing of this product, please recycle the paper responsibly.

IntechOpen Book Series

Materials Science

Volume 10

Aims and Scope of the Series

Materials science has always occupied an extremely high position in the human development process. As we explore the oceans of stars, various industries have put forward more stringent requirements for the performance of materials, forcing us to pay more and more attention to the development of new materials. At the same time, the formation of a data-driven scientific paradigm is dramatically shortening the development cycle of new materials. The huge data generated by synergistically combining theories, high-throughput experiments, high-throughput computation, and artificial intelligence is greatly contributing to our ability to utilize materials science to solve real-world problems. The three topics of this book series - Metals and Nonmetals; Composite Materials; and Surface Science - will address important areas of advancement in materials science. There will be a range of interesting works published under these topics.

Meet the Series Editor

Prof. Chonghe Li received his Ph.D. from the Chinese Academy of Sciences in 1995. From 1995 to 2000, he worked as a researcher at the Shanghai Institute of Metallurgy, Chinese Academy of Sciences, where he also served as director of the research laboratory. In 2000, he was appointed Professor at the Institute of High-Performance Computing in Singapore, where he worked on computation and simulation in materials science until 2004. Since then, he has been a professor at the School of Materials Science and Engineering, Shanghai University, China, as well as the director of the Shanghai Specialty Casting Engineering and Technology Research Center. Prof. Li's research focuses on titanium alloys, titanium-aluminum single crystals, intermetallic compounds, theoretical calculations, alloy design, and special refractory materials. His broad scientific expertise is well recognized by the scientific community around the world. He is a member of the editorial board of the journal Metals. As an author, he has published more than 200 peer-reviewed papers, 2 books, and over 40 patents.

Meet the Volume Editor

Petrică Vizureanu obtained an MSc and Ph.D. in Heating Equipment at the „Gheorghe Asachi" Technical University, Iasi, Romania, in 1992 and 1999, respectively. Dr. Vizureanu is currently a Full Professor and scientific supervisor in Materials Engineering at the same university, and has been appointed Director of the Department of Technologies and Equipment for Materials Processing. He serves as an editor and guest editor for numerous journals and publishing houses. His research focuses on expert systems for heating system programming, computer-assisted design for heating equipment, and the application of heating equipment in materials processing, as well as heat transfer, biomaterials, and geopolymers. He has published more than 250 papers in international journals and conference proceedings, as well as over 40 books.

Contents

Preface

Composite materials are a dynamic and interdisciplinary field at the intersection of physics, chemistry, materials science, and engineering, with extensive applications in various modern industries. For engineering, composite materials represent an essential solution for designing structures that require an optimal balance between low weight, high mechanical strength and functional performance under extreme conditions. This book provides a comprehensive overview of these advanced materials, highlighting their production methods, internal structure, physicochemical properties and multiple uses.

Unlike traditional materials, composites are made by combining two or more distinct components, resulting in materials with properties superior to those of the individual elements. Well-known examples include reinforced concrete, laminated wood panels (such as plywood), fiber-reinforced plastics (e.g., carbon or glass fiber reinforced polymers), metal-ceramic composites or metal matrix composites.

Their use is increasingly widespread, covering fields such as construction, shipbuilding, automotive, aeronautics and aerospace. From race car bodies and pool panels to structural components for spacecraft, composite materials offer unique combinations of lightweight, high strength and structural adaptability.

This book explores both thermoplastic composites, with short, long, or mixed fibers, and thermoset composites, including those with aramid or carbon fibers in an epoxy matrix. It also examines advanced materials, such as shape-memory polymer composites, which retain their properties after multiple cycles of thermal deformation, and high-strain composites, designed for flexible applications and folding structures.

The structural testing and analysis component is not neglected, either—from numerical simulations using finite element analysis (FEA) to non-destructive investigation methods such as ultrasound, X-ray radiography, or laser inspection. These techniques play a crucial role in ensuring the reliability and safety of composite materials in real-world operating conditions.

With its rich and well-documented content, this work is addressed not only to researchers and engineers but also to students or practitioners interested in the latest developments in the field of composite materials. We are confident that this volume will represent a valuable contribution to the understanding and promotion of composite materials-based technologies in a global context of innovation and sustainability.

Petrică Vizureanu
Gheorghe Asachi Technical University of Iași,
Iași, Romania

Section 1

Metallic Composites

Chapter 1

Biocomposites: Materials, Properties, and Applications

Mădălina Simona Bălţatu, Petrică Vizureanu, Andrei Victor Sandu, Dragos Cristian Achitei, Dumitru Doru Burduhos Nergis, Manuela Cristina Perju and Marcin Nabiałek

Abstract

The rapid evolution of technology necessitates the continuous development of materials that can fulfill a broad spectrum of industrial and technical requirements. Composite materials, while ancient in concept, have become increasingly significant in modern contexts due to their ability to integrate the favorable properties of their constituent materials. This chapter delves into biocomposites, a sustainable subcategory of composite materials that use natural fibers and biodegradable matrices to address current environmental challenges while reducing carbon footprints. Notably, biocomposites are utilized extensively in areas such as bioceramics for medical applications, including bone regeneration and cardiovascular implants. These materials are not only pivotal for reducing environmental impact but also play a critical role in the medical field by enhancing biointegration and promoting tissue regeneration. The utilization of biocomposites spans various industries including construction, automotive, and packaging, underscoring their importance due to their lightweight, durability, and sustainable properties. This chapter aims to provide a comprehensive overview of biocomposites, exploring their key components, manufacturing techniques, applications, and their significant role in advancing sustainable and innovative industrial practices.

Keywords: biocomposites, composite materials, sustainable materials, biodegradable matrices, bioceramics, bone regeneration, medical implants

1. Introduction

The creation of new materials that can meet an increasingly wide range of technical and industrial requirements is a constant necessity, driven by the rapid development of various fields of application. The concept of composite material, although known since antiquity, has gained major relevance in the current technological context. In the past, combining different materials was used to achieve superior properties compared to individual materials; however, spatial development and new technological requirements have brought this concept to a new level [1].

Composite materials are distinguished by their ability to combine the favorable properties of their components, thus providing innovative solutions for designers who can customize materials according to the specific needs of applications. These materials not only allow for the achievement of established performance parameters but also enable the creation of structures with unique properties, oriented toward clearly defined outcomes [2, 3].

Composite materials are multiphase systems obtained by the artificial combination of at least two chemically distinct materials, separated by a well-defined interface, in order to achieve superior performance compared to the individual component materials [4]. Alternatively, a composite material can be defined as a material composed of multiple components, whose organization and formulation allow for the exploitation of the best characteristics of each component, creating a material with final properties that exceed those of its base components.

Through the concept of composite materials, new perspectives are opened in utilizing materials with known properties to create materials with superior and controllable properties, offering new possibilities for innovation and adaptation to modern technological demands.

Biocomposites, a subcategory of composite materials, represent a modern and sustainable solution. They utilize natural fibers and biodegradable matrices to produce durable materials that address contemporary challenges related to sustainability and reducing carbon footprints. Natural fibers, such as flax, hemp, or bamboo, provide biocomposites with mechanical strength and rigidity, while biodegradable matrices (natural or synthetic polymers) ensure the cohesion of the material and contribute to its ecological properties. Unlike traditional composites made from synthetic fibers and petroleum-based matrices, biocomposites can be integrated into a circular life cycle, being recyclable or compostable [5].

A remarkable application of these materials is in the field of bioceramics, where ceramic composites, used for millennia, have begun to be utilized for medical and surgical purposes. Calcium phosphate-based bioceramics have demonstrated remarkable biointegration capabilities, favorably interacting with biological tissue and stimulating bone regeneration. At the same time, carbon-based bioceramics offer superior biocompatibility and hemocompatibility, making them suitable for cardiovascular implants and prosthetics. By combining these ceramic materials with biodegradable matrices, advanced biocomposites are created, ideal for critical medical applications such as orthopedics and dental implantology, while simultaneously reducing environmental impact [6].

Thus, biocomposites have become a central focus in the research of sustainable materials, recognized for their ability to combine exceptional mechanical properties with ecological compatibility. They represent a promising alternative to conventional materials, offering innovative solutions to global challenges such as climate change and the depletion of petroleum resources.

In recent years, biocomposites have found applications in various industries, including construction, automotive, medical equipment, and consumer products. These materials are valued for their unique combination of lightness, durability, and sustainability, attracting the interest of researchers and developers aiming to expand their use and technological innovation in production [7].

This chapter will provide a detailed analysis of biocomposites, discussing their fundamental components, manufacturing methods, mechanical and chemical properties, as well as their ecological and economic advantages. The aim of this chapter is to provide a comprehensive perspective on the importance of biocomposites, encouraging their integration into sustainable and innovative industrial practices.

2. Constituent phases of composite material

A composite material is made up of several distinct components, which are combined to improve the mechanical, chemical, or thermal properties of the final product. These phases are the discontinuous phase, the continuous phase, and the technological additives (**Figure 1**) [8].

2.1 Discontinuous phase (reinforcement)

The discontinuous phase, also known as reinforcement, is responsible for increasing the stiffness and strength of the composite material. The reinforcement can consist of:

- *Fibers* are the most commonly used reinforcing elements, typically unidimensional in orientation, and significantly contribute to the increase in strength and stiffness in the direction of the fibers.

- *Sheets and scales*, these forms of reinforcement provide better stress distribution across the surface, enhancing the structural behavior of the composite.

- *Particles* are used in various shapes and sizes, and reinforcing particles are beneficial for improving resistance to abrasion and wear. The discontinuous phase makes part in defining the mechanical properties of the composite, and the choice of reinforcing material must be made carefully to avoid destructive effects such as corrosion or unwanted chemical reactions at the interface with the matrix [9, 10].

2.2 Continuous phase (matrix)

The matrix, also referred to as the continuous phase, forms the base component of the composite. Make part in maintaining the structural integrity of the material, ensuring cohesion between the reinforcing fibers, and facilitating the transfer of stresses between them. The matrix (**Figure 2**) can be made from various materials, such as:

- *Polymers* widely used due to their ease of processing and insulating properties.

- *Metals* provide high resistance to elevated temperatures and durability.

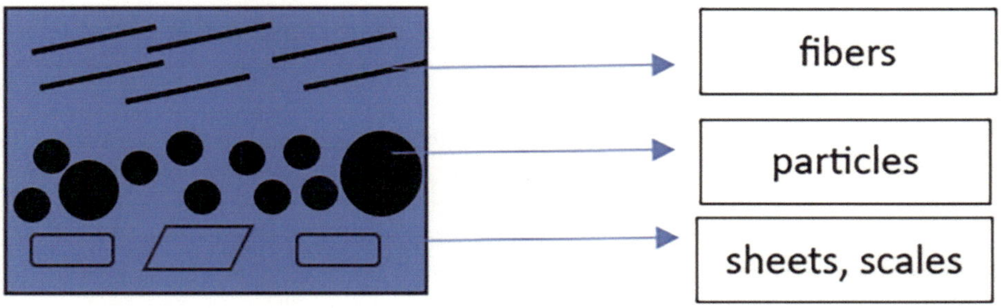

Figure 1.
Component phases of composites.

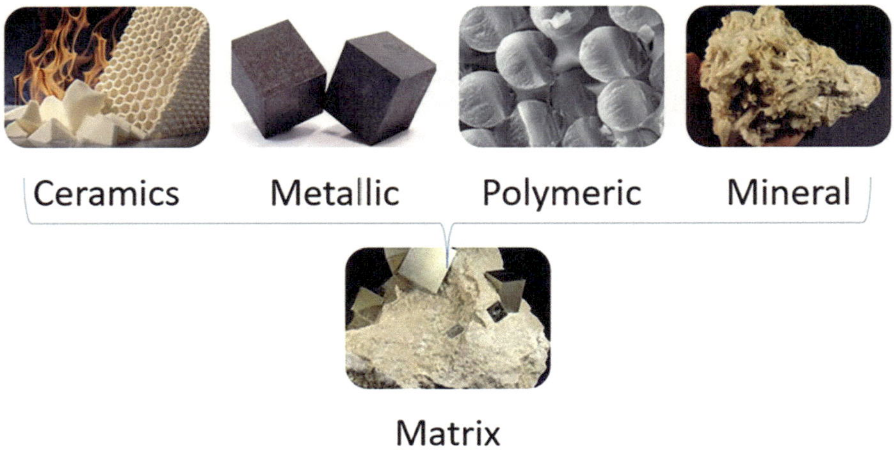

Figure 2.
Nature of the matrix.

- *Ceramics* are used when resistance to extreme temperatures or chemical stability is required [11].

The main functions of the matrix include:

- Protecting the reinforcement during both the manufacturing process and usage;
- Preventing fiber buckling, keeping them at appropriate distances to efficiently transfer loads;
- Redistributing stresses and preventing crack propagation, thus contributing to the durability of the composite.

2.3 Technological additives

Technological Additives are substances added during the manufacturing process of composites to enhance specific characteristics, such as:

- *Strengthening additives* increase the material's resistance to external factors, such as high temperatures or UV radiation.
- *Stabilizers* ensure the chemical stability of the composite under harsh environmental conditions.
- *Pigments and dyes* added to improve the esthetic appearance or to provide certain visible properties for technical purposes (e.g., defect detection) [12, 13].

3. Parameters influencing the properties

The general properties of composite materials depend on several important factors (**Figure 3**), such as:

DOI: http://dx.doi.org/10.5772/intechopen.1010197

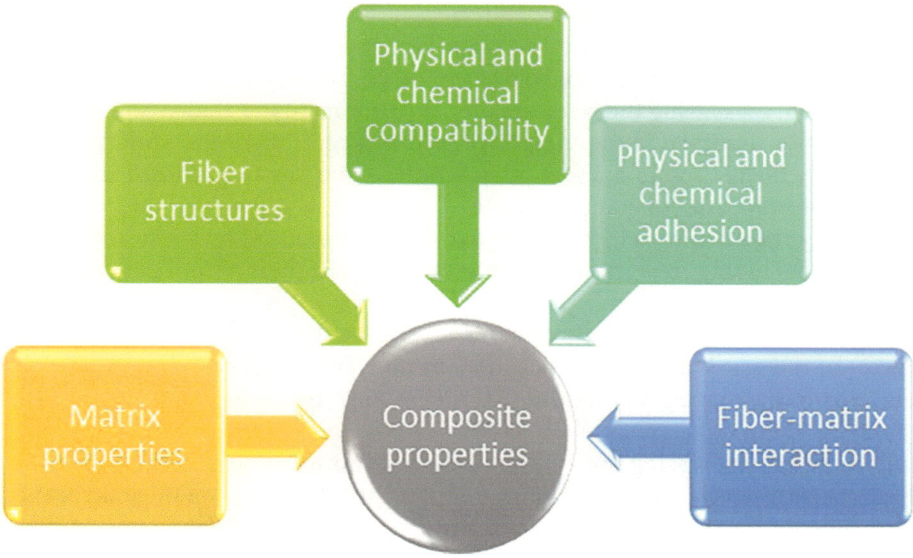

Figure 3.
Composite properties.

- Each component of the composite, whether matrix, reinforcement, or technological additive, must be carefully selected based on the specific requirements of the application.
- The uniformity of the distribution of the constituent phases directly influences the material's ability to absorb and redistribute mechanical loads.
- Molecular contact and cohesion at the fiber-matrix interface are essential for the overall performance of the composite material. Therefore, it is important to select material combinations that prevent destructive effects, such as unwanted chemical reactions [14].

The properties of composite materials are profoundly influenced by the characteristics of the constituent materials, their distribution, and their interactions.

In describing a composite material as a system, besides identifying the component materials and their properties, it is essential to mention the geometry of the reinforcement. This can be described by shape, dimensions (proportions), and distribution (size and concentration). Systems containing reinforcements with identical geometry can be distinguished by concentration, distribution of concentrations, and orientation of the reinforcements. The shape of the discrete unit of the discontinuous phase can be approximated by spheres or cylinders, and in general, reinforcements, whether natural or industrial, resemble prismatic sections [15].

The orientation of the reinforcement influences the isotropy of the system. When the reinforcement is represented by particles of approximately equal size, the composite behaves like an isotropic material, with properties independent of direction. If the particles are of unequal size, the composite becomes quasi-isotropic, especially in the case of reinforcement with randomly oriented short fibers. In composites reinforced with long and continuous fibers, such as unidirectional ones, anisotropy becomes a key characteristic. A major advantage of these composites is the ability to control anisotropy through design and manufacturing.

The distribution of particle concentration refers to the spatial relationship between them, measuring the homogeneity and uniformity of the system. Particles must be evenly dispersed in the composite to form a continuous network that connects them, which is possible at low concentrations and has a significant impact on the electrical properties of composites [16].

The properties of biocomposites (**Figure 4**) are influenced by a series of key parameters, including the type of constituent materials, their proportions, and the distribution of fibers. The biodegradable matrix, such as PLA or PHA polymers, together with natural fiber reinforcements determine the mechanical and thermal behavior of the material. Additionally, the size and shape of the fibers, as well as their orientation in the matrix, has an essential role in the isotropy and anisotropy of the biocomposite.

Moisture and processing temperature influence both the final properties and the stability of the material, while the degree of compatibility between the matrix and the reinforcement affects adhesion and stress transfer. Therefore, a deep understanding of these parameters is essential for developing biocomposites with optimal performance for various applications.

The particularities of composition, design, manufacturing, and use of composite materials provide this category of materials with significant advantages (**Table 1**) compared to traditional ones [17–19].

Biocomposite materials offer a number of significant advantages over traditional materials, especially in the context of sustainability and environmental protection.

Firstly, they are biodegradable, meaning they naturally decompose in biological environments, reducing the impact on pollution. Additionally, biocomposites are made from renewable resources, contributing to a circular economy. The mechanical properties of biocomposites can be superior to those of conventional plastics, offering good strength and flexibility, ideal for a wide range of applications, from packaging to

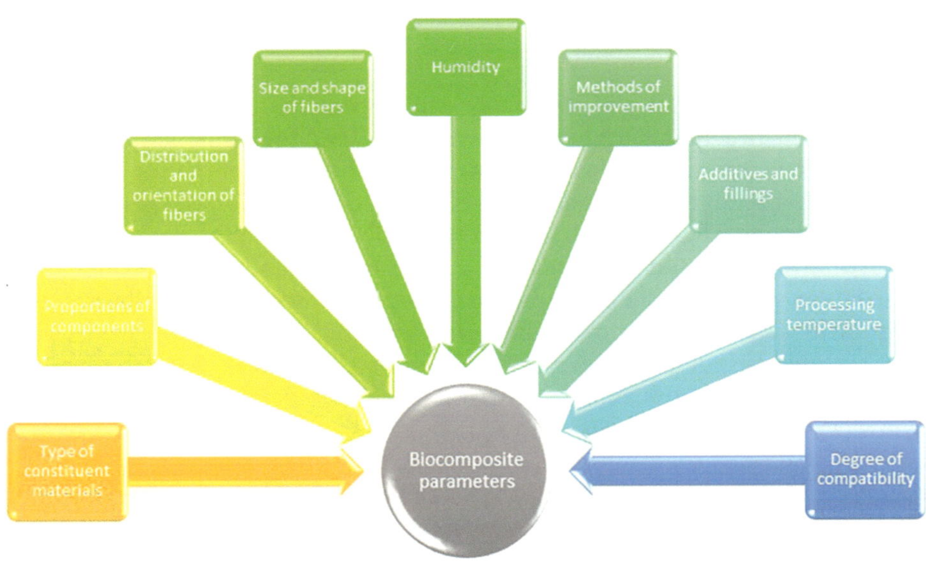

Figure 4.
Biocomposite parameters.

Advantages of composite materials:
• Enable the simultaneous design and optimization of both the material properties and the structural configuration of components.
• Offer a broad spectrum of adjustable variables, facilitating multi-objective optimization through comprehensive and advanced analytical methods.
• Components fabricated from composite materials exhibit enhanced operational safety and reliability.
• Exhibit high mechanical performance under both static and dynamic loading conditions, coupled with an advantageous strength-to-weight ratio that surpasses most conventional materials.
• Certain composite materials can be engineered to maintain their integrity and performance in aggressive environments over extended periods.
• Demonstrate excellent vibration damping capabilities, contributing to noise reduction and improved structural longevity.
• Although the reinforcing fibers typically exhibit low toughness, this limitation is offset by the ductility of the matrix and the energy dissipation at the fiber-matrix interface.
• Possess notable chemical stability and retain high strength at elevated temperatures (e.g., Kevlar fibers withstand temperatures up to 500°C, while ceramic fibers such as Al_2O_3 and SiC maintain structural integrity between 1400 and 2000°C).
• Characterized by low density in comparison with metallic materials, contributing to overall weight reduction in applications.
• Offer high resistance to crack propagation and wear, ensuring long service life in mechanically demanding conditions.
• Provide a tunable coefficient of thermal expansion, allowing compatibility with various materials and environmental conditions.
• Exhibit superior resistance to atmospheric degradation, making them suitable for long-term outdoor use.
• Can be fabricated into complex geometries and customized to meet specific functional requirements.
• Show exceptional durability in harsh environments, including marine, chemical, and civil infrastructure applications.

Table 1.
Advantages of composite materials.

structural components. Furthermore, biocomposites can be customized to meet the specific requirements of various applications, with excellent capabilities for vibration damping and resistance to different atmospheric conditions. These advantages make them an increasingly popular choice in modern industries where innovation and sustainability are key priorities.

4. Classification

The classification of these materials, initially carried out by Heimke and his collaborators, divides bioceramics into four main groups based on their specific behavior in interaction with biological environments.

This approach classifies bioceramics not only by chemical or physical criteria but also according to how they interact with biological tissues, how they integrate into the body, and how they maintain or modify their properties over time. Thus, bioceramics are divided into bioinert, bioactive, biodegradable, and carbon-based ceramics, each with unique characteristics that determine their use in medical applications such as implants, prostheses, and dental restorations. This classification highlights the diversity of behaviors that ceramic materials can exhibit when in contact with biological environments, allowing for optimal selection based on the specific needs of clinical applications.

- *The first group includes BIOINERT ceramics*, which consist of oxides, nitrides, and metal carbides; materials used in both dense and porous forms, such as Al_2O_3, MgO, ZrO_2, Si_3N_4, and SiC. The ceramic materials in this group are characterized by high values of density, Young's modulus, hardness, and certain mechanical properties such as compressive strength. Among this group, the most widely used ceramic material is aluminum trioxide, in fields such as bone substitutes, hip joint components, dental implants, and maxillofacial surgery.

- *The second group represents BIOACTIVE ceramics*, which includes various glass ceramics and materials containing calcium phosphate with a composition and structure similar to the inorganic components of bone. These materials are characterized by a certain solubility, which allows the biological tissue in contact with them to form direct bonds with the implant, capable of transferring shear and stress along the implant-tissue interface, thus covering a larger area. The main drawback of these ceramic materials is their weak mechanical properties, and consequently, they can only be used for devices that are not subjected to mechanical stress. Experimental trials have shown that these ceramics can be used as surface coatings for orthopedic devices and in dental surgery.

- *The third group of bioceramics includes the so-called BIODEGRADABLE ceramics*, which contain calcium phosphates as their primary component in various structural forms. These ceramics differ from bioactive ones by their high degree of solubility, which leads to gradual dissolution and controlled resorption by the adjacent biological tissue, stimulating bone growth on and within the pores of the material, thereby transforming the ceramic material into real bone tissue. These ceramics offer new possibilities in reconstructive surgery by filling or bonding defective bones, which is why biodegradable ceramics have garnered significant interest in recent years, especially in research aimed at improving their mechanical properties by using variable phosphate structures and special processing technologies. Attempts have been made to use these ceramics as surface coatings on stronger materials in medical fields such as dental implants and maxillofacial surgery.

- *The fourth group includes CARBON-BASED MATERIALS* in various structural forms, determined by the processing technology. The main advantages of carbon materials are that they are completely inert to living tissue and exhibit high biocompatibility when in contact with tissue and blood (hemocompatibility), which is why they can be used in the cardiovascular system. This group includes carbon, carbides, and nitrides with carbon [7].

The classification of biocomposites follows a similar principle, focusing on the type of natural fibers and matrices used, as well as their behavior in biological and environmental conditions. While bioceramics are classified based on their interaction with biological tissues, biocomposites are differentiated according to the type of natural fibers (such as flax, hemp, or sisal) and biodegradable matrices (such as PLA or PHA), with the aim of creating eco-friendly and sustainable materials that offer both good mechanical performance and environmental compatibility.

Natural fibers represent an essential component in the structure of biocomposites, providing mechanical strength, flexibility, and stability. These fibers are derived from plants, animals, or minerals and have gained popularity due to their availability, low cost, and reduced environmental impact compared to synthetic fibers. The most commonly used natural fibers in biocomposites include:

1. *Flax* is one of the strongest natural fibers used in biocomposites. It is lightweight, has excellent tensile strength, and a good moisture absorption capacity. Due to its structure, flax offers significant rigidity, making it ideal for applications requiring high mechanical strength, such as automotive components and construction. Flax is sensitive to moisture, and if not properly treated, it may degrade in humid environments, limiting its use in extreme conditions.

2. *Hemp* is a durable and strong fiber with good tensile and impact resistance. It is easy to grow, with a low environmental impact due to its minimal need for pesticides and fertilizers. Hemp is widely used in automotive applications, furniture parts, and construction due to its ecological and mechanical properties. Hemp may have variable consistency in terms of quality, which can require additional treatments to standardize the fiber's properties.

3. *Sisal* is a strong and durable fiber with good absorbency and excellent mechanical strength. It is commonly used in applications requiring tensile and friction resistance, such as in the automotive industry (vehicle interiors), rope making, and insulation materials. Like other natural fibers, sisal is susceptible to degradation when exposed to high moisture levels and may require special treatments to prevent biological degradation.

4. *Bamboo* is an extremely strong and lightweight fiber, being one of the fastest-growing natural materials. It is durable, flexible, and biodegradable, making it ideal for applications in construction and furniture industries. Bamboo is known for its excellent strength-to-weight ratio and its ability to absorb impact. Bamboo may have varying mechanical properties due to growing and harvesting conditions, which can affect its performance in biocomposites. It may also have lower moisture resistance if not treated.

5. *Jute* is an affordable and biodegradable natural fiber widely used for packaging, furniture parts, and technical textiles. It has good tensile strength and is considered one of the lightest natural fibers. Jute has lower mechanical strength compared to other natural fibers such as hemp or flax and may be more sensitive to moisture [20, 21].

In biocomposites, biodegradable matrices are essential for bonding natural fibers and ensuring the material's structural cohesion. The matrices transfer stresses between fibers and protect the fibers from external damage and moisture. Biodegradable matrices are preferred in biocomposites due to their ability to decompose in the environment, reducing environmental impact after use. These matrices can be natural or synthetic polymers with biodegradable properties. Some common examples include:

- *Polylactic Acid (PLA)* is a synthetic biodegradable polymer derived from renewable resources such as corn starch or sugarcane. It is one of the most widely used biodegradable polymers in biocomposites due to its good mechanical properties and ability to break down in the environment. PLA is biocompatible, easy to process, and has a relatively low melting point, making it ideal for use in plastic molding processes. It is also biodegradable and compostable in controlled environments. PLA has lower impact resistance compared to other polymers and may degrade more quickly in conditions of heat and humidity, which can limit its use in outdoor applications.

- *Polyhydroxyalkanoates (PHA)* is a biodegradable polymer produced by bacteria through fermentation. It is fully biodegradable and can decompose in natural environments without leaving toxic residues. PHA has mechanical properties similar to polypropylene and is very stable in biological environments, making it ideal for medical and packaging applications. It is biodegradable in soil and water, making it environmentally friendly. However, the production cost of PHA is higher compared to other biodegradable matrices, limiting its large-scale use.

- *Polybutylene Succinate (PBS)* is another biodegradable polymer produced from petroleum or biological resources. It has excellent processing properties and good resistance to heat and oils, making it suitable for packaging and industrial products. PBS has good tensile strength and excellent biodegradability in soil and composting conditions. It is compatible with natural fibers and can be used in packaging applications and disposable products. Compared to PLA, PBS has a higher production cost and may require stricter conditions for biodegradation.

- *Mater-Bi* is a starch-based biodegradable matrix developed to provide an eco-friendly alternative to traditional plastics. It is used in packaging, biodegradable garbage bags, and agricultural products. It is compostable in both industrial and home environments, making it an excellent eco-friendly solution for disposable products. Mater-Bi can be combined with natural fibers to produce efficient biocomposites. Mater-Bi may have lower mechanical strength and a shorter lifespan in harsh environments, such as high moisture exposure [22, 23].

5. Manufacturing techniques for biocomposites

The manufacturing of biocomposites involves both traditional composite material processing methods and innovative techniques that have evolved due to the specific sustainability and performance requirements of these materials. The choice of manufacturing technique depends on the final application of the biocomposites, the desired properties, and the costs associated with the production process.

Biocomposite manufacturing techniques include both traditional methods such as compression molding, resin injection, and lamination, as well as innovative techniques like 3D printing, nanotechnology, and low-temperature processing. Each contributes to optimizing the performance, sustainability, and energy efficiency of these materials.

- *Compression molding* involves compressing a layer of natural fibers and matrix in a mold under pressure and heat to create a compact part. It is one of the most common methods for producing biocomposites and is used for large parts such as automotive panels or structural elements in construction. Adapting this method for biocomposites required optimizing temperature and pressure parameters to match the characteristics of biodegradable matrices, which may have lower melting points than traditional synthetic materials.

- *Resin injection* is a method that involves injecting a biodegradable resin into a mold filled with natural fibers, after which the material is allowed to solidify. The process is fast and allows the production of complex parts. For biocomposites, injecting biodegradable resins (such as PLA or PHA) requires strict control of temperature and pressure to ensure uniform distribution and avoid thermal degradation of the material.

- *Lamination* involves the successive application of layers of natural fibers and matrix, which are then subjected to pressure and temperature to create a layered and consolidated material. This technique is widely used to obtain lightweight but strong biocomposites, used in the automotive and aerospace industries. Manual or automated lamination is often used for large and complex components.

Traditional methods have been adapted for biocomposites by adjusting processing temperatures and cycle times to match the thermal behavior of biodegradable matrices, which are more sensitive to high temperatures compared to conventional synthetic resins. Adjustments have also been made to integrate natural fibers, which may have greater variability in properties compared to synthetic fibers.

With the development of biocomposites, modern manufacturing methods have emerged, aimed at improving the efficiency, precision, and sustainability of these materials.

- *3D Printing (Additive manufacturing)* is an innovative method that allows the creation of biocomposite parts by adding material layer by layer. This technique offers great design flexibility and is used for rapid prototyping or customized components, particularly in fields such as the medical industry or custom furniture. The use of 3D printing for biocomposites involves extruding a mixture of natural fibers and biodegradable polymers (such as PLA) to create complex structures, with minimal material waste and increased energy efficiency.

- *Nanotechnology* is used in biocomposites to improve the mechanical, thermal, and barrier properties of materials by integrating nanofibers or nanoparticles into the matrix. For example, adding clay or carbon nanoparticles to biodegradable polymer matrices can improve the mechanical strength and thermal stability of biocomposites. These improvements are essential in advanced applications, such as in the automotive or aerospace industries, where materials need to be both strong and lightweight.

- *Low-temperature processing techniques.* Many biodegradable matrices, such as PLA, have lower melting points than conventional plastics, leading to the development of manufacturing technologies that operate at lower temperatures, reducing energy consumption. These processes are essential to ensuring minimal degradation of polymers and protecting the integrity of natural fibers.

- *Advanced biolamination* has enabled the production of more uniform and durable biocomposites. These techniques ensure better matrix distribution and the elimination of air between layers, improving the mechanical and esthetic performance of the final product [24, 25].

The manufacturing of biocomposites can have a higher initial cost compared to traditional composites due to the higher cost of natural fibers and biodegradable matrices. For example, biodegradable polymers such as PLA and PHA are often more expensive than petroleum-derived polymers. Additionally, the need for adjustments to processing equipment to handle these materials can add extra costs. However, in the long term, biocomposites offer economic advantages due to their recycling and composting potential, as well as reducing waste management fees.

One of the major advantages of biocomposites is the energy efficiency in the manufacturing process. Many biodegradable matrices, such as PLA, have lower melting and processing temperatures than conventional synthetic polymers, resulting in reduced energy consumption during production. Moreover, natural fibers are generally lighter than synthetic fibers, reducing energy costs associated with transportation and material handling. In addition, innovative processes such as 3D printing and advanced lamination allow for more efficient use of materials, reducing waste and minimizing by-products. This contributes to a decrease in overall energy consumption and environmental impact.

6. Applications in industry

Biocomposites with unique properties find their applications in many fields (**Figure 5**) thanks to their sustainability, versatility, and performance benefits [7].

Biocomposites have become widely used in the *automotive sector* due to their sustainability and performance benefits. They are applied in various vehicle components, including door panels, dashboards, center consoles, and seats. By reducing the vehicle's weight, biocomposites help improve energy efficiency and lower CO_2

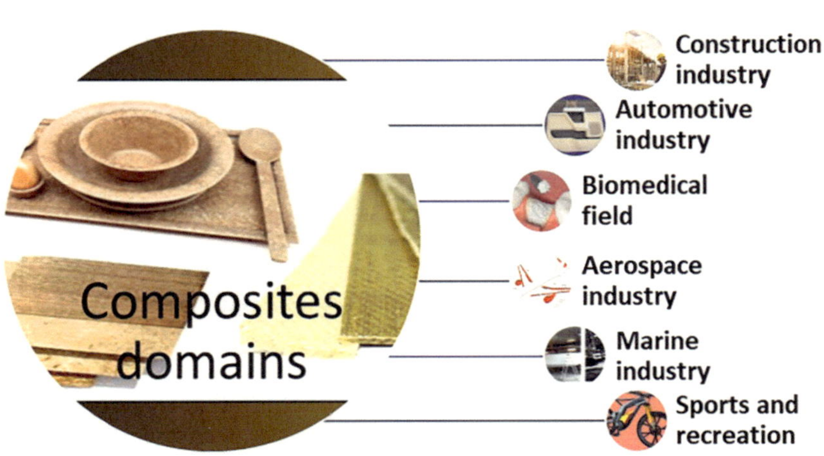

Figure 5.
Composites domain.

emissions. Natural fibers like hemp are integrated into automotive interiors, as seen in BMW's i3 series, offering both strength and eco-friendliness.

In the *construction field*, biocomposites are utilized for their excellent insulating properties and reduced environmental impact. They are commonly used for facade panels, roofing, and structural elements, providing a sustainable alternative to conventional cement and steel materials. Green building projects often use biocomposites made from natural fibers like hemp and flax to improve energy efficiency and thermal comfort.

Biocomposites are increasingly replacing conventional plastics in the production of *consumer goods*. In packaging, biodegradable biocomposites made from materials like starch, PLA, or PHA are used to create compostable packaging, reducing plastic pollution. In the furniture industry, biocomposites are used to manufacture durable, esthetically pleasing pieces such as chairs and tables, while sports equipment like snowboards and bicycles benefit from the lightweight and durable nature of these materials.

In medicine, biocomposites are used in medical devices, implants, and prostheses due to their biocompatibility and controlled biodegradability. They support tissue regeneration when used in implants that are absorbed by the body. For example, biocomposites made from calcium phosphate and PHA are employed in bone implants, providing mechanical support during healing and being gradually resorbed by the body.

In the *aerospace sector*, biocomposites are valued for their lightweight and ability to withstand extreme conditions. They are used in the production of interior panels, body elements, and lightweight structural components, helping improve fuel efficiency by reducing the weight of aircraft. Companies like Airbus are incorporating natural fiber-based biocomposites into aircraft interiors to enhance sustainability.

The *marine industry* utilizes biocomposites for constructing boats and marine equipment, thanks to their resistance to corrosion and excellent performance in wet environments. Biocomposites provide a sustainable alternative to traditional materials like fiberglass, reducing the environmental impact and the overall weight of vessels.

Biocomposites are widely used in *sports equipment*, including tennis rackets, bicycles, and skis. Their combination of strength, flexibility, and lightweight makes them ideal for producing durable and high-performance products that also reduce environmental impact.

7. Applications of biocomposites in living tissues

Biocomposites, which combine ceramics such as hydroxyapatite with polymers, are crucial in bone regeneration through the formation of synthetic scaffolds that facilitate cell growth and communication, as well as nutrient flow. These synthetic scaffolds are designed to be highly porous to allow the incorporation of specific bone cells, such as mesenchymal stem cells. In the biomedical field, biocomposites offer innovative solutions for a wide range of applications due to their remarkable properties. Here is a detailed list [26] of the different applications and compositions of these materials:

Bone regeneration: Nanohydroxyapatite combined with glycol chitosan for improved scaffold integration; blend of hydroxyapatite with a polymeric mixture (silk fibroin, chitosan, and agarose) enhancing mechanical strength and biocompatibility; innovative mixture of calcium silicate, zinc silicate, and graphene oxide to support osteoconductivity and osteoinductivity; composite of collagen, silk fibroin, and decellularized extracellular matrix (dECM) tailored for natural bone mimicry;

boron nitride and boron trioxide incorporated for their osteogenic and antibacterial properties; nanohydroxyapatite with calcium sulfate and bioactive molecules to facilitate rapid bone repair and growth.

Orthopedic implants: PEEK reinforced with graphene oxide to improve wear resistance and mechanical properties; CFRPEEK blended with nanohydroxyapatite, carboxymethyl chitosan, and bone forming peptides for enhanced bioactivity and integration; polyphenylene sulfide combined with nanohydroxyapatite, offering high mechanical performance and stability; polyimide fused with tantalum pentaoxide for superior durability and biocompatibility in demanding load-bearing applications; hydroxyapatite mixed with ceria nanoparticles and silver nanoparticles to provide antimicrobial properties and promote osteointegration.

Wound healing: Polycaprolactone and gelatin for creating moisture-maintaining yet breathable dressings; chitosan blended with polyethylene oxide and fibrinogen to accelerate clotting and promote rapid healing; collagen and alginate infused with silver nanoparticles for enhanced antimicrobial efficacy; polyurethane combined with keratin and silver nanoparticles to create scaffolds that support healing while preventing infection; collagen and dextran for dressings that promote cellular attachment and proliferation.

Tissue engineering: Fibrin, alginate, and genipin for creating highly biocompatible scaffolds that support cell growth; conductive PEDOT, chitosan, and gelatin composite for applications requiring electrical stimulation; polycaprolactone blended with silk fibroin and carbon nanotubes for enhanced structural integrity and functionality; silk fibroin and melanin for biocompatible and photoprotective tissue engineering applications; polycaprolactone and collagen optimized for versatility in scaffold fabrication and natural integration; gelatin, alginate, and fibrinogen providing a customizable matrix for diverse tissue engineering needs; collagen type I and gelatin methacryloyl, engineered for photo-crosslinking capabilities in precise scaffold modeling.

In addition to replacing a tissue or damaged area, the scaffolds must be biocompatible and have a controllable degradation rate that matches the tissue cell growth, allowing attachment, proliferation, and ultimately replacement of the scaffold by tissue growth. Moreover, the mechanical properties of the scaffold, such as stiffness, should be similar to the characteristics of the tissue being repaired [27–30].

For example, in one study, Dumont et al. [31] selected nanohydroxyapatite and glycol chitosan as the ceramic and polymeric components, respectively, of the developed biocomposite. Nanohydroxyapatite particles were synthesized using glycol chitosan as the ligand. Membranes of the composite material were created by pouring the polymer/calcium phosphate mixture into plastic molds and allowing them to dry. Scanning electron microscope (SEM) analysis of the membranes showed a homogeneous distribution of the nanohydroxyapatite particles, resulting in uniform dispersion within the polymeric matrix. To verify the biocompatibility of the composite membranes, live/dead and MTT assays were performed using human osteosarcoma (Saos-2) and human bone marrow mesenchymal stem cells (HBMS), displaying high cell viability (~100%) and, consequently, non-toxic behavior. Additionally, HBMS cells showed good adhesion, spreading, and proliferation on the surface of the biocomposite, demonstrating the valuable potential of the developed scaffolds for applications such as moldable membranes for orthopedic surgeries. These positive results obtained in the study demonstrate that the developed biocomposite offers valuable applications in the field of tissue regeneration.

In tissue engineering, biocomposites are applied to create artificial scaffolds that support cellular growth and tissue regeneration, adapting to the regeneration rate of the affected area. In wound healing, biocomposites are used to develop dressings that balance the degradation rate with skin repair, maintaining an essential moist environment for regeneration. In orthopedic implants, biocomposites [32] offer alternatives to traditional orthopedic implants, reducing infection risk and ensuring better mechanical compatibility with bone tissues. At bone regeneration, biocomposites form porous synthetic scaffolds that facilitate the integration of specific bone cells and are crucial in treating bone defects caused by trauma or disease [33, 34].

8. Trends in research and development of biocomposites

Globally, research and development trends in biocomposites are influenced by national and regional economic strategies, focusing on the development of innovative and sustainable materials. Among the main trends is the creation of new types of multiphase biocomposites that meet the diverse needs of industries, from construction to medicine. Research is also focused on theoretical and experimental studies on the behavior of materials under deformation or radiation, mathematical modeling of properties based on processing parameters, as well as analysis of the physicochemical phenomena at the phase separation boundaries [35, 36]. Non-destructive testing techniques are being developed to assess the integrity of biocomposites, and studies on reinforcement with innovative textile structures are expanding.

Another important trend is the integration of nanotechnologies into the creation of multifunctional materials that combine advanced properties with sustainability. This includes not only the development of nanomaterials but also the implementation of innovative processes that integrate material science at the nanoscale with biotechnologies and information technologies. These promising directions highlight the potential of biocomposites to revolutionize traditional industries while contributing to environmental protection.

9. Conclusions

Biocomposites represent a promising solution in the context of sustainable development, with the potential to replace traditional fossil-based materials. Due to their biodegradability and the use of renewable materials, they contribute significantly to reducing environmental impact. Their superior mechanical properties, customization for specific applications, and ability to be used in various industries, from construction to medicine, give them a competitive advantage. Furthermore, current research and development directions aim at innovation in multiphase structures, the integration of nanotechnologies, and the development of advanced processing techniques, which further improve their versatility and efficiency. These advances emphasize the importance of continued investment in research and development to fully harness the potential of biocomposites and contribute to a greener, more sustainable future.

Acknowledgements

This work was supported by a grant of from the Ministry of Research, Innovation and Digitization, CNCS/CCCDI—UEFISCDI, project number ERA-NET-ERAMIN-3-Cool&SmartTit-1, contract no. 8/2024 within PNCDI IV.

Conflict of interest

The authors declare no conflict of interest.

Author details

Mădălina Simona Bălțatu[1], Petrică Vizureanu[1,2,3]*, Andrei Victor Sandu[1,2,3], Dragos Cristian Achitei[1], Dumitru Doru Burduhos Nergis[1], Manuela Cristina Perju[1] and Marcin Nabiałek[4]

1 Faculty of Materials Science and Engineering, "Gheorghe Asachi" Technical University of Iasi, Iasi, Romania

2 Academy of Romanian Scientists, Bucharest, Romania

3 Romanian Inventors Forum, Iasi, Romania

4 Faculty of Production Engineering and Materials Technology, Department of Physics, Częstochowa University of Technology, Częstochowa, Poland

*Address all correspondence to: petrica.vizureanu@academic.tuiasi.ro

References

[1] Bibu M, Nemeş T, Studiul materialelor. Materiale utilizate în construcţii de maşini. Sibiu: Universităţii Lucian Blaga; 2004. ISBN 973-651-825-6

[2] Ispas SC. Materiale compozite. Bucureşti: Ed. Tehnică; 1987

[3] Chiara G, Letizia F, Lorenzo F, Edoardo S, Diego S, Stefano S, et al. Nanostructured biomaterials for tissue engineered bone tissue reconstruction. International Journal of Molecular Sciences. 2012;**13**:737-757

[4] Agarwal BD, Broutman LJ. Analysis and Performance of Fiber Composites. New York: John Wiley & Sons, Inc.; 1990

[5] Hench LL. Bioceramics: From concept to clinic. Journal of the American Ceramic Society. 2002;**81**(7):1705-1728

[6] Tajbakhsh S, Hajiali F. A comprehensive study on the fabrication and properties of biocomposites of poly (lactic acid)/ ceramics for bone tissue engineering. Materials Science and Engineering: C. 2017;**70**:897-912

[7] Hench LL, Wilson J. An Introduction to Bioceramics. Singapore: World Scientific; 1993. ISBN 978-981-02-1180-1

[8] Drăgulescu D, Popescu M. Materiale compozite metalice. Prezent cu perspective. Timişoara: Orizonturi Universitare, Timişoara; 2001. ISBN 973-9400-48-5

[9] Goanţă V. Materiale ceramice tehnice. Studiul proprietăţilor mecanice. Iaşi: Performantica; 2003. ISBN 973-7994-15-9

[10] Pickering KL, Efendy MA, Le TM. A review of recent developments in natural fibre composites and their mechanical performance. Composites. Part A, Applied Science and Manufacturing. 2016;**83**:98-112

[11] Ţăranu N, Secu A, Decher E, Isopescu D. Structuri din materiale compozite şi asociate. Iaşi: Rotaprint; 1992

[12] Ţăranu N. Elemente portante din materiale plastice. Institutul Politehnic Iaşi: Teza de doctorat; 1978

[13] Zgură G, Moga V. Bazele proiectării materialelor compozite. Bucureşti: Ed. Bren; 1999. ISBN 973-9493-01-7

[14] Bejan L. Introducere in micromecanica materialelor composite armate cu tesaturi. Iasi: Ed. Gheorghe Asachi; 2000. ISBN 973-8050-34-0

[15] Alsuwait RB, Souiyah M, Momohjimoh I, Ganiyu SA, Bakare AO. Recent development in the processing, properties, and applications of epoxy-based natural fiber polymer biocomposites. Polymers. 2023;**15**:145. DOI: 10.3390/polym15010145

[16] Hubcă G, Iovu H, Tomescu M, Roşca ID, Novac O, Gh I. Materiale compozite. Bucureşti: Tehnică; 2000. ISBN 973-31-1126-0

[17] Baley C, Gomina M, Breard J, Bourmaud A, Davies P. Variability of mechanical properties of flax Fibres for composite reinforcement. A review. Industrial Crops and Products. 2020;**145**:111984

[18] Dorozhkin SV. Biocomposites and hybrid biomaterials based on calcium orthophosphates. Biomatter. 2011;**1**:3-56

[19] Haraguchi K. Biocomposites. In: Encyclopedia of Polymeric Nanomaterials. Berlin/Heidelberg, Germany: Springer; 2015. pp. 124-130

[20] Vaiani L, Boccaccio A, Uva AE, Palumbo G, Piccininni A, Guglielmi P, et al. Ceramic materials for biomedical applications: An overview on properties and fabrication processes. Journal of Functional Biomaterials. 2023;**14**:146. DOI: 10.3390/jfb14030146

[21] LeGeros RZ. Properties of Osteoconductive biomaterials: Calcium phosphates. Clinical Orthopaedics and Related Research. 2002;**395**:81-98

[22] Vallet-Regí M. Ceramics for medical applications. Journal of Chemical Society. 2001;**31**(3):290-295

[23] Yunus Basha R, Sampath SK, Doble M. Design of Biocomposite Materials for bone tissue regeneration. Materials Science and Engineering: C. 2015;**57**:452-463

[24] Piconi C, Maccauro G. Zirconia as a ceramic biomaterial. Biomaterials. 1999;**20**(1):1-25

[25] Bružauskaitė I, Bironaitė D, Bagdonas E, Bernotienė E. Scaffolds and cells for tissue regeneration: Different scaffold pore sizes—Different cell effects. Cytotechnology. 2016;**68**:355-369

[26] Valente KP, Brolo A, Suleman A. From dermal patch to implants—Applications of biocomposites in living tissues. Molecules. 2020;**25**:507

[27] Bose S, Bandyopadhyay A. Bone Substitutes. Berlin, Heidelberg: Springer; 2013. ISBN 978-3-642-38542-9

[28] Kokubo T, Takadama H. How useful is SBF in predicting in vivo bone bioactivity? Biomaterials. 2006;**27**(15):2907-2915

[29] Swetha M, Sahithi K, Moorthi A, Srinivasan N, Ramasamy K, SelvamuruganN.Biocompositescontaining natural polymers and hydroxyapatite for bone tissue engineering. International Journal of Biological Macromolecules. 2010;**47**:1-4

[30] Regí MV, Navarrete DA. Biological Apatites in bone and teeth. In: Nanoceramics in Clinical Use: From Materials to Applications; RSC Nanoscience & Nanotechnology. Cambridge, UK: Royal Society of Chemistry; 2015. pp. 1-29 ISBN 978-1-78262-104-1

[31] Dumont VC, Mansur HS, Mansur AAP, Carvalho SM, Capanema NSV, Barrioni BR. Glycol chitosan/Nanohydroxyapatite biocomposites for potential bone tissue engineering and regenerative medicine. International Journal of Biological Macromolecules. 2016;**93**:1465-1478

[32] Sanchez C, Arribart H, Guille MMG. Biomimetism and bioinspiration as tools for the design of innovative materials and systems. Nature Materials. 2005;**4**:277-288

[33] Akter M, Uddin MH, Tania IS. Biocomposites based on natural fibers and polymers: A review on properties and potential applications. Journal of Reinforced Plastics and Composites. 2022;**41**(17-18):705-742. DOI: 10.1177/07316844211070609

[34] Dumitraş D, Opran C. Prelucrarea materialelor composite, ceramice şi minerale. Bucureşti: Ed. Tehnică; 1994. ISBN 973-31-0602-X

[35] Aydoğmuş E, Yanen C, Kıstak C. Advancing sustainable

materials: Synthesis and analysis of polyurethane biocomposites from hydrogenated safflower oil. Applied Sciences. 2025;**15**(3):1017. DOI: 10.3390/app15031017

[36] Ismail I, Marni A, Yufita E, Rahwanto A, Fadzullah S. Effect of composition on physical, mechanical, and thermal properties of oil palm empty fruit bunch epoxy resin biocomposite. Key Engineering Materials. 2024;**1001**:99-109. DOI: 10.4028/p-82tfuv

Chapter 2

Composite Materials Used to Repair Industrial Equipment: Analytical, Numerical and Experimental Studies

Alin Dinita and Maria Tanase

Abstract

Composite materials offer superior strength-weight ratios, corrosion resistance, and durability, making them ideal for restoring and reinforcing equipment. Analytical approaches employ mathematical models, while numerical analysis, particularly finite element analysis (FEA), simulates stress, strain, and failure, experimental analysis evaluates repair effectiveness. This research emphasizes fiber-reinforced polymers (FRP) for pipeline defect repairs, showcasing high tensile strength, corrosion resistance, and lightweight properties. For a case study involving a natural gas pipeline repair, the chapter details the calculation of the minimum required thickness of the GFRP composite. Depending on the material used, the number of layers required to meet strength demands varies, with examples showing 91 layers for MCM composite material and 30 layers for TDW composite, demonstrating the relationship between material properties and repair design. Numerical simulations using FEA predict FRP repair performance, aiding decision-making and ensuring safety standards. Results show that composite repairs can maintain stress states below 290 MPa in pipelines (considering a steel with a yield strength of 360 MPa) with varying defect depths (d/t ratios). Furthermore, experimental tests validate the quality of GFRP composite repairs, demonstrating failure at pressures exceeding design limits, with composite coatings functioning as rupture stoppers at pressures of approximately 31.5 MPa. The chapter details design steps for Defect Type A repairs according to ISO 24817, including risk assessment, material selection, and design calculations for minimum repair laminate thickness. Relevant ISO, ASTM, BS, and ASME standards are discussed to ensure repair reliability and safety.

Keywords: GFRP, pipeline, mechanical properties, yield strength, FEA

1. Introduction

Composite materials have become increasingly popular in the repair of industrial equipment due to their excellent strength-to-weight ratio, corrosion resistance, and versatility [1–4]:

- Fiber-reinforced plastics: This includes materials like fiberglass and carbon fiber. They offer high tensile strength and are often used to reinforce structures or repair cracks in machinery.
- Epoxy resins: Often used as a matrix by combining them with reinforcements like carbon or glass fibers. Epoxy resins provide strong adhesion and chemical resistance, making them suitable for a variety of industrial repairs.
- Polymer-matrix composites: These utilize thermosetting or thermoplastic polymers as the matrix material. They are lightweight and resistant to environmental factors, making them suitable for repairing equipment exposed to harsh conditions.
- Metal matrix composites: Combining metal with ceramic fibers or particles, these composites offer improved thermal and wear resistance, making them ideal for high-stress applications.
- Concrete composites: For structural repairs, fiber-reinforced concrete can be used to strengthen and repair concrete structures in industrial settings.
- Bamboo and natural fiber composites: These are gaining attention for their sustainability and can be used in less critical applications or as environmentally friendly repair options.

2. Standards used to define composite materials

Utilizing the relevant BS and ASME standards, alongside ISO and ASTM standards, ensures comprehensive coverage of requirements and performance assessments for composite materials used in repairs. Adhering to these standards enhances the reliability, safety, and longevity of the repaired equipment, making sure all repairs meet industrial quality expectations [5–9]. **Table 1** presents the main standards and regulations containing information regarding repair with composite materials.

Standard	Usage	Importance
ISO 527 Plastics — Determination of tensile properties	Specifies methods for determining the tensile properties of polymer materials, including composites.	Essential for assessing the strength and durability of composite materials used in repairs.
ISO 16773 Plastics — Determination of the resistance to solvent action of plastic materials	This standard assesses the resistance of plastic materials, including composites, to the action of solvents.	Critical for ensuring that repair materials can withstand exposure to industrial chemicals.

Standard	Usage	Importance
ISO 527-4 Plastics — Determination of tensile properties — Part 4: Test conditions for isotropic and anisotropic fiber-reinforced plastics	Specifies test conditions relevant to fiber-reinforced composites.	Helps in understanding the mechanical performance of composite materials under tensile stress.
ISO 14129 Fiber-reinforced plastic composites — Determination of interlaminar shear strength by short-beam method	This standard provides a method for assessing the shear strength of laminated composite materials.	Useful for evaluating the bonding quality in laminate repairs.
ISO/IEC 17025 General requirements for the competence of testing and calibration laboratories	This standard pertains to laboratories that test the properties of composite materials, ensuring accuracy and reliability in testing.	Important for ensuring that tests of materials used in repairs are performed to recognized quality standards.
ISO 9060 Solar energy — Specification and classification of instruments for measuring shortwave solar radiation	While not directly related to composite repairs, it offers insights into materials used in solar panels, which sometimes utilize composites.	Relevant for industries integrating composite technology with renewable energy applications.
ISO 14130 Fiber-reinforced plastic composites — Determination of the durability of fiber-reinforced plastic composites at elevated temperatures	This standard evaluates how composites behave when exposed to high temperatures over time.	Important for repairing components that operate under thermal stress.
ISO 24817:2017 Petroleum, petrochemical, and natural gas industries — Composite repairs for pipework — Qualification and design, installation, testing, and inspection	Requirements and recommendations for the qualification and design, installation, testing and inspection for the external application of composite repair systems.	Important for petroleum, petrochemical, and natural gas industries.
	British Standards (BS) for composite materials	
BS EN 1372 Fiber-reinforced plastics — Definitions, designation, and classification.	Provides definitions and classification of different types of fiber-reinforced plastics.	Essential for ensuring clarity in the materials used for repairs.
BS EN 14573 Plastics — Determination of the tensile properties of polymers and polymer composites.	This standard specifies methods for tensile testing in fiber-reinforced plastics.	Helps assess the mechanical properties of materials used in repair applications.
BS EN ISO 14129 Fiber-reinforced plastic composites — Determination of interlaminar shear strength by the short-beam method.	Like its ISO counterpart, this standard helps determine the shear strength of composite .	Crucial for evaluating the performance of repaired composite structures.
BS EN ISO 527 Plastics. Determination of tensile properties.	Specifies the methods to determine tensile properties for all types of polymers, including composites.	Critical for understanding the strength and durability of repair materials.

Standard	Usage	Importance
BS 4970 Design, manufacture, and testing of composite materials.	This standard provides guidelines for the design and manufacturing processes of composite.	Useful for ensuring that repair techniques align with best practices in composite manufacturing.
	ASME standards for composite materials	
ASME BPVC Section III Rules for construction of nuclear facility components, subsection NE - Core support structures.	Discusses requirements for materials in nuclear applications, including composites.	Ensures safety and reliability of composite materials in critical applications.
ASME B31.3 Process piping code.	While primarily focused on piping systems, it provides guidelines on the use of materials, including composites.	Offers essential data for ensuring the integrity of piping repairs in industrial settings.
ASME B16.20 +Metallic gaskets for pipe flanges.	This standard includes guidelines on the application of composite materials within gaskets.	Helps in utilizing composites effectively in sealing solutions in piping systems.
ASME Y14.5 Dimensioning and tolerancing.	Although not specific to composites, this standard is essential for defining and communicating design requirements in composite repair and manufacturing.	Ensures drawings and specifications for composite repairs are clear and standardized.
ASME PTC 19.1 Test codes for piping.	Provides guidelines on testing methods that may involve composite.	Crucial for assessing the performance of repairs in piping systems.
	ASTM standards for composite materials	
ASTM D638 Standard test method for tensile properties of plastics	This standard specifies the method for determining the tensile of plastic materials.	Critical for evaluating the strength and elastic properties of composites used in repair applications.
ASTM D3039 Standard test method for tensile properties of fiber-reinforced polymer matrix composites	Focuses specifically on the tensile testing of fiber-reinforced composites.	Essential for assessing the mechanical performance of materials used in structural repairs.
ASTM D256 Standard test methods for determining the izod pendulum impact resistance of plastics	This test method evaluates the impact resistance of plastics, including composite materials.	Important for determining how well materials can withstand sudden impacts during service.
ASTM D2290 Standard test method for tensile properties of adhesive bonds	Specifies a method for the tensile strength of adhesive bonds used between composite.	Helps ensure the integrity of bonded composite repairs.
ASTM D4862 Standard guide for the evaluation of the effect of moisture on the physical and mechanical properties of fiber-reinforced polymer matrix composites	Evaluates how moisture absorption affects composite materials.	Vital for understanding the durability of composites in humid or wet environments.
ASTM D5400 Standard guide for the evaluation of polymer matrix composites	Provides guidelines for characterizing polymer matrix composites, including composite repairs.	Offers insight into material selection and performance criteria for repairs.

Standard	Usage	Importance
ASTM D1980 Standard guide for determining material properties of reinforced thermosetting resin composites	Focuses on determining the mechanical properties of thermosetting resin composites.	Useful for selecting appropriate composite materials based on their performance in repairs.
ASTM D794 Standard test method for flexural properties of unreinforced and reinforced plastics and electrical insulating materials	Assesses the flexural properties of unreinforced and reinforced plastics, which can include composite repairs.	Helps in evaluating how well materials will perform under bending loads.
ASTM C1355 Standard test method for the determination of the interlaminar shear strength of composite materials	Specifies a method for measuring the shear strength between layers in composite materials.	Important for evaluating lamination quality in repaired structures.
ASTM E694 Standard guide for package design and testing of composite containers	Provides guidelines for testing composite packaging materials, which can also be relevant for industrial components.	Ensures that composite repairs maintain the integrity of enclosed materials.

Table 1.
ISO standards for composite materials.

3. Composite materials used to repair industrial equipment

When selecting a composite material for repair, consider factors such as the environment (temperature, chemicals), the type of stress (tensile, compressive), and the required durability. Proper surface preparation and application techniques are also crucial to ensure the effectiveness of the repair (*Patch Repairs*: Composite materials can be used to apply patches to damaged surfaces; *Reinforcement*: They can add strength to weakened areas of equipment; *Sealant Applications*: Composites can seal and protect interfaces to prevent leakage or corrosion). Fiber-reinforced polymers (FRP) can be classified in several ways based on the type of fiber used, the nature of the polymer matrix, and their applications [10–13].

- Fiber type: Firstly, FRPs can be categorized by the type of fibers used. Glass Fiber-Reinforced Polymer (GFRP) is common due to its low cost and good mechanical properties, making it suitable for construction, automotive, and marine applications. Carbon Fiber-Reinforced Polymer (CFRP) is known for its high strength-to-weight ratio and stiffness, finding applications in aerospace, high-performance sports equipment, and the automotive industry. Aramid Fiber-Reinforced Polymer (AFRP) is noted for its impact resistance and toughness, often used in ballistic applications, aerospace, and protective gear. Basalt Fiber-Reinforced Polymer (BFRP) offers high temperature and chemical resistance, making it suitable for construction and infrastructure uses.

- Polymer matrix type: Another classification is based on the type of polymer matrix. Thermosetting resins, such as epoxy, polyester, and vinyl ester, are widely used. Epoxy offers excellent mechanical properties and moisture resistance and is favored in combination with carbon fibers. Polyester is cost-effective and commonly used with glass fibers. Vinyl ester provides better chemical resistance than polyester, making it ideal for use in corrosive environments. On the other hand, thermoplastic resins like polypropylene, polycarbonate, and polyether ether ketone (PEEK) allow for reshaping and reprocessing. These offer impact resistance and are utilized in automotive and consumer goods.

- Structural form: FRPs can also be differentiated by structural form. Unidirectional tapes, where fibers are aligned in a single direction, maximize strength in that direction. Woven fabrics, with interlaced fibers in varying patterns, provide balanced properties. Multiaxial fabrics consist of layers of fibers in multiple directions, offering tailored mechanical properties for specific applications.

- Application: Finally, classifying FRPs by application helps identify their diverse uses. In construction and infrastructure, they reinforce concrete, bridges, and repair materials. In the automotive and transportation sectors, they contribute to lightweight panels, structural components, and interior parts. Aerospace and defense applications include structural components, fuselages, and bulletproof vests. In the marine industry, FRPs are used for boats, hulls, and marine structures due to their corrosion resistance.

This detailed classification helps in understanding the versatility and application-specific design of FRPs, aiding in the selection of the appropriate type for desired performance characteristics.

Glass fiber-reinforced polymer (GFRP) and carbon fiber-reinforced polymer (CFRP) are two types of composite materials that are widely used due to their advantageous properties.

3.1 Glass fiber-reinforced polymer (GFRP)

- *Mechanical properties*: *Tensile strength* - Typically ranges from 350 to 620 MPa, depending on the type of glass fiber and the resin used. *Modulus of Elasticity* - Generally, around 20 to 40 GPa, providing a good balance of stiffness and flexibility. *Impact resistance* - Good, making GFRP suitable for applications where moderate impact resistance is needed.

- *Physical properties*: *Density* - Approximately 1.8 to 2.0 g/cm^3, which is lower than metals, contributing to its lightweight nature. *Thermal properties* - Exhibits good thermal insulation properties but has a relatively low maximum service temperature, usually up to 200°C (around 392°F).

- *Chemical resistance*: Offers excellent resistance to a wide range of chemicals and environmental conditions, including corrosion.

- *Cost*: Generally lower cost than CFRP, making it an affordable option for a variety of applications.

- *Applications*: Commonly used in construction (rebars, bridge decks), automotive parts, water and sewage piping, and marine applications.

3.2 Carbon fiber-reinforced polymer (CFRP)

- *Mechanical properties*: *Tensile strength*: Ranges from 600 to 8000 MPa, significantly higher than GFRP, making it suitable for high-performance applications. *Modulus of elasticity* - Typically, between 70 to 700 GPa, which provides exceptional stiffness and is tailored based on fiber composition and orientation. *Impact resistance* - Lower than GFRP because of the material's stiffness, which can make it more brittle in some applications.

- *Physical properties*: *Density* - Around 1.6 g/cm^3, lighter than GFRP and metals, contributing to its high strength-to-weight ratio. *Thermal Properties* - Better thermal resistance than GFRP, with service temperatures up to 400°C (around 752°F) or higher with specialized resins.

- *Chemical resistance*: Excellent resistance to chemicals and environmental degradation, similar to GFRP, but can be prone to galvanic corrosion when used in contact with metals.

- *Cost*: Generally, more expensive than GFRP due to the cost of carbon fibers, but this is often justified by its superior mechanical properties and performance benefits.

- *Applications*: Widely used in aerospace (aircraft components), automotive (racing and performance cars), sports equipment (bicycles, golf clubs), and high-end construction projects.

In summary, GFRP is cost-effective and has good mechanical and chemical resistance properties, making it suitable for a wide range of general applications. CFRP, on the other hand, offers superior strength and stiffness, making it highly desirable for applications where performance is critical, despite its higher cost. The selection between the two often depends on the specific requirements of the application, including factors like budget, mechanical demands, and environmental conditions.

4. The determination of the thickness of a coating made of GFRP composite materials used for repairing a natural gas pipeline

ISO 24817 addresses the requirements for composite repairs to pipework and pipelines within the petroleum, petrochemical, and natural gas sectors [14–17]. The standard specifies criteria for:

- *Qualification*: Establishing procedures to assess the durability and performance of composite materials used in repairs, ensuring they meet specific needs and conditions of the pipework they are intended to repair.

- *Design*: Guidelines for designing repairs that consider operational pressures, temperatures, and environmental factors to ensure compatibility and reliability of the repair over its intended lifespan.

- *Installation*: Defining best practices for the application of composite repair systems, including surface preparation, material application methods, and curing processes necessary to achieve a successful repair.

- *Testing*: Providing methodologies for testing repaired sections to confirm their integrity and performance, including pressure testing, non-destructive examination, and other appropriate tests.

- *Inspection*: Outlining procedures for inspecting the repair during and after installation to ensure compliance with safety and performance standards, and to confirm the repair's effectiveness in restoring the pipe's original function.

By covering these areas, ISO 24817 helps ensure that composite repairs are properly executed and maintain the safety and functionality of critical infrastructure in demanding industrial environments.

Alright, based on the document, here's a detailed breakdown of the design steps for a Defect Type A repair, according to ISO 24817.

- *Defect Type A*: Defect is within the substrate, not through-wall, and not expected to become through-wall within the lifetime of the repair system, requiring structural reinforcement only.

4.1 Phase 1: Preliminary assessment and data gathering

- Risk assessment: Conduct a thorough risk assessment considering the nature and location of the defect, operating conditions, repair lifetime, geometry of the substrate, hazards, availability of skilled personnel, ease of surface preparation, potential upset conditions, operational measures, failure modes, and repair system materials.

- Determine repair class: Based on the risk assessment, determine the appropriate repair class (Class 1, 2, or 3) according to the design pressure and temperature requirements.

- Define repair lifetime: Establish the required lifetime of the repair system, considering the defect type and service conditions. The minimum lifetime should be 2 years.

- Gather required data: Original equipment design data: Collect piping line lists, isometric drawings, operating mechanical loads, and original design calculations. Maintenance and Operational Histories: Document changes in service conditions, past service conditions, alterations, past repairs, and inspection reports. Service Condition Data: Define lifetime requirements, design and operating pressures/temperatures, expected future service conditions, and *MAWP* (if applicable). Repair System Qualification data: Obtain material documentation, design capability, surface preparation documentation, and short-term/long-term test data from the repair system supplier.

4.2 Phase 2: Design calculation and material selection

Select design method: For Defect Type A, choose *one* of the design methods:

- Include allowance for the substrate: Design based on substrate-allowable stress.

- Exclude allowance for the substrate: Design based on repair laminate allowable strains.

- Long-term performance test data: Design based on repair-allowable stresses determined by performance testing.

Environmental compatibility: Verify the suitability of the repair system for the service environment, considering external and internal factors. Address resistance to UV degradation and weathering. If compatibility is unknown, obtain supporting data from the repair system supplier or conduct specific environmental testing.

Design temperature effects: Consider the design temperature (T_d) and its effect on the repair system.

Perform design calculations:

If using substrate-allowable stress:

- Calculate the minimum repair laminate thickness ($t_{min,c}$) in the circumferential direction.

- Calculate the minimum repair laminate thickness ($t_{min,a}$) in the axial direction.

- The design repair thickness (t_{design}) is the maximum of $t_{min,c}$ and $t_{min,a}$.

If using repair laminate allowable strains:

- Calculate the minimum repair laminate thickness ($t_{min,c}$) in the circumferential direction.

- Calculate the minimum repair laminate thickness ($t_{min,a}$) in the axial direction.

- The design repair thickness (t_{design}) is the maximum of $t_{min,c}$ and $t_{min,a}$.

- Calculate allowable strains, considering thermal expansion coefficient differences.

- If using performance testing:

- Calculate the minimum repair laminate thickness (t_{min}) if the substrate allowance is not included.

- Calculate the design repair laminate thickness (t_{design}) if the substrate allowance is included.

Axial extent of repair: Determine the required overlay length for slot-type defects or for circular-type defects. Calculate the total axial length of the repair (l), considering the defect length and taper lengths. Verify that the axial extent is sufficient to transfer the applied axial load. If the required axial extent cannot be achieved, follow the procedures for limited axial extent, increasing the design repair thickness.

Optional design considerations: Consider impact resistance and cyclic loading if applicable. For cyclic loading, de-rate the allowable strains.

External loads: If external loads are a concern, calculate the minimum repair thickness to resist external pressure or vacuum and to resist soil loads.

4.3 Important considerations

- Material properties: Use qualified material properties from the repair system supplier.
- Safety factors: Ensure appropriate safety factors are applied in the design calculations.
- Limitations: Be aware of the limitations of the design methods and the repair system.
- Expertise: Consult with experienced composite repair specialists.
- Specific requirements: Always adhere to any specific requirements or recommendations from the repair system supplier.

Procedure for calculating the wall thickness for repairing a gas transmission pipeline:

- In circumferential direction, the minimum thickness of the GFRP, (in mm) due to internal pressure

$$t_{min,c} = \frac{D_e}{2\sigma_a} \bullet \left(\frac{E_s}{E_c}\right) \bullet \left(p_{eq} - p_s\right); \tag{1}$$

- In the axial direction, the minimum thickness of the GFRP, (in mm) due to internal pressure, bending and axial thrust

$$t_{min,a} = \frac{D_e}{2\sigma_a} \bullet \left(\frac{E_s}{E_a}\right) \bullet \left(\frac{2F_{eq}}{\pi D^2} - p_s\right); \tag{2}$$

- E_a is the axial modulus of the composite material in MPa;
- E_c is the circumferential modulus of the composite material, expressed in MPa;
- E_s is the modulus of elasticity of the steel, expressed in MPa;
- D_e is the outside diameter in mm;
- F_{eq} is the equivalent axial loading in N;
- σ_a is the allowable steel stress in MPa;
- p_{eq} is the internal working pressure in MPa;
- p_s is the maximum operating pressure in the presence of the defect, MPa

Case study for evaluating the thickness of GRRP composite materials used for repairing natural gas pipelines, depending on the mechanical characteristics, especially related to the modulus of elasticity of GFRP (Young modulus). Considering that the axial and bending forces are negligible compared to the action given by the

Pipe material	**Steel X52N *(API 5 L)* L360NB *(EN 10208–2)***					
Chemical composition	$\%C_{max}$	$\%Mn_{max}$	$\%P_{max}$	$\%S_{max}$	$\%Ti_{max}$	—
	0.22	1.40	0.025	0.015	0.04	Nb,V,Ti
Mechanical properties $R_{t0,5min}$ – yield stress $R_{m\ min}$ – ultimate tensile stress E_s – young's modulus	$R_{t0,5min}$ [MPa]	$R_{t0,5max}$ [MPa]	$R_{m\ min}$ [MPa]	$R_{m\ max}$ [MPa]	E_s [MPa]	
	360	531	455	758	$2.1 \cdot 10^5$	
Pipe dimensions D_e – diameter t – thickness	D_e [mm]			t [mm]		
	508			9.5		

Table 2.
Parameters of the pipeline under repair.

internal pressure (the natural gas transmission pipeline is characterized by the parameters specified in **Table 2** and **Figures 1 and 2**), formula (1) will be used to calculate the thickness of the repair shell. The wall thickness for different types of composite materials is shown in **Table 3**, and the wall thickness is calculated using a computer product made with a software called MathCad, presented below.

Input data

$$D_e := 508\ mm\ t := 9.5\ mm\ MAOP := 2\ MPa\ P_{eq} := 7.8\ MPa$$
$$\downarrow$$
$$R_{t0.5} := 360\frac{N}{mm^2}$$
$$\downarrow$$
$$\sigma_a := 0.72 \cdot R_{t0.5} = 259.20\frac{N}{mm^2}$$
$$\downarrow$$
$$E_S := 2.05 \cdot 10^5\frac{N}{mm^2}$$
$$\downarrow$$
$$d = \text{maximum depth of the corroded area}$$
$$\downarrow$$
$$\mathbf{Lm} = \text{measured axial length of the corroded area}$$
$$d := 7\ mm\ L_m := 150\ mm \tag{3}$$
$$\downarrow$$
$$E_{MCM} := \mathbf{22700}\ MPa$$
$$\downarrow$$
$$E_{Perma} := \mathbf{38000}\ MPa$$
$$\downarrow$$
$$E_{Fiba} := \mathbf{8700}\ MPa$$
$$\downarrow$$
$$E_{Clock} := \mathbf{34500}\ MPA$$
$$\downarrow$$
$$E_{TDW} := \mathbf{68000}\ MPa$$
$$\downarrow$$
$$E_{composite} := (\mathbf{E_{MCM}}\ \mathbf{E_{Perma}}\ \mathbf{E_{Fiba}}\ \mathbf{E_{Clock}}\ \mathbf{E_{TDW}})$$

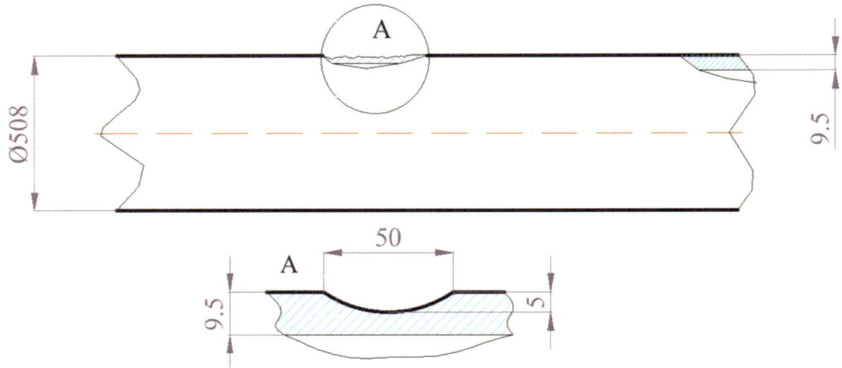

Figure 1.
Dimensional characteristics of the pipe and the defect to be repaired.

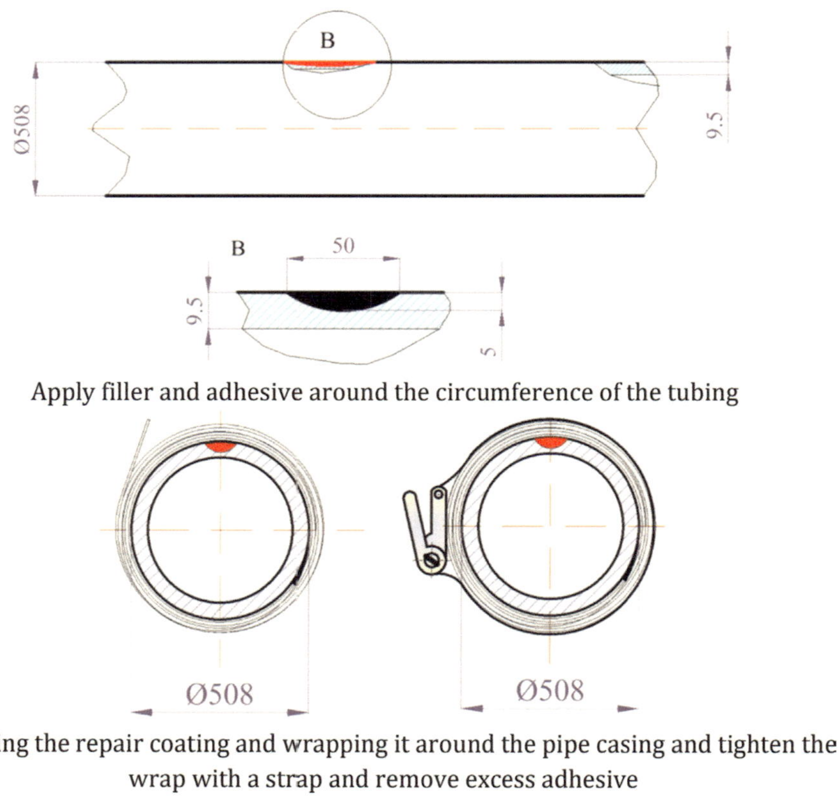

Figure 2.
Methodology for repairing a natural gas transmission pipeline with GFRP.

Formula for determining the thickness of the composite material

$$t_{\text{min\_stress}} = \frac{D_e}{2 \cdot \sigma_a} \cdot \frac{E_s}{E_c} \cdot \left(P_{eq} - p_s\right) \; t_{\text{min\_strain}} = \frac{1}{\varepsilon_c} \frac{P_{eq} \cdot D_e}{2 \cdot E_c} \tag{4}$$

Composite materials	Fiber type	Mechanical characteristics		
		E_C, GPa	R_{mC}, MPa	A_C, %
MCM	Fiberglass	17.5 ... 22.7	265 ... 315	1.32 ... 1.60
Perma wrap	Fiberglass	34.0 ... 38.0	580 ... 620	1.00 ... 1.10
Fiba roll	Fiberglass	7.9 ... 8.7	86 ... 72	2.60 ... 3.10
Clock spring	Fiberglass	33.8 ... 34.5	630 ... 650	1.06 ... 1.36
TDW RES-Q Wrap	Carbon fiber	68.8	1028	—

Table 3.
Characteristics of the composite material used in the repair piping.

Determination of the maximum operating pressure of the pipeline with defect – ASME B31.G

$$A := 0.893 \cdot \frac{L_m}{\sqrt{d \cdot t}} (A) = 16.43$$

$$P_{S\_1} := \text{if}\left[A \le 4{,}1.1\text{MAOP} \cdot \left[\frac{1-\frac{2}{3}\cdot\left(\frac{d}{t}\right)}{1-\frac{2}{3}\cdot\left(\frac{d}{t\cdot\sqrt{A^2+1}}\right)}\right], 1.1\cdot\text{MAOP}\cdot\left(1-\frac{d}{t}\right)\right] P_{S\_} = 0.58\ \text{MPa} \tag{5}$$

$$\lambda := \frac{1.2855 \cdot L_m}{\sqrt{2\cdot\frac{\mathrm{D_e}-2\cdot \mathrm{t}}{2}\cdot \mathrm{t}}} \lambda = 2.83$$

$$\left(M_t := \sqrt{\frac{1.02 + 0.441\cdot\lambda^2 + 0.006124\lambda^4}{1+0.02642\cdot\lambda^2 + 1.533\cdot 10^{-6}\cdot\lambda^4}}\right)$$

$$(M_t) = 2.02\text{RSF} := \frac{1-\frac{d}{t}}{1-\frac{1}{M_t}\cdot\frac{d}{t}} RSF = 0.41$$

$$\mathrm{P_{S\_2}} := \mathrm{RSF}\cdot\mathrm{MAOPP_{S\_2}} = 0.83\ \mathrm{MPa}$$

$$\mathbf{p_S} := min\left(\mathbf{p_{S\_1}}, \mathbf{p_{S\_2}}\right)\mathbf{p_S} = 0.58\ \mathrm{MPa}$$

$$t_{\text{min\_stress}}\left(\mathbf{E}_{\text{composite}}\right) := \frac{D_e}{2\cdot\sigma_a}\cdot\frac{\mathbf{E}_s}{\mathbf{E}_{\text{composite}}}\cdot\left(\mathbf{P}_{\text{eq}} - \mathbf{p}_{\text{S}}\right)$$

$$t_{\text{min\_stress}}\left(\mathbf{E}_{\text{composite}}\right)^T = \begin{pmatrix} 63.90 \\ 38.17 \\ 166.74 \\ 42.05 \\ 21.33 \end{pmatrix}$$

$t_{\text{min\_stress}}(\mathbf{E_{MCM}}) = 63.90$ mm
...
$t_{min\_stress}(\mathbf{E_{Perma}}) = 38.17$ mm
...
$t_{\text{min\_stress}}(\mathbf{E_{Fiba}}) = 166.74$ mm
...
$t_{\text{min\_stress}}(\mathbf{E_{Clock}}) = 42.05$ mm
...
$t_{\text{min\_stress}}(\mathbf{E_{TDW}}) = 21.33$ mm

$$t_{\text{min\_strain}}(E_{\text{composite}}) := \frac{1}{\varepsilon_c} \frac{P_{eq} \cdot D_e}{2 \cdot E_{\text{composite}}}$$

$$t_{\text{min\_strain}}(E_{\text{composite}})^T = \begin{pmatrix} 21.82 \\ 13.03 \\ 56.93 \\ 14.36 \\ 7.28 \end{pmatrix} \quad \begin{matrix} t_{\text{min\_strain}}(E_{MCM}) = 21.82 \text{ mm} \\ \ldots \\ t_{\text{min\_strain}}(E_{\text{Perma}}) = 13.03 \text{ mm} \\ \ldots \\ t_{\text{min\_strain}}(E_{\text{Fiba}}) = 56.93 \text{ mm} \\ \ldots \\ t_{\text{min\_strain}}(E_{\text{Clock}}) = 14.36 \text{ mm} \\ \ldots \\ t_{\text{min\_strain}}(E_{\text{TDW}}) = 7.28 \text{ mm} \end{matrix}$$

$$t_{min}(E_{MCM}) := max\,(t_{\text{min\_stress}}(E_{MCM}), t_{\text{min\_strain}}(E_{MCM}))$$

$$\downarrow$$

$$t_{min}(E_{MCM}) = 63.90 \text{ mm}$$

$$\ldots$$

$$t_{min}(E_{\text{Perma}}) := max\,(t_{\text{min\_stress}}(E_{\text{Perma}}), t_{\text{min\_strain}}(E_{\text{Perma}}))$$

$$\downarrow$$

$$t_{\text{min}}(E_{\text{Perma}}) = 38.17 \text{ mm}$$

$$\ldots$$

$$t_{min}(E_{\text{Fiba}}) := max\,(t_{\text{min\_stress}}(E_{\text{Fiba}}), t_{\text{min\_strain}}(E_{\text{Fiba}}))$$

$$\downarrow$$

$$t_{min}(E_{\text{Fiba}}) = 166.74 \text{ mm}$$

$$\ldots$$

$$t_{\text{min}}(E_{\text{Clock}}) := max\,(t_{\text{min\_stress}}(E_{\text{Clock}}), t_{\text{min\_strain}}(E_{\text{Clock}}))$$

$$\downarrow$$

$$t_{\text{min}}(E_{\text{Clock}}) = 42.05 \text{ mm}$$

$$\ldots$$

$$t_{\text{min}}(E_{\text{TDW}}) := max\,(t_{\text{min\_stress}}(E_{\text{TDW}}), t_{\text{min\_strain}}(E_{\text{TDW}}))$$

$$\downarrow$$

$$t_{\text{min}}(E_{\text{TDW}}) = 21.33 \text{ mm}$$

Minimum width of the GFRP material

$$L_{c\_\text{cover}} := 2.5 \cdot \sqrt{\frac{D_e \cdot t}{2}} \quad L_{c\_cover} = 122.81 \text{ mm} \tag{6}$$

$$L_{-\text{defect}} := L_m = 150.00 \text{ mm}$$

$$L_{\text{composite}} := 2 \cdot L_{c\_}\text{cover} + L_m$$

$$\mathbf{L_{composite} = 395.61\ mm}$$

Number of layers of GFRP composite material thickness of a layer material $t_{\text{layer}} := 0.7\ mm$

$$
\begin{array}{ll}
N_{layer}(E_{MCM}) := \dfrac{t_{min}(E_{MCM})}{t_{layer}} & \\
\downarrow & N_{layem}(E_{TDW}) := \dfrac{t_{min}(E_{TDW})}{t_{layer}} \\
\mathbf{N_{layer}(E_{MCM}) = 91.29} & \\
\ldots & \downarrow \\
N_{layer}(E_{Perma}) := \dfrac{t_{min}(E_{Perma})}{t_{layer}} & \mathbf{N_{layer}(E_{TDW}) = 30.48} \\
\downarrow & \ldots \\
\mathbf{N_{layer}(E_{Perma}) = 54.53} & N_{layepl}(E_{Clock}) := \dfrac{t_{min}(E_{Clock})}{t_{layer}} \\
\ldots & \\
N_{layem}(E_{Fiba}) := \dfrac{t_{min}(E_{Fiba})}{t_{layer}} & \downarrow \\
\downarrow & \mathbf{N_{layer}(E_{Clock}) = 60.07} \\
\mathbf{N_{layer}(E_{Fiba}) = 238.20} &
\end{array}
\tag{7}
$$

5. Numerical simulation of a GFRP repair of a defective pipe

- *Finite element analysis (FEA)* is a crucial tool in evaluating the performance and integrity of pipelines, especially when assessing repairs using advanced materials like Glass Fiber Reinforced Polymer (GFRP) composites. When a pipeline has a defect—such as corrosion, dents, or cracks—it is essential to ensure that any repair method restores the pipeline's structural integrity and extends its service life. GFRP composites are increasingly popular for pipeline repairs due to their strong mechanical properties, corrosion resistance, and ease of application. In an FEA analysis involving a pipeline with a defect repaired using GFRP, the primary objectives are to assess the effectiveness of the composite repair in restoring the pipeline's load-bearing capacity and to predict the repaired system's behavior under various operational conditions.

- *Modeling geometry – see* **Figure 3**: Creating a detailed 3D model of the pipeline section, including the defect and the composite repair layer. This model helps simulate real-life conditions accurately.

- *Material properties*: Defining the mechanical properties of both the pipeline material and the GFRP composite. This includes properties like modulus of elasticity, Poisson's ratio, and tensile strength, which are crucial for accurate simulation results.

- *Load application*: Applying operational loads and conditions to the model, such as internal pressure, temperature variations, and external forces, to understand their effects on both the pipeline and the repair.

- *Boundary conditions*: Setting appropriate constraints and interactions within the model, such as the contact between the pipeline and the composite layer.

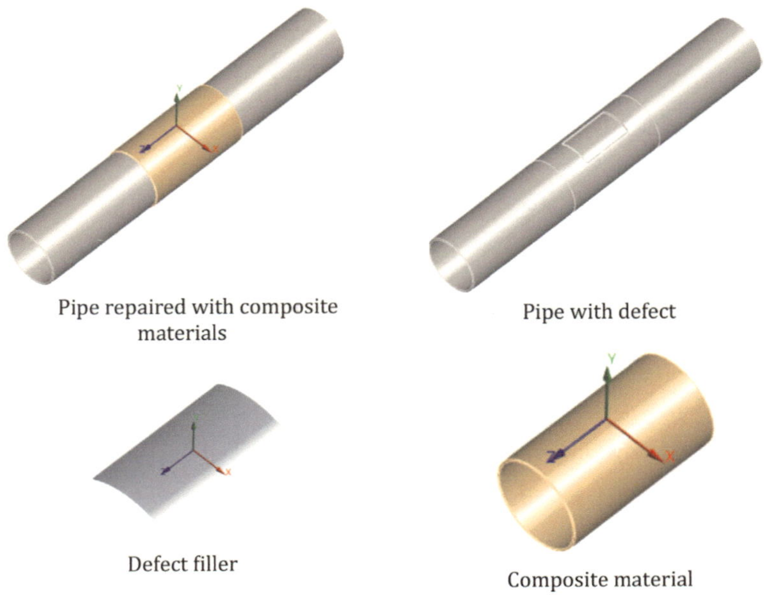

Figure 3.
Geometric modeling used in FEA analysis.

- *Simulation and analysis – see* **Figure 4**: Running the FEA to observe stress distribution, deformation patterns, and potential failure points. This helps in understanding how well the composite repair can withstand operational demands.

Through FEA, we can effectively predict how the GFRP repair will perform, aiding in decision-making and ensuring that the repaired pipeline section meets safety and reliability standards. This predictive capability is invaluable for extending the lifespan of pipelines and minimizing the risk of failure, ultimately protecting both infrastructure and the environment. The use of FEA allows successive simulations so that the quality of the repair of a natural gas pipeline with GFRP composite materials (see **Table 3**) can be evaluated, and the results are presented in **Table 4**.

Table 4 presents the results obtained by varying the depth of the defect *($d/t = 0.3$; $d/t = 0.4$; $d/t = 0.5$; $d/t = 0.6$; $d/t = 0.7$)* so that the stress state in the repaired pipeline is below 290 MPa, considering a steel with a yield strength of 360 MPa, but taking into account the safety factor and the reduction coefficient related to the fact that the pipeline is longitudinally welded.

6. Carrying out an experimental test for a pipe repaired with GFRP material

Carrying out an experimental test on a pipe repaired with GFRP material involves several important steps. First, the type of pipe should be selected, determining its original condition before repairing. Next, the appropriate GFRP repair technique, such as patching or wrapping, should be chosen based on the observed damage. After

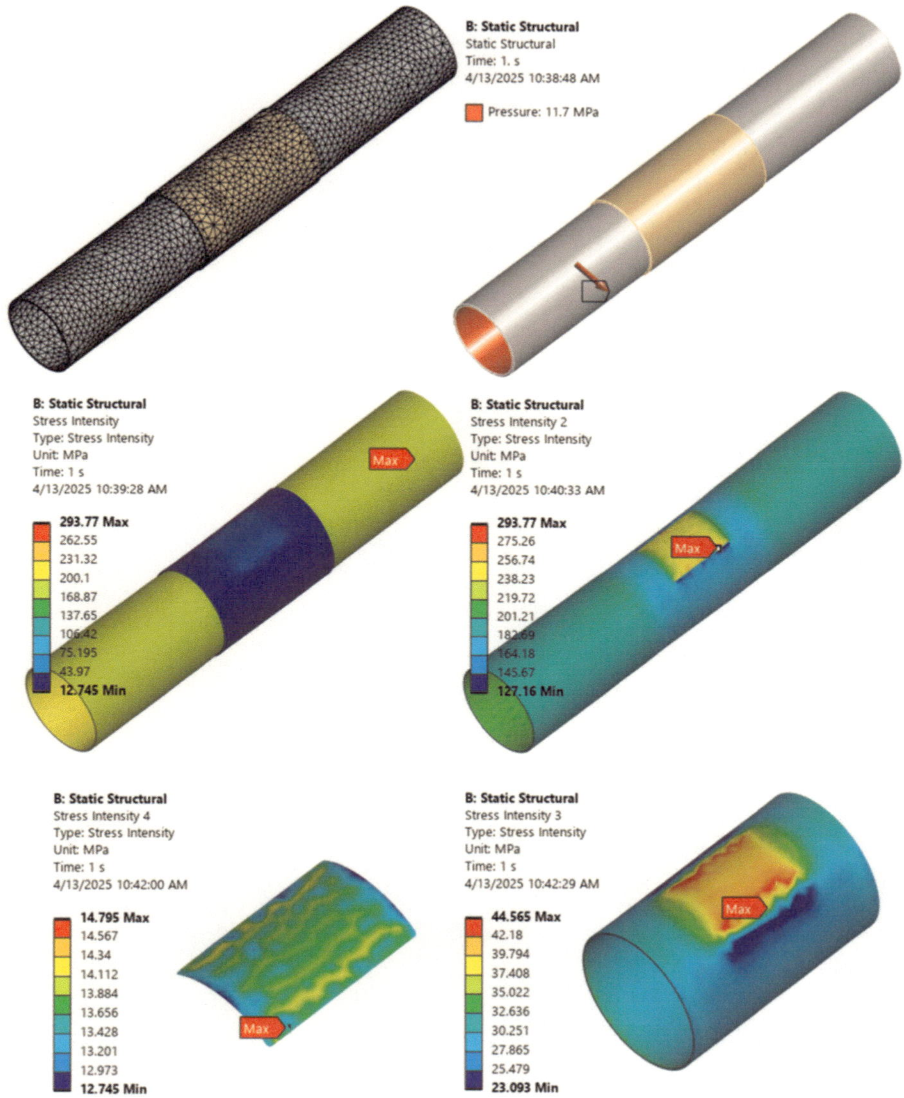

Figure 4.
Result obtained though FEA.

preparation, the GFRP material can be applied over the damaged area according to the manufacturer's instructions, which may involve layers of fabric and resin followed by a curing process (the pipe subjected to analysis is shown in **Figure 5**). With the GFRP applied, the next step is to set up the testing environment. The pipe should be equipped with the necessary sensors and gauges to measure key parameters such as pressure, temperature, and strain. When ready to conduct the test, pressure within the pipe should be gradually increased to a predetermined level while monitoring signs of leaks or failures (see **Figure 6**).

The probe was subjected to testing on a stand using a pump which can achieve pressures of up to 100 MPa, the failure of pipe occurred due to the rupture of the pipe in the area of defect, accompanied by the bursting of the composite coating applied to

		MCM									
d	d/t	t_{MCM} min	Stress intensity	t_{MCM} min	Stress intensity	t_{MCM} min	Stress intensity	t_{MCM} min	Stress intensity	t_{MCM} min	Stress intensity
mm	%	mm	MPa	mm	MPa	mm	MPa	mm	MPa	mm	MPa
2.85	30%	**10**	**244**	**10**	**244**	**10**	**244**	**10**	**244**	**10**	**244**
3.80	40%	**10**	**275**	**10**	**275**	**10**	**275**	**10**	**275**	**10**	**275**
4.75	50%	10	319 > 290	**15**	**283**	**15**	**283**	**15**	**283**	**20**	**281**
5.70	60%	10	376 > 290	15	327 > 290	20	293 > 290	**25**	**268**	**25**	**268**
6.65	70%	10	420 > 290	15	364 > 290	20	324 > 290	25	292 > 290	**30**	**265**

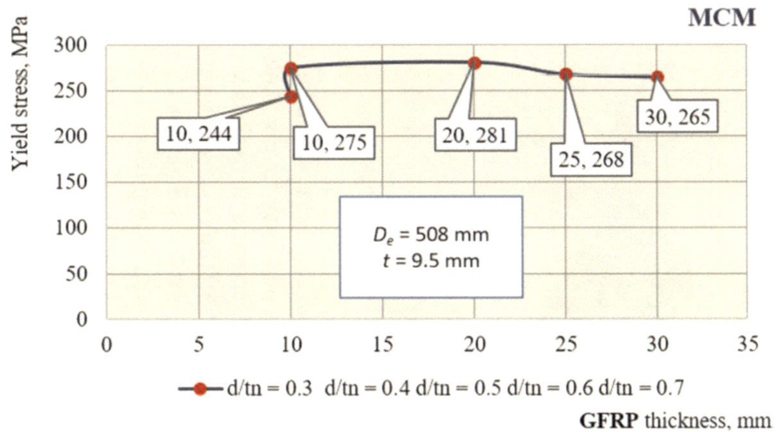

Perma wrap

d	d/t	t_{perma} min	Stress intensity	t_{perma} min	Stress intensity	t_{perma} min	Stress intensity	t_{perma} min	Stress intensity	t_{perma} min	Stress intensity
mm	%	mm	MPa	mm	MPa	mm	MPa	mm	MPa	mm	MPa
2.85	30%	**10**	**260**	**10**	**260**	**10**	**260**	**10**	**260**	**10**	**260**
3.80	40%	10	294 > 290	**15**	**269**	**15**	**269**	**15**	**269**	**15**	**269**
4.75	50%	10	346 > 290	15	298 > 290	**20**	**281**	**20**	**281**	**20**	**281**
5.70	60%	10	409 > 290	15	350 > 290	20	330 > 290	**30**	**285**	**30**	**285**
6.65	70%	10	460 > 290	15	433 > 290	20	368 > 290	30	330 > 290	**40**	**274**

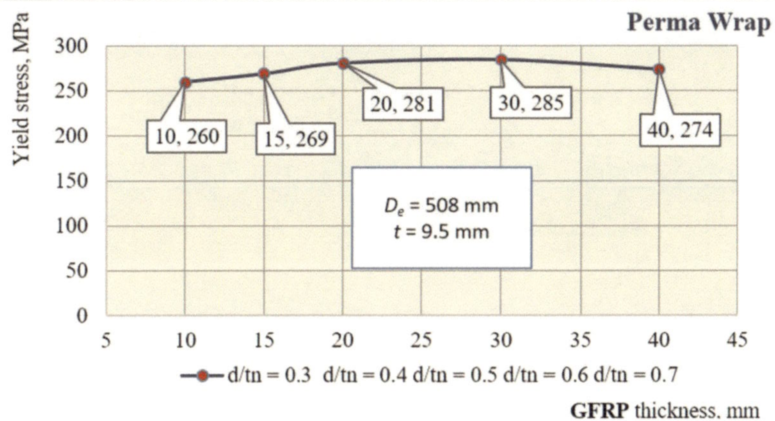

Fiba roll

d	d/t	t_{fiba} min	Stress intensity	t_{fiba} min	Stress intensity	t_{fiba} min	Stress intensity	t_{fiba} min	Stress intensity
mm	%	mm	MPa	mm	MPa	mm	MPa	mm	MPa
2.85	30%	**70**	**265**	**70**	**265**	**70**	**265**	**70**	**265**
3.80	40%	**80**	**287**	**80**	**287**	**80**	**287**	**80**	**287**
4.75	50%	80	364 > 290	**120**	**276**	**120**	**276**	**120**	**276**
5.70	60%	80	472 > 290	120	365 > 290	**150**	**280**	**150**	**280**
6.65	70%	80	531 > 290	120	416 > 290	150	341 > 290	**180**	**272**

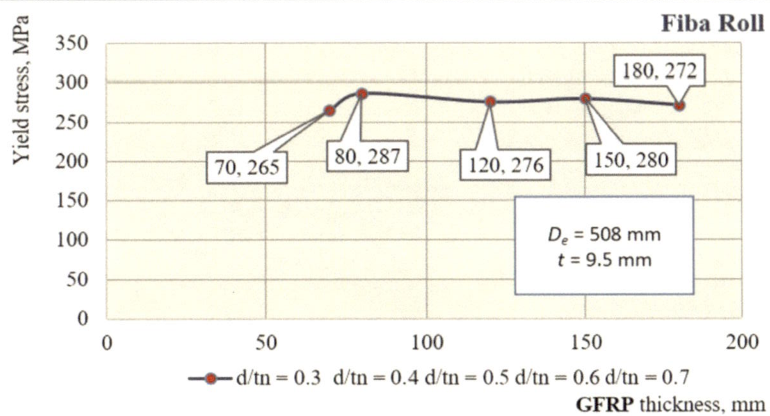

Clock spring

d	d/t	t_{clock} min	Stress intensity	t_{clock} min	Stress intensity	t_{clock} min	Stress intensity	t_{clock} min	Stress intensity	t_{clock} min	Stress intensity
mm	%	mm	MPa	mm	MPa	mm	MPa	mm	MPa	mm	MPa
2.85	30%	**10**	**264**	**10**	**264**	**10**	**264**	**10**	**264**	**10**	**264**
3.80	40%	10	301 > 290	**15**	**275**	**15**	**275**	**15**	**275**	**15**	**275**
4.75	50%	10	353 > 290	15	320 > 290	20	295 > 290	**30**	**259**	**30**	**259**
5.70	60%	10	418 > 290	15	372 > 290	20	339 > 290	30	294 > 290	**40**	**263**
6.65	70%	10	472 > 290	15	415 > 290	20	378 > 290	30	330 > 290	**40**	**282**

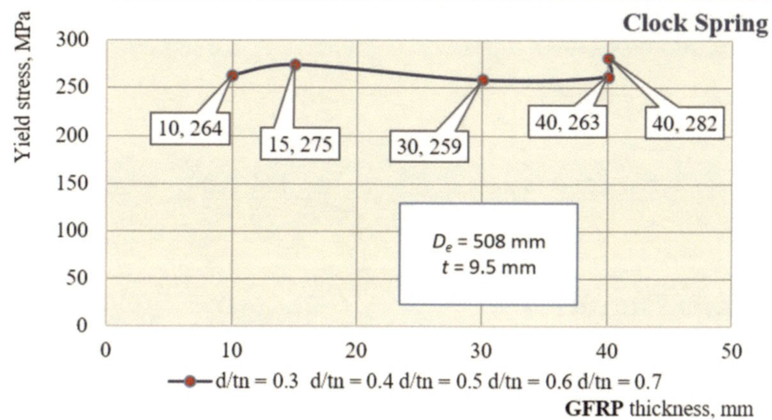

		TDW RES-Q wrap					
d	d/t	t_{TDW} min	Stress intensity	t_{TDW} min	Stress intensity	t_{TDW} min	Stress intensity
mm	%	mm	MPa	mm	MPa	mm	MPa
2.85	30%	**10**	**220**	**10**	**220**	**10**	**220**
3.80	40%	**10**	**244**	**10**	**244**	**10**	**244**
4.75	50%	**10**	**280**	**10**	**280**	**10**	**280**
5.70	60%	10	328 > 290	**15**	**279**	**15**	**279**
6.65	70%	10	361 > 290	15	308 > 290	**20**	**267**

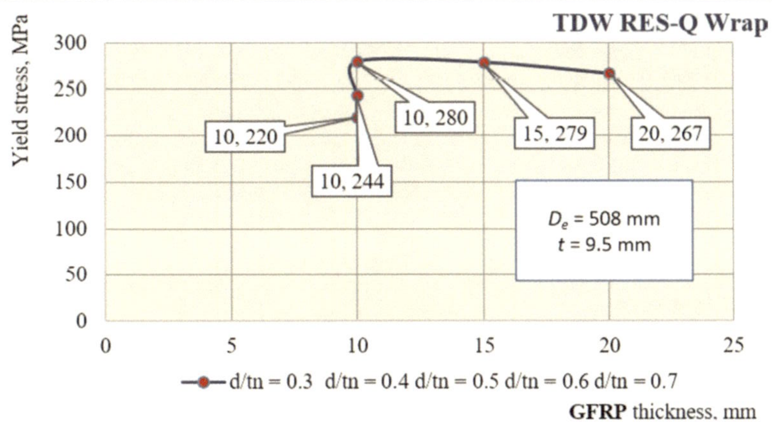

The bolding is necessary to identify the crossing of the limit imposed by the value of 290 MPa, it can be omitted because, in the other cases, it was passed as a mathematical relationship (example: 319 > 290)

Table 4.
FEA results obtained by varying the depth of the defect and different composite thickness.

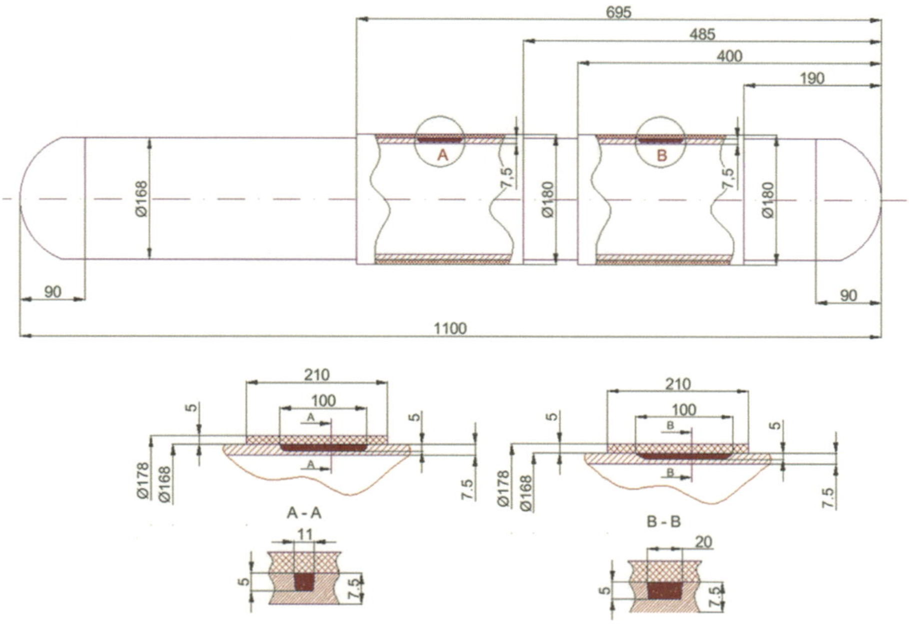

Figure 5.
Pipe with local defects repaired with composite material coatings.

repair this defect, with the propagation of the rupture reaching the area of the second composite coating (applied to repair defect 2) and stopping in that area.

The pressure at which failure occurred was approximately 31.5 MPa. The examination of the appearance of probe after failure suggested that the composite coatings used for repairing local surface defects of the "material loss" type can also function as rupture stoppers.

7. Discussion

The integration of analytical calculations, FEA simulations, and experimental testing provides a comprehensive understanding of composite material repairs.

7.1 Analytical calculations

Guided by standards like ISO 24817, these calculations relieve a combination of material and operational data. Some common input parameters include pipe diameter (D_e = 508 mm), wall thickness (t = 9.5 mm), maximum allowable operating pressure (MAOP = 2 MPa), steel yield strength, allowable stress in the steel, maximum depth of corroded area (d = 7 mm), and measured axial length of corroded area (L_m = 150 mm). Typical outputs of the composite analytical include the following:

- *MCM*: Requires 91 layers, resulting in an approximate thickness of 63.9 mm.

- *Perma Wrap*: Requires 54 layers, resulting in an approximate thickness of 38.17 mm.

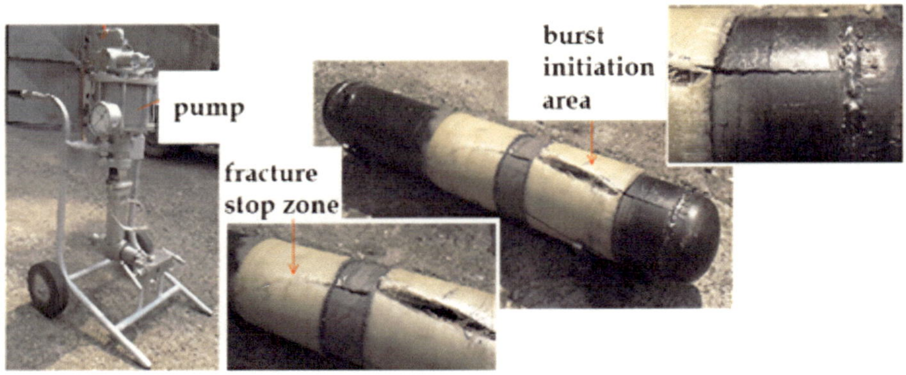

Figure 6.
Experimental results.

- *Fiba Roll*: Requires 238 layers, resulting in an approximate thickness of 166.74 mm.
- *Clock Spring*: Requires 60 layers, resulting in an approximate thickness of 42.05 mm.
- *TDW*: Requires 30 layers, resulting in an approximate thickness of 21.33 mm.

7.2 FEA results

Finite element analysis provides a detailed visualization of stress distribution and deformation patterns within the repaired structure. The FEA reveals what would be the minimum thickness for each material to reach or exceed the stress level of 290 MPa (considering a steel with a yield strength of 360 MPa).

7.2.1 MCM composite

- For a *d/t* of 30% and 40%, a composite thickness of 10 mm is sufficient to bring the stress state of the pipe into the acceptable range (244 MPa and 275 MPa).
- For a *d/t* of 50%, a composite thickness of 15 mm is sufficient to bring the stress state of the pipe into the acceptable range (283 MPa).
- For a *d/t* of 60%, a composite thickness of 25 mm is sufficient to bring the stress state of the pipe into the acceptable range (268 MPa).
- For a *d/t* of 70%, a composite thickness of 30 mm is sufficient to bring the stress state of the pipe into the acceptable range (265 MPa).

7.2.2 Perma wrap composite

- For a *d/t* of 30%, a composite thickness of 10 mm is sufficient to bring the stress state of the pipe into the acceptable range (260 MPa).

- For a *d/t* of 40%, a composite thickness of 15 mm is sufficient to bring the stress state of the pipe into the acceptable range (269 MPa).
- For a *d/t* of 50%, a composite thickness of 20 mm is sufficient to bring the stress state of the pipe into the acceptable range (281 MPa).
- For a *d/t* of 60%, a composite thickness of 30 mm is sufficient to bring the stress state of the pipe into the acceptable range (285 MPa).
- For a *d/t* of 70%, a composite thickness of 40 mm is sufficient to bring the stress state of the pipe into the acceptable range (274 MPa).

7.2.3 Fiba roll composite

- For a *d/t* of 30%, a composite thickness of 70 mm is sufficient to bring the stress state of the pipe into the acceptable range (265 MPa).
- For a *d/t* of 40%, a composite thickness of 80 mm is sufficient to bring the stress state of the pipe into the acceptable range (287 MPa).
- For a *d/t* of 50%, a composite thickness of 120 mm is sufficient to bring the stress state of the pipe into the acceptable range (276 MPa).
- For a *d/t* of 60%, a composite thickness of 150 mm is sufficient to bring the stress state of the pipe into the acceptable range (280 MPa).
- For a *d/t* of 70%, a composite thickness of 180 mm is sufficient to bring the stress state of the pipe into the acceptable range (272 MPa).

7.2.4 Clock spring composite

- For a *d/t* of 30%, a composite thickness of 10 mm is sufficient to bring the stress state of the pipe into the acceptable range (264 MPa).
- For a *d/t* of 40%, a composite thickness of 15 mm is sufficient to bring the stress state of the pipe into the acceptable range (275 MPa).
- For a *d/t* of 50%, a composite thickness of 30 mm is sufficient to bring the stress state of the pipe into the acceptable range (259 MPa).
- For a *d/t* of 60% and 70%, a composite thickness of 40 mm is sufficient to bring the stress state of the pipe into the acceptable range (263 MPa and 282 MPa).

7.2.5 TDW composite

- For a *d/t* of 30%, 40% and 50%, a composite thickness of 10 mm is sufficient to bring the stress state of the pipe into the acceptable range (220 MPa, 244 MPa and 280 MPa).

- For a *d/t* of 60%, a composite thickness of 15 mm is sufficient to bring the stress state of the pipe into the acceptable range (279 MPa).
- For a *d/t* of 70%, a composite thickness of 20 mm is sufficient to bring the stress state of the pipe into the acceptable range (267 MPa).

7.3 Experimental results

Experimental tests serve as the ultimate validation of the repair's effectiveness. The experimental tests showed that the composite repairs withstood pressures significantly exceeding the design limits. The "rupture stopper" effect was observed in experimental tests, with failures occurring at pressures of approximately 31.5 MPa (315 bar), which is significantly higher than the maximum operating pressure.

The convergence of these three methodologies presents a robust approach to designing and implementing composite repairs for industrial equipment, ensuring both performance and safety.

8. Conclusions

Glass fiber-reinforced polymer (GFRP) is cost-effective with good mechanical and chemical resistance, while Carbon Fiber-Reinforced Polymer (CFRP) offers superior strength and stiffness but at a higher cost. The ISO 24817 standard provides requirements for composite repairs to pipework in the petroleum, petrochemical, and natural gas sectors, covering qualification, design, installation, testing, and inspection.

Design calculations for composite repairs include considerations for substrate-allowable stress, repair laminate allowable strains, and long-term performance test data. The minimum thickness of the GFRP composite material is calculated based on circumferential and axial directions, considering factors like modulus of elasticity, allowable steel stress, and internal working pressure.

Finite element analysis (FEA) is crucial for evaluating the performance and integrity of pipelines, especially when assessing repairs using GFRP composites. FEA involves modeling geometry, defining material properties, applying loads, setting boundary conditions, and analyzing stress distribution and deformation patterns. FEA simulations help evaluate the stress state in the repaired pipeline by varying the depth of the defect, ensuring it remains below a certain threshold (e.g., 290 MPa).

The number of layers of composite material needed for repair is determined based on the minimum thickness required and the thickness of each layer, with specific calculations provided for different composite materials.

The effective repair of pipelines with defects using GFRP composites requires a comprehensive approach that adheres to established standards (like ISO 24817) and involves detailed engineering analysis. The selection of the composite material depends on the specific application requirements, balancing cost with necessary mechanical properties.

Accurate calculations, often aided by FEA simulations, are essential to determine the appropriate repair design, including the number of composite layers and their dimensions, ensuring the repaired pipeline meets safety and operational standards for its intended lifespan.

The minimum required thickness of GFRP composite repairs varies significantly based on material properties. For instance, MCM requires 91 layers (∼63.9 mm), while TDW needs only 30 layers (∼21.33 mm) for the same pipeline defect.

Design calculations (guided by ISO 24817) account for operational pressures, defect geometry (e.g., depth-to-thickness ratio, d/t), and material properties (e.g., circumferential vs. axial elasticity). Key formulas determine thickness in both circumferential and axial directions.

Material selection (GFRP vs. CFRP) balances cost, mechanical properties (tensile strength, corrosion resistance), and environmental compatibility, with GFRP being cost-effective for general applications and CFRP preferred for high-performance demands.

Finite element analysis validates that composite repairs restore pipeline integrity by maintaining stress levels below the yield threshold (<290 MPa for steel with *yield stress* = 360 MPa). For example, a 15 mm MCM repair reduces stress to 283 MPa for a 50% d/t defect.

The required repair thickness increases with defect severity (Fiba Roll needs 180 mm for d/t = 70% vs. 70 mm for d/t = 30%). TDW (carbon fiber) performs best, requiring minimal thickness (10–20 mm) across all defect depths.

FEA simulations highlight stress concentration zones and predict failure modes, ensuring compliance with safety standards (e.g., ISO 24817) and optimizing repair designs for diverse operational conditions.

Experimental tests confirm GFRP repairs withstand pressures far exceeding design limits (failure at ∼31.5 MPa vs. MAOP of 2 MPa), demonstrating their effectiveness as "rupture stoppers" to prevent catastrophic pipeline failures.

Composite coatings not only reinforce defects but also arrest crack propagation. For example, a rupture initiated at Defect 1 stopped at Defect 2's GFRP coating, proving their dual role as reinforcements and failure inhibitors.

The alignment of experimental results with analytical and FEA predictions (e.g., stress thresholds, layer counts) validates the reliability of composite repair systems for industrial applications, ensuring long-term safety and performance.

The integration of *analytical* (ISO 24817-guided calculations), *numerical* (FEA stress simulations), and *experimental* (pressure testing) methodologies provide a robust framework for designing and validating composite repairs, ensuring they meet mechanical, safety, and operational requirements.

Author details

Alin Dinita* and Maria Tanase
Petroleum-Gas University of Ploiesti, Ploiesti, Romania

*Address all correspondence to: adinita@upg-ploiesti.ro

References

[1] Zhang H, Lan HQ. A review of internal corrosion mechanisms and experimental study for pipelines based on multiphase flow. Corrosion Reviews. 2017;**35**:425-444

[2] Mirshamsi M, Rafeeyan M. Speed control of inspection pig in gas pipelines using sliding mode control. Journal of Process Control. 2019;**77**: 134-140

[3] Xie M, Tian Z. A review on pipeline integrity management utilizing in-line inspection data. Engineering Failure Analysis. 2018;**92**:222-239

[4] Bhadran V, Shukla A, Karki H. Non-contact flaw detection and condition monitoring of subsurface metallic pipelines using magnetometric method. Materials Today Proceedings. 2020;**28**: 860-864

[5] ASME. B31G Manual for Determining the Remaining Strength of Corroded Pipelines. New York, NY, USA: The American Society of Mechanical Engineers;

[6] DNV. RP-F101 Corroded Pipelines. Recommended Practice. Hovik, Norway: Det Norske Veritas A.S;

[7] ISO. TS 24817 Petroleum, Petrochemical and Natural Gas Industries—Composite Repairs for Pipework—Qualification and Design, Installation, Testing and Inspection. Geneva, Switzerland: Technical Specification; International Organization for Standardization;

[8] ISO. ISO 3183 Petroleum and Natural Gas Industries—Steel Pipe for Pipeline Transportation Systems. Geneva, Switzerland: International Organization for Standardization;

[9] ASME. B31.8 Gas Transmission and Distribution Piping Systems. New York, NY, USA: The American Society of Mechanical Engineers;

[10] Lowea PS, Sandersonb R, Pedrama SK, Boulgourisb NV, Mudgeb P. Inspection of pipelines using the first longitudinal guided wave mode. Physics Procedia. 2015;**70**:338-342

[11] El Mountassir M, Mourot G, Yaacoubi S, Maquin D. Damage detection and localization in pipeline using sparse estimation of ultrasonic guided waves signals. IFAC PapersOnLine. 2018;**51–24**:941-948

[12] Engineering (OMAE 2019), Glasgow, Scotland, UK, 9–14 June 2019. Paper No. 96279

[13] Valadi Z, Bayesteh H, Mohammadi S. XFEM fracture analysis of cracked pipeline with and without FRP composite repairs. Mechanics of Advanced Materials and Structures. 2020;**27**:1888-1899

[14] Abd-Elhady A, Sallam HE, Alarifi I, Malik R, El-Bagory T. Investigations of fatigue crack propagation in steel pipeline repaired by glass fiber reinforced polymer. Composite Structures. 2020;**242**:123-130

[15] Hocine A, Achira FSK, Habbar G, Levent A, Medjdoub SM, Maizia A, et al. Structural integrity assessment of corroded pipelines repaired with composite materials – Literature review. International Journal of Pressure Vessels and Piping. 2024;**210**:105253

[16] Arifin HH, Noor NM, Yahya N, Lim KS, Shah MSA, Umar S, et al. Influence of defect geometry on putty

performance in pipeline composite repair assessments. International Journal of Pressure Vessels and Piping. 2024; **209**:105190

[17] Andrade DM, Areias BS, Costa MLM. Felipe Bastos de Freitas Rachid, Heraldo da Costa Mattos, failure of composite repair systems used in corroded pipelines due to pressure transients. Engineering Failure Analysis. 2025;**171**: 109368

Chapter 3

Damage Mechanisms and Mechanical Behavior of Epoxy Composites after Low Velocity Impact

Marina Bunea

Abstract

This study provides a comprehensive review of existing research on the low velocity impact response, post-impact mechanical strength, and damage mechanisms of epoxy hybrid composites reinforced with fibers or fabrics. It focuses on the investigation of the influence of various factors on the impact damage mechanisms, such as matrix cracking, delamination, fiber breakage, and fiber pullout, which significantly reduce the residual mechanical properties of these composites. Herein, it is discussed the ways of optimizing impact response, post-impact mechanical properties, and resistance to damage by hybridization, incorporation of fillers into epoxy matrix, and insertion of metal wires between certain layers. This study reveals that the impact energy level, impact angle, impactor size, and shape significantly affect damage severity and energy absorption. The hybrid composites with specific fiber combinations and ply orientations can enhance impact resistance and post-impact mechanical strength. The combination of different fibers can offer tailored properties and improved impact behavior compared to pure fabric composites. By adding fillers to epoxy matrix and metal wires between certain plies, it can reduce residual deformations and damage degree. The natural fiber-reinforced epoxy composites exhibit high impact strength due to their fiber structure and can be a sustainable alternative to synthetic fibers.

Keywords: epoxy composites, hybrid composites, woven fabrics, impact response, damage mechanisms, residual mechanical properties

1. Introduction

Recently, the composite materials have gained widespread use in all industrial sectors. Among these, the composite materials with epoxy matrix are the most popular due to their high strength and favorable weight-to-cost ratio as high-performance materials, which allow the construction of lightweight and resistant structures and can be adapted to a wide range of applications [1]. As a result, they have surpassed the metallic materials in applications such as automotive, aircraft, shipbuilding, sports equipment, wind turbines, printed circuit boards, and so on [2]. Other advantages of using epoxy composite materials include their high adhesion to all types of fibers and fillers [3, 4], as well as simple formation techniques. For example, the composites with fillers can be blended using a mixer, while materials reinforced with long fibers or fabrics can be formed using the wet-hand lay-up technique in comfortable room conditions [5]. However, the epoxy composites, compared to thermoplastic composites, have a longer polymerization time, are not recyclable, and are difficult to repair.

Generally, the epoxy composites reinforced with fibers demonstrate an exceptional balance of fiber flexibility and tensile strength with matrix rigidity and mechanical resistance, which significantly contribute to their low velocity impact response and residual mechanical properties. The properties of these types of composites can be customized in dependence of functional fibers used as reinforcement. The most commonly used reinforcements for composites are synthetic fibers such as carbon, glass, and aramid. So, the carbon fibers exhibit high tensile strength, high modulus, high rigidity, and good thermal and electrical conductivity, but unfortunately low impact resistance due to high brittleness and low energy absorption [6]. Also, the rigidity of these fibers can lead to the occurrence of voids, leading to a poor quality of matrix-fiber interfaces and interfaces between adjacent plies, especially when the adjacent plies are oriented at different angles. In **Figure 1** are plotted the microscopic images of a carbon fabric-reinforced epoxy composite with ply orientation at various angles, where it can be observed the voids occurring between plies (**Figure 1a**) and their removal by addition of fillers into the epoxy matrix (**Figure 1b**).

The glass fibers show higher flexibility as compared to carbon fibers, which is important for impact energy absorption, high tensile strength, high corrosion resistance, electrical and thermal insulation, and low cost [7]. The aramid fibers present the highest flexibility, impact performance, very high tensile strength, non-flammability, and electrical and thermal insulation, but exhibit low compression strength and sensibility to UV rays and moisture [8].

So, in dependence of exploitation conditions (temperature, humidity, acidic environment, cleanliness, vibration, mechanical shock, etc.), the suitable fibers or combination of fibers are selected for composite materials. For high-performance applications in the aeronautical industry, the carbon fibers can be used together with glass or aramid fibers to manufacture epoxy composites with higher impact performance. Pincheira et al. [9] investigated how aramid fibers affect the strength and durability of hybrid aramid-carbon woven fabrics when they are used as

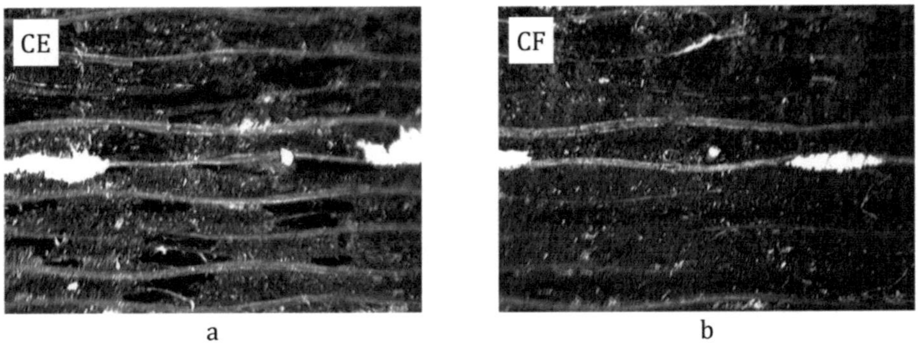

Figure 1.
Microscopic images of the cross-sectional surfaces of carbon fabric-reinforced epoxy composite materials with fiber orientation at different angles: (a) pure epoxy matrix; (b) filled epoxy matrix.

reinforcement for epoxy composites. The obtained results showed that adding aramid fibers significantly improved impact energy absorption. Although the overall stiffness and strength decreased slightly, these fibers made the composite more resistant to impact and fracture because they are more flexible than carbon fibers. Also, Lei et al. [10] found that adding glass fiber effectively enhanced both the impact resistance and compression performance after impact. The hybrid laminates showed a smaller area of damage and a higher CAI strength by 9–16% at impact energy levels of 10–30 J.

The investigation of the low velocity impact response of epoxy composite materials and their residual mechanical properties after impact is very important to ensure the safety and reliability of structures made of these materials. In the case of many applications, the components made of these materials may be subject to impact or shock loadings, either during transport or during operation, such as collisions or accidental impacts [11, 12]. Impact at low velocity can create internal damage that is not immediately visible, such as delamination, cracks, or fiber debonding, which can later significantly affect mechanical performance and compromise structural integrity [13].

The study of post-impact behavior helps to assess the degree of damage and to determine how internal damage influences their mechanical properties. If the material no longer has the ability to withstand mechanical stress after an apparently minor impact, the ultimate failure may occur.

Therefore, understanding how these materials react and deteriorate is essential for risk assessment and accident prevention.

This chapter briefly outlines the damage mechanisms and residual mechanical properties of fiber/fabric-reinforced epoxy composites subjected to low velocity impact loadings. Herein, it is discussed the effects of fiber type, fabric weave, ply orientation, arrangement of different fabric sheets, embedding of metallic wires between certain layers, and filler addition into epoxy matrix on these damage mechanisms developed under impact and post-impact mechanical loadings.

2. Damage mechanisms of fiber-reinforced epoxy composites

The fiber-reinforced epoxy composite materials exhibit damage mechanisms, such as matrix cracks, delamination, fiber debonding, fiber pullout, and fabric splitting in the fiber directions, under impact and mechanical loadings [14]. Commonly, the damage of laminates initiates with matrix cracking, which occurs within the epoxy matrix that binds the fabric fibers together in material structure.

The matrix cracking of fiber-reinforced epoxy composites is the key damage mechanism that can significantly affect their performance. When they are subjected to impact and mechanical loadings, the epoxy matrix can develop microcracks, which are the first signs of material degradation. These microcracks can be formed at stress concentration points, such as defects in the matrix (voids, pores, impurities, and poor polymerization) at fiber-matrix interfaces and between adjacent layers made of the different types of fibers and/or their different orientations (±15°, ±30°, ±45°, 0°/90°).

The matrix cracking of composite materials may occur by three different damage modes, usually denoted by Mode I, Mode II, and Mode III (**Figure 2**).

The damage mode I is named also "opening mode" and occurs in tensile stresses perpendicular to the crack plane, separating the layers. The modes II and III generate matrix cracking due to shear-dominated stresses. The mode II, also called "sliding

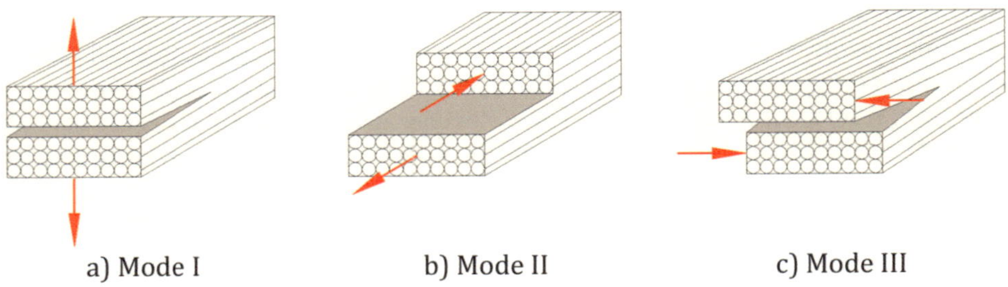

Figure 2.
Damage modes of matrix cracking of composite materials reinforced with unidirectional fibers.

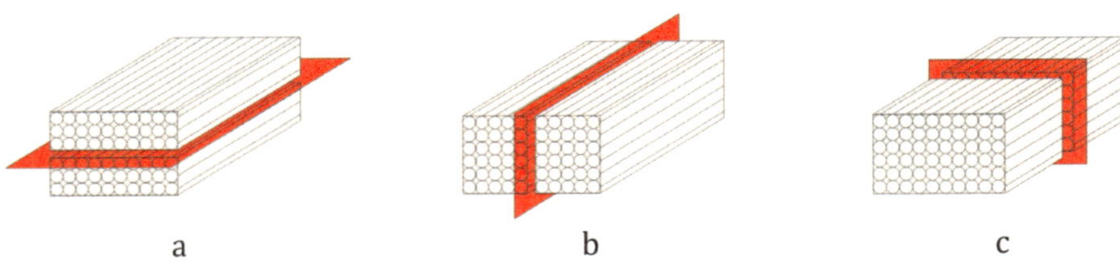

Figure 3.
Types of delamination: (a) interlaminar direction; (b) intralaminar direction; (c) translaminar direction.

mode" or "in-plane shear," leads to the sliding of crack surfaces over each other, and the crack propagates in a parallel direction to the load. The damage mode III, also known as "anti-plane shear," refers to the type of matrix cracking when shear stresses act perpendicular to the crack front and parallel to the crack surfaces. This mode involves tearing or sliding action in the plane of the material, typically characterized by out-of-plane shear deformations.

In the case of fiber-reinforced epoxy composites, the matrix properties dominate, and the degradation of composite materials begins at an early stage with the appearance of microcracks, which can be observed in the force-time and force-displacement curves of mechanical and impact tests before the force reaches maximum value.

As the loading force increases, the microcracks grow, extend between plies, and develop delamination. The delamination in the unidirectional fiber-reinforced epoxy composites can propagate in interlaminar, intralaminar, and translaminar directions (**Figure 3**). In contrast, the delamination in epoxy composites reinforced with fabrics tends to propagate primarily in the interlaminar direction and through fiber breakage.

As the applied load increases, delamination progresses through the thickness of the composite laminate and leads to fiber breakage and fiber pullout. In the case of fabric-reinforced composites, the fabric splits in the warp and weft yarn directions on the bottom layers.

2.1 Damage mechanisms of fiber-reinforced epoxy composites subjected to low velocity impact

During their service life, the fiber-reinforced epoxy composites can be subjected to low velocity impacts such as tool drops during maintenance or encounters with birds, hail, and debris during aircraft takeoff or landing. Generally, the fiber-reinforced epoxy composites have greater damage resistance in the longitudinal and lateral directions compared to the transverse direction. Their impact resistance is largely determined by their ability to absorb impact energy. The impact energy is absorbed by materials in three ways: elastic deformation, plastic deformation, and the creation of new surfaces [15]. Composites primarily absorb energy through elastic deformation until they reach a threshold. Beyond this threshold, they absorb energy through both plastic deformation and damage. The type of damage depends on the material properties and geometry of the impactor and target.

Common damage mechanisms include matrix cracking, delamination, fiber breakage, and fiber pullout discussed above. The composites reinforced with plain weave fabric, generally, show better low velocity impact response than the composites reinforced with unidirectional fibers due to their interwoven fibers, which confine delaminations to a smaller area [16]. The laminates reinforced with unidirectional fibers are prone to large delaminations, especially in the intralaminar direction. A crack on the back surface can split the ply and lead to multiple damages.

Mostly, the low velocity impact loadings with reduced energy can cause barely visible impact damages, which are difficult to detect and can compromise the structural integrity and the strength of composite, potentially leading to failure over time if not addressed.

As an example, **Figure 4** presents images of damages on the impacted (Frontal-F) and non-impacted (Back-B) surfaces of a fabric-reinforced epoxy hybrid composite with outer layers made of carbon fabric, subjected to energy levels of 15 and 30 J. By visual inspection, it can be observed on the impacted surfaces that at an impact energy level of 15 J, the hybrid material suffered a visible bending deformation accompanied by damage mechanisms such as matrix crazing and small, barely visible cracks around the contact area between the impactor and composite surface. On the back surface, the material has suffered damages such as matrix cracks, fiber breakage, fabric splitting in the fiber directions, and fiber peeling due to flexural waves that resulted after the contact with the impactor transferred through the thickness of the plate. At the impact energy level of 30 J, it can be seen more accentuated damages as matrix cracks, fiber breakage, and fabric splitting in the fiber directions on a non-impacted surface. On the contact area of the impactor with the composite plate, a long crack of the matrix and carbon outer layer has occurred.

Figure 5 illustrates tomographic images of the cross-sectional area damaged under impact loading at an energy level of 45 J in the case of a hybrid composite reinforced with four different types of fabrics (carbon, aramid, glass, and hybrid) and filled epoxy matrix with structure architecture $\left[0_{3C}/0_{3A}/0_{2G}/\overline{90_{H}}\right]_{S}$.

It can be easily remarked that the cracks propagate in both the translaminar direction breaking the carbon outer layers and the interlaminar direction between them and their adjacent aramid plies. On the back surface, the translaminar

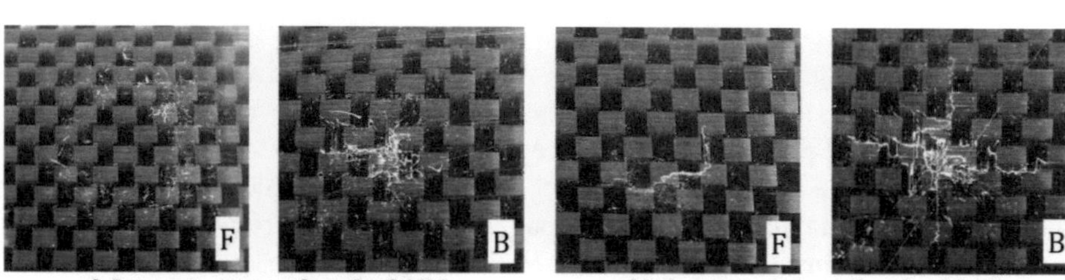

a) Impact energy level of 15 J   b) Impact energy level of 30 J

Figure 4.
Images of impacted (F) and non-impacted (B) surfaces of a hybrid composite with outer layers made of carbon fabric.

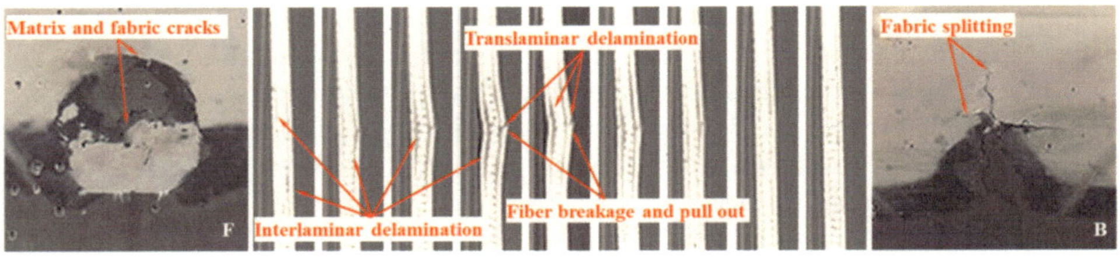

Figure 5.
Tomographic images of the cross-sectional area damaged on impact loading at an energy level of 45 J of a hybrid epoxy composite [17].

delamination produced the breakage and pullout of the carbon fibers and fabric splitting in the fiber directions. Also, on the impacted surface, a circular crack of matrix and outer plies is visible around the contact area between the impactor and specimen surface, with a horizontal crack crossing the center of this area.

2.2 Damage mechanisms of fiber-reinforced epoxy composites subjected to mechanical stresses after impact

When the epoxy composite materials reinforced with fibers or fabrics are subjected to mechanical stresses after low velocity impact loadings, the damage mechanisms become more complex and weaken their structure stability and integrity, contributing to crushing and ultimate failure.

Under tensile after impact, the existing matrix cracks and delaminations propagate and contribute to ply debonding and separation [18]. Also, the pre-existing broken fibers lead to poor residual tensile strength and faster ultimate failure depending on how high the applied impact energy level was and how severe the material was damaged.

The compression after impact stresses amplifies and propagates the damage mechanisms of fiber-reinforced epoxy composites initiated by low velocity impact loadings, often leading to ultimate failure. When the material is subjected to compressive loading, it is focused on the effects of pre-existing damage caused by low velocity impact (matrix cracking, delamination, fiber breakage, fiber pullout, and surface indentation), and it involves fiber kinking, buckling, and delamination propagation, which reduce the composite residual strength [19].

The low velocity impact damages usually contribute to stress concentrations, especially around defects like cracks, delaminations, and broken fibers. These stress concentrations increase the likelihood of localized failure of the material under compression and accelerate it, as localized regions experience much higher stresses than the undamaged material.

Initially, the damages produced by low velocity impact loadings may not be identified by visual inspection, but the compression after impact testing can reveal how they affect the overall performance of the epoxy composite material.

Under compressive loadings, the matrix cracks and delamination initiated during low velocity impact may open up and propagate, contributing to premature buckling and structural instability since it acts as stress concentrators. It leads to composite weakening by initiating further damage mechanisms during compression.

The local buckling in epoxy composites can be caused by delaminated plies due to the lack of linkage support between the layers. Also, the pre-existing impact damage, such as delamination or fiber breakage, leads to the kinking or buckling of the fibers within laminate. The fiber kinking or microbuckling occurs when the fibers deform plastically or elastically in a localized region under compressive stresses. This often happens after delamination, where load redistribution leads to higher stress concentrations in certain regions.

In severe cases, the entire structure may buckle or collapse due to the compounding effects of fiber damage, matrix cracking, and delamination.

Another damage mechanism generated by compression after impact is the interlaminar shear, which occurs when the layers of the composite slide relative to each other. This can be typically observed at the edges of delaminated regions. The interlaminar shear stresses exacerbate delamination and reduce the overall stiffness of the structure, leading also to structural instability.

Out-of-plane deformations are caused by delaminated regions, which may buckle. This buckling can drive the growth of existing delaminations, making the problem progressively worse. Buckle-driven delamination growth causes rapid loss of structural integrity under compression, leading to failure much sooner than if the delamination was stable. In some composite materials, the fibers and matrix may experience debonding at the interface due to the weak fiber-matrix adhesion. The debonding reduces the effectiveness of load transfer between the fibers and the matrix, leading to poor compressive strength and failure under post-impact loads.

High compressive forces that lead to localized damages, typically near the point of the impact or along the edges of delaminated areas, contribute to the composite material crushing by matrix failure, fiber breakage, and even local fiber bundle fragmentation.

In the case of bending after impact stresses are applied to the fabric-reinforced epoxy composite materials, both the compression stresses in the top layers and the tensile stresses in the bottom layers are involved [20]. Under the action of the compressive stresses, the specimens are deformed so that the upper layers are kinked, and the existing damage caused by the impact loadings accelerates the linear expansion of the matrix and fabric cracks along the entire specimen width at the impacted side. The interlaminar delamination propagates between the adjacent layers, and the fibers of the bottom plies break and pull due to the action of tensile stresses. The ultimate failure of the specimens is achieved by propagation of translaminar delamination through their thickness.

3. Factors influencing damage mechanisms

The factors as impact energy level, shape and size of impactor, direction of loading, environmental conditions, properties of epoxy matrix and fibers, type and orientation of fibers, quality of fiber-matrix interface, and plate thickness influence damage mechanisms of the fiber-reinforced epoxy composites subjected to low velocity impact loadings and mechanical stresses after impact [21].

3.1 Parameters of impact tests

As the impact energy level increases, the damage can progress from matrix cracking to fiber-matrix debonding. The delamination becomes more widespread and severe, and it extends in the interlaminar, intralaminar, and translaminar directions. The impact energy level influences also the energy absorption capacity of the composites because the energy is dissipated from the impactor by plastic deformation through composite thickness [22].

The low velocity impact response and resulted damage mechanisms of fiber-reinforced epoxy composites are also influenced by the frontal geometry and size of the impactor [23, 24]. The size and level of damaged area on the impacted surface depends on the shape of the impactor [25]. It was found that round-nosed impactors produce more damage to epoxy composite reinforced with carbon unidirectional fibers at low energy levels compared to flat-ended impactors [26]. This is due to deeper indentation created by the round-nosed impactor, which leads to variations in the size and shape of delaminations within the plate. Delaminations grew larger as the impact energy increased from 10 to 30 J, but flat-ended impactors only cause delaminations at energies above 20 J. The flat-ended impactor's larger contact area produces a circular damage zone without delaminations directly under the impact point. At higher impact energies, the flat-ended impactor leads to kinking-type fractures around the edge of the impactor, while the round-nosed impactor contributes to fiber breakage and pullout.

The maximum penetration of epoxy composite reinforced with carbon fabrics can be achieved with a conical impactor shape at an energy level of 45 J, followed by ogival, hemispherical, and flat shapes, as presented in Ref. [27]. The largest delamination was recorded using a flat-ended impactor shape, followed by hemispherical, ogival, and conical shapes.

Regarding the effects of impactor diameter size on impact behavior, Icten et al. [28] investigated the effects of hemispherical impactors with different sizes of diameters (12.7 mm, 20.0 mm, 25.4 mm, and 38.1 mm) on woven glass epoxy composites. The specimens were impacted until they were penetrated or perforated, and then compressed to measure their strength. The impact results obtained at an energy level of 60 J showed that the size of the impactor significantly influences the perforation threshold of the composite. It was determined that the 38.1 mm diameter impactor did not perforate the composite, while the composite impacted by the impactor with a diameter of 12.7 mm was completely perforated. The area of damage produced by the 38.1 mm impactor was smaller compared to the other impactor diameters. The composite was perforated by the 12.7 mm diameter impactor at 60 J, the 20.0 mm diameter impactor at 100 J, the 25.4 mm diameter impactor at 125 J, and the 38.1 mm diameter impactor at 210 J. It results in the larger impactors requiring more energy to penetrate or perforate the plates. So, the increasing of impactor diameter generally leads to an increase in CAI strength. The investigation of damages reveals that the composite suffered fabric splitting in the transversal direction on the impacted area.

Also, the angle at which the impactor strikes affects the force distribution and energy transfer within the composite [29]. A more oblique angle might lead to a more gradual load application, reducing the likelihood of immediate damage. Perpendicular impact directly targets the composite's thickness, resulting in more severe damage. Oblique impact might distribute the force more evenly, reducing damage severity. The strength to impact damage can be improved in this case by interlacing composite layers with z-fibers that allow for more efficient energy dissipation through deformation and fracture. This is especially beneficial at higher impact angles where the z-fibers can absorb a significant portion of the energy and reduce damage level.

3.2 Characteristics of epoxy composites

Epoxy's high crosslinking density creates a rigid and inflexible structure. This rigidity makes the epoxy matrix brittle and less tolerant to impact or mechanical stresses. When subjected to loads that exceed its elastic limit, epoxy matrix is more vulnerable to crack or fracture than to deform or absorb the energy [30, 31]. This susceptibility to microcracks can significantly reduce the material's overall toughness and impact resistance. Over time, these cracks can grow and propagate, leading to ultimate failure. As a result, the limitations imposed by epoxies' high crosslinking density often restrict their use in applications that require exceptional durability, impact resistance, or the ability to withstand cyclic loading. These properties can be improved by reinforcing with woven or non-woven fibers.

The properties of fibers significantly influence how materials respond to impact and the damage mechanisms that occur. Materials with more complex structures are harder to understand when it comes to impact effects and resulting damage. In fiber-reinforced composites, the failure theory divides damage into two types: matrix failure and fiber failure. Matrix failure happens when fibers are stronger than the surrounding material, while fiber failure occurs when the matrix is stronger. Usually, plain simple woven fabrics are used as reinforcement for epoxy composite laminate due to their structural stability, delamination resistance [28], mechanical resistance on both warp and weft directions [32], and better impact response [33] compared to non-woven fabrics [34] due to the yarns interlacing, exhibiting thus the same properties in both directions of fibers [35].

The carbon fabrics are often preferred in the aerospace industry for their stiffness, while glass fabrics are favored in the automotive sector for their mechanical strength. Aramid fabrics are commonly used in military applications due to their high impact resistance and ability to absorb energy. Epoxy composites reinforced with carbon fibers offer exceptional mechanical properties, but they are prone to impact damages due to their rigidity and brittleness [36].

The ply orientation within a fabric-reinforced composite significantly influences its deformation mechanisms under impact loading [37]. It was found that the composites with greater angular differences between adjacent plies exhibit a higher tendency to delamination, which propagates in a manner dependent on the fiber direction [38]. The ply orientation at 0° contributes to maximum impact load and minimum deflection, and conversely for composites with ply orientation at 90° [39]. Liu et al. [40] found that the inclusion of plies with orientation at 90° in composite structures led to a reduction in response time and an increase in delamination along the transverse direction. The plies with this orientation introduce a complex interplay between competing deformation mechanisms associated with unidirectional and woven fiber reinforcements. These layers can alter the stress distribution within the composite, potentially leading to localized stress concentrations that initiate and propagate delamination. Additionally, the transverse stiffness provided by the plies with orientation at 90° can influence the overall deformation behavior of composite, potentially affecting its impact resistance and energy absorption capabilities. The laminates made of plies with orientations at 45° contribute to more severe impact damages such as indentation, cracks on the surface, delamination, and a fabric splitting in the direction of the fibers on the back side [15]. Liu [41] determined that the highest resistance to delamination and highest thresholds of penetration and perforation under low velocity impact loadings are exhibited by composites containing layers with orientation at 15°.

4. Optimizing impact strength and mechanical properties

4.1 Hybridization of epoxy composites

In order to improve the impact strength and mechanical response of the epoxy composite materials, it can be used different types of fabrics as reinforcement for a laminate, especially in the case of composites made of carbon fabrics. The hybridizing of composites allows to obtain materials with desired properties and to monitor their impact behavior. Also, the properties of the hybrid composites depend on the careful arrangement of layers using different types of fabrics.

Randjbaran et al. [42] found that the fabrics of the outer layers made of glass fabric of the composites led to a lower absorption of impact energy as compared to those made of aramid outer plies. The carbon outer plies are not suitable for composites used in applications that involve impact loadings, but it is better to use them for medial layers. Also, Sayer et al. [43] obtained that the perforation threshold of carbon/glass hybrid laminate was higher by 30% in comparison with glass/carbon hybrid composite. With the increase of impact energy to 35 J, the impactor rebounds without causing significant damage. It was observed that minor cracks and delaminations occurred on the glass surface, while the carbon ply only showed dents and cracks. As the impact energy got higher, there was fiber breakage and fiber pullout.

The effects of fabric types (carbon, aramid, and glass) used for outer layers and the specific density of glass fabric on the low velocity impact behavior of hybrid epoxy composites at an energy level of 90 J were analyzed in Ref. [44]. It was recorded that the composites with outer layers made of glass fabric with higher specific density (390 g/m^2) exhibited a higher impact resistance than those made of aramid outer layers with lower specific density (173 g/m^2). The arrangement of the layers made of carbon and aramid fabrics in the structure of the composites with glass outer layers also influenced their impact performance. So, the material with glass outer layers preceded by carbon layers and aramid central layers showed lower impact strength and a higher damage level as compared to the one with preceded glass outer layers by aramid layers and carbon central layers. By tomographic investigation, it was found that the matrix cracking and delaminations occurred between adjacent plies made of different types of fabrics. The damage mechanisms observed were matrix cracking, interlaminar and translaminar delaminations, fiber breakage, fiber pullout, and the splitting of back faces in the fiber directions. The hybrid epoxy composites with carbon outer plies suffered the highest level of damage, resulting that the carbon fabric is not suitable to be used for outer plies, as determined by Randjbaran et al. [42]. However, the carbon fabrics are more appropriate for the medial layers.

Ni et al. [45] obtained that the hybrid carbon-glass epoxy laminates used in yachts with a carbon fiber layer in the middle were found to have the smallest damage area under low velocity impact and post-impact compression loadings. The laminate with a carbon fiber bottom layer was effective in distributing the impact load evenly among the layers, preventing stress concentration and maintaining structural integrity. This type of laminate showed the highest resistance to deformation and compressive strength after impact, making it suitable for areas of the yacht that may experience low energy impacts. It can help reduce damage caused by impacts and prevent deformation.

The carbon layers lead to a larger area of interlaminar delamination between them and adjacent glass plies due to differences in flexibility, reducing thus the extension of damage through thickness [46].

Also, the stacking sequence of the hybrid epoxy composites significantly influences their damage response under low velocity impact and compression after impact loadings. In Ref. [47], it was found that the laminates with medial plies made of glass fibers with orientation at $\pm 45°$ have more strength to impact and compression after impact than those with medial plies oriented at 0° and 90°, reducing damage extension through laminate thickness. The ultimate failure of the specimens was produced by delamination buckling propagated in line from the center (indentation point) and toward edges at the impacted side.

Wang et al. [48] studied the effects of hybrid ratio and stacking sequence on damage mechanisms of hybrid epoxy laminates reinforced with carbon and Kevlar fabrics subjected to bending after impact stresses. The obtained results revealed that the bending properties were improved by hybridization. The changing of fiber arrangement in the epoxy composite material improved its performance. When the composite was subjected to bending stress, the Kevlar fibers on the compressed side buckled, preventing cracks from spreading in the carbon fibers. This resulted in minimal damage to the intermediate carbon fiber. Overall, the composite with the Kevlar-carbon-carbon stacking sequence had the best resistance to bending. Also, the impact response and post-impact flexural strength of a hybrid epoxy composite reinforced with carbon and aramid fibers were investigated in Ref. [49]. The aramid layer was positioned between two carbon layers in the epoxy laminate structure. This arrangement of the layers significantly enhanced the composite's ability to withstand loads, particularly after impact, due to its energy dissipating capabilities. This effect was evident in the preservation of the lower carbon fibers, which remained undamaged after the post-impact bending test. Tian et al. [6] also studied the damage mechanisms of carbon-aramid epoxy hybrid composites under post-impact flexural. It was found that the damage pattern of laminates CCAAA and CCCCC was similar. The compressive stress under bending loading has led to the fracture of carbon fibers, and the cracks followed a straight path. In the case of epoxy laminates AAAAA and ACACA, the hybrid one was least damaged, but it was observed for both that the aramid fibers of the bottom layer were broken and pulled out. The composites CCAAA and ACACA showed significant brittle damage under bending loads, but nevertheless the laminate ACACA exhibited the best damage resistance. When the upper aramid layer was kinked under compression stress, it was not recorded the crack extension, but it was observed a mild delamination and a low degree of bottom aramid layer damage by tensile stress.

4.2 Addition of fillers into epoxy matrix

In order to improve the low velocity impact resistance and mechanical strength after impact, different types of fillers can be added to the pre-polymeric blend of epoxy matrix components (resin and hardener). In this case, the impact and post-impact mechanical performance of the composite materials depends on the properties of fillers, their compatibility with epoxy matrix, the quality of the filler-matrix interface, and their size and shape. The nanoclay [50, 51], nano-alumina [52], multi-walled carbon nanotubes [53, 54], carbon nanofibers [55], graphene [56], PA66 fibrous nanomats [57], milled glass fibers [58], and many others are often used as fillers in epoxy matrix.

Hoseinlaghab et al. [18] found that the nanoparticles of nanoclay and carboxy-modified multi-walled carbon nanotubes increased the impact performance of glass fiber-reinforced epoxy laminates, reduced residual deformations, identified damages (matrix cracking, delamination, fiber breakage and pullout), and improved tensile strength after impact response. The carbon nanotubes showed 66% reduction of deformation and 21–39% enhanced residual tensile strength versus nanoparticles of clay, which exhibited 42% reduction and 9–37% improved residual strength. The addition of nanoparticles to epoxy composites made them stronger to impact and post-impact mechanical loadings. The damage mechanisms observed in cases of specimens with nanoparticles were matrix cracking and delamination. The carbon nanotubes reduced these two types of damage by 20–27% and nanoparticles of clay reduced them by 8–11%.

The results presented in Ref. [59] showed that adding nanoclay to unidirectional glass fiber-reinforced epoxy composites improved their impact strength and hardness. The best results were achieved with 3.5% nanoclay, which increased impact strength and hardness by up to 24%. The modified epoxy in the glass fiber-reinforced epoxy composite improved the adhesion between the matrix and the fiber, thus resulting in a stronger interface. Additionally, the SEM investigation revealed minimal fiber pull-out, indicating the effectiveness of the matrix modification in creating a strong bond with the glass fibers. Even on the fractured surface, the fibers were well-aligned in the correct direction, demonstrating the composite's high degree of fiber orientation under stress and leading to superior fiber efficiency. Singh et al. [60] also determined that the nanoclay and compatibilized ethylene propylene diene monomer (EPDM) rubber improved the low velocity impact strength of the epoxy composites reinforced with glass fibers and their resistance to damage. The highest impact resistance was recorded for a 5 wt.% concentration of rubber because it enhanced interfacial adhesion with glass fibers due to silane treatment.

The electrospun PA66 nanofibrous mats also improve low velocity impact behavior, mechanical after-impact strength, and damage mechanisms of the epoxy composites. Their effects on impact response and tensile residual strength of laminates reinforced with plain woven aramid fabric were investigated by Goodarz et al. [61]. They also analyzed the influence of nanomats position in the laminate interlaminar regions with three different configurations: back-side, central, and two-side interlayering. The pulled off nanofibers of nanofibrous mats occurred in zones of plastic deformation that behaved as links for the adjacent layers. These links can prevent the spreading of microcracks and delamination in interlaminar regions. Thus, the breakage of bridged nanofibers improves interlaminar toughness due to their capacity to absorb a large energy amount.

Also, the fillers, such as potato starch, aramid powder, carbon black, and barium ferrite, added into the epoxy matrix between certain layers improve the resistance of epoxy hybrid composites to impact damages such as matrix cracking and delamination propagation in the interlaminar direction and through the thickness, especially in the case of the composite materials with glass outer plies. Stefanescu et al. [17] studied the influence of these fillers, fabric type of outer plies, and different fabric layer arrangement on the low velocity impact behavior of the epoxy hybrid composites at energy levels of 45 and 90 J. The effect of outer layer fabrics was similar to that presented in Ref. [44]. Regarding the influence of the arrangement of plies made of different fabrics, it was determined that, in this case, the laminate with glass outer layers preceded by carbon layers and aramid central layers showed better impact resistance as compared to the one with preceded outer layers by aramid layers and carbon central layers due to the presence of fillers. In this case the damage level at energy of 45 J was reduced, so as it was observed by tomographic investigation, a delamination between the bottom glass and carbon adjacent layers.

Further, Bunea et al. [62] used these fillers for epoxy matrix between certain layers of hybrid composites reinforced with different fabric types and ply orientation at various angles (0°, ±15°, ±30°, 45°, and 90°), which exhibited a good behavior under low velocity impact at an energy level of 45 J. These fillers improved the impact strength of the aramid fabric-reinforced epoxy laminate with various ply orientations at an energy level of 90 J [5]. Also, they examined how different factors, such as the amount of carbon and aramid sheets, fiber orientation at 0°, ±15°, ±30°, and ± 45°, and the properties of filled epoxy matrix affected the impact strength of hybrid epoxy composites with stratified filled matrix and reinforced with different types of fabrics (carbon, aramid, glass, and hybrid) at an energy level of 90 J [14]. Various types of fillers, such as potato starch, carbon black, aramid powder, barium ferrite, carbon whiskers, and glass whiskers, were added into the pre-polymeric epoxy blend between certain layers. They also varied the number of carbon and aramid sheets in the laminates and the ply orientation. The obtained results revealed that all specimens were not perforated, and the hybrid composites with fiber orientation at 0° and a greater number of carbon layers exhibited the highest impact resistance. It was observed that the properties of the epoxy matrix dominated and its failure significantly influenced the damage impact responses of the composites. The damaged areas presented matrix cracks, delaminations in the interlaminar and translaminar directions, fiber breakage, fiber pullout, and fabric splitting in the fiber directions. The orientation of plies at various angles contributed to more severe damages.

The addition of milled glass fiber filler to glass/epoxy laminates reduces the area of impact damage and improves overall impact performance compared to unfilled laminates, as it was found in Ref. [63]. These fillers also decrease the degree of damage, indicating enhanced impact resistance. Additionally, the filler improves the adhesion between the fiber and matrix, leading to more efficient load transfer and reduced damage size. Thus, it improves the quality of the fiber-matrix interface and damages the toughness behavior of the epoxy matrix by decreasing crack propagation.

4.3 Insertion of metal wires

One potential approach is to incorporate metal wires, stainless steel (SS) wire meshes [64], and shape memory alloys (SMAs) wires [65] into the composite structures. These materials can absorb impact energy, reduce the propagation of matrix cracking and delamination, and exhibit improved compression after impact strength [66].

In Ref. [5], it was investigated the low velocity impact behavior of epoxy composite laminates reinforced with 17 hybrid fabric layers at an energy level of 90 J. A simple plain woven carbon/aramid fabric in geometry 2:1 in the warp direction and 1:2 in the weft direction was used to manufacture this hybrid fabric. Each second aramid fiber in the weft direction was replaced by a glass fiber together with a tinned copper wire. The impact results showed that the composite was not perforated by the impactor at an energy level of 90 J, but it suffered damage mechanisms such as matrix cracking, delamination in the interlaminar and translaminar directions, fiber breakage, fiber pullout and, fabric splitting in the fiber directions. The addition of fillers (potato starch, aramid powder, carbon black, and barium ferrite) into the epoxy matrix between certain layers improved the impact response of the composite and reduced its damage degree.

The embedding of 304 SS (stainless steel) wire mesh improves significantly mechanical resistance (tensile and bending), impact strength, and energy absorption capacity of epoxy composites reinforced with glass fibers, as determined by Wan et al. [67]. The results of their investigation presented that the reference specimen and the one with a single layer of wire net were perforated by an impactor at an impact energy level of 24 J. The other specimens embedded with wire nets from two to five layers have withstood impact loadings. The inspection of damages reveals that these composites suffered breakage of fiber/wires and delaminations. Liu et al. [68] also used this type of wire net in order to improve the strength of glass fiber-reinforced epoxy composites under compression after double impact. It was found that the wire nets enhanced the mechanical resistance of the composite and reduced out-of-plane deformation and internal damage. However, the compression after impact at high energy level (60 J) revealed that the resistance of composite embedded with wire nets was lower than that without wire nets.

The damage tolerance of the glass fiber-reinforced epoxy laminates under low velocity impact can be significantly reduced by incorporating shape memory alloy (SMA) wires [69]. Rajendran et al. [70] determined that the embedding with SMA wires improves also the flexural strength of the carbon fiber-reinforced epoxy composite by 29.9% and its interlaminar fracture toughness by 39.53%. The matrix cracking in damage Mode I was reduced due to solid SMA wire-matrix bonding, so as to increase the resistance of material to fracture.

5. The potential of natural fibers

The natural fiber plain woven fabric-reinforced epoxy composites offer a significant advantage in terms of impact resistance, as the fibers, moving in the warp and weft directions, respectively, block the propagation of cracks and delamination due to the presence of lateral filaments within these fibers. The natural fiber fabrics are an excellent alternative to be used in the automotive and aeronautical industries and to replace synthetic fibers in many applications. The most common natural fabrics for epoxy composites as reinforcements are hemp [71], kenaf [72], flax [73], banana [74], sisal [75], jute [76], etc.

Shaik et al. [77] performed a comparative investigation on the low velocity impact behavior of abaca/epoxy composite and abaca/rubber hybrid epoxy composite at an impact energy of 10.24 J. The primary cause of failure in the abaca/epoxy composite was a combination of fiber breakage and debonding. In contrast, the abaca/rubber hybrid composites experienced puncture-type damage due to the tearing of the rubber layer. This tearing happened after the fiber breakage in the abaca layer. The fiber breakage and matrix microcracks were the dominant damage for the abaca/epoxy composite in the impact area. The abaca/rubber hybrid composite, however, suffered more severe damage, with a hole penetrating the bottom layers.

In Ref. [78], low velocity impact and post-impact tensile and compressive responses of epoxy composites reinforced with unidirectional flax fibers were analyzed. The low velocity impact tests were performed using conical impactors with angles of 15°, 30°, 45°, and 60° and an impact energy range of 2–5 J. The increase in impactor angle led to propagation of shear matrix cracks, delaminations, and fiber breakage; thus, the larger damage areas were recorded under impact tests with a flattened impactor. The tensile after impact results presented a significant influence of induced damages on residual strength, which was reduced by 41% at an energy level of 5 J. The residual compressive strength of the composite was not significantly affected by low velocity impacts with energy less than 3 J. However, it suffered local buckling and delamination due to existing impact damage. With the increasing of impact energy to 5 J and impactor angle to 60°, the compressive strength decreased by 40%, and more severe damage mechanisms occurred as local buckling and extension of existing cracks.

The damage modes of the composite reinforced with untreated raw jute fibers, such as matrix cracking, delamination, and fiber debonding, are not significantly severe under impact loadings at energy levels of 3 and 6 J, according to a study [79]. The chemical treatment using alkali for jute fibers contributes to a toughened fiber-matrix interface that reduces the impact performance due to low energy absorption ability. Also, the usage of PVA binders also leads to a decrease in impact strength due to their brittleness and contributes to ultimate matrix cracking and fiber debonding in the impacted area.

Green epoxy bio-composites reinforced with palm leaflet fabric exhibit double impact strength compared to those reinforced with palm leaflet tree at an impact energy of 3 J, as found by Bezazi et al. [80]. However, in the case of the bending investigation, the results showed higher bending resistance (76.6%) and higher modulus (149.8%) for bio-composites reinforced with palm leaflets tree compared to those reinforced with palm leaflets fabric. These natural materials are suitable to be used in many industries, including automotive for door panels or seat backs, while their mechanical strength does not permit them to be used for important car parts due to unguaranteed structural stability.

6. Conclusions

This chapter delves into a review of existing research on the low velocity impact response, post-impact mechanical strength, and damage mechanisms of epoxy hybrid composites reinforced with fibers/fabrics. There are discussed damage mechanisms (matrix cracking, delaminations, fiber breakage, and fiber pullout) involved by impact loading that act as stress concentration points and reduce residual mechanical properties. These existing impact damages reduce significantly the post-impact mechanical resistance and contribute to matrix cracks and delamination propagations between adjacent layers and through the specimen thickness, leading to the ultimate failure.

Based on the bibliographic study, several conclusions can be drawn regarding the impact and post-impact mechanical responses of hybrid epoxy composites:

- The impact test parameters influence the behavior of materials under impact and mechanical loadings and their resistance to damage. It was found that as impact energy increases, damage severity progresses from matrix cracking to fiber-matrix debonding and delamination. The delamination becomes more widespread and severe, extending in the interlaminar, intralaminar, and translaminar directions. Also, the increasing impact energy level leads to a decrease in energy absorption capacity as energy is dissipated through plastic deformation.

- The round-nosed impactors produce more damage at low energy levels compared to flat-ended impactors due to deeper indentation and variations in delamination size and shape. The impactors with larger diameters require more energy to penetrate or perforate composites, leading to increased CAI strength.

- The angle at which the impactor strikes affects force distribution and energy transfer within the composite. Oblique impact may lead to more gradual load application, reducing immediate damage. Perpendicular impact directly targets composite thickness, resulting in more severe damage.

- The impact damage resistance can be improved by interlacing with z-fibers, especially at higher impact angles, by providing more efficient energy dissipation.

- The epoxy's high crosslinking density makes it rigid and inflexible, leading to susceptibility to cracking and fracture under impact loads. Due to these limitations, epoxy matrixes are often unsuitable for applications requiring exceptional durability, impact resistance, or cyclic loading. Reinforcing epoxy matrixes with fibers can enhance their properties, such as impact resistance and durability.

- The type and orientation of fibers significantly influence the composite's response to impact and damage mechanisms. Composites with greater angular differences between adjacent plies tend to delaminate more easily. The ply orientations at 0° generally provide maximum impact load and minimum deflection, while orientations at 90° can reduce response time and increase delamination. The orientation of fibers at 45° may lead to severe impact damage, including indentation, cracks, and delamination. The epoxy composites with layers at 15° often exhibit the highest resistance to delamination and penetration under low velocity impact.

- The hybridization of epoxy composites can offer desired properties and improved impact behavior compared to those reinforced with pure fabrics. The choice of fabric for outer layers significantly influences impact resistance. It was found that the glass outer layers generally exhibit lower impact energy absorption compared to aramid outer layers. Carbon outer layers are not suitable for impact-loaded applications but are better used for medial layers. The carbon plies can lead to larger interlaminar delamination areas between them and adjacent glass or aramid layers due to differences in flexibility. Also, the hybrid composites can exhibit good post-impact flexural strength and damage resistance, especially when aramid layers are strategically placed.

- Composites with higher specific density glass outer layers and aramid-carbon-glass stacking sequences often demonstrate better impact performance. The hybrid ratio of carbon and aramid fibers can enhance bending properties and damage resistance. An aramid-carbon-carbon stacking sequence can provide optimal resistance to bending.

- The carbon-glass epoxy laminates with a carbon fiber layer in the middle can provide high resistance to deformations and compressive strength under impact, making it suitable for areas of the yacht. They can reduce damages caused by impacts and prevent deformation.

- The addition of various fillers to epoxy composites significantly enhances their low velocity impact resistance and post-impact mechanical strength. The properties of fillers, their compatibility with the epoxy matrix, the quality of the filler-matrix interface, and their size and shape all influence the composite's performance. Thus, nanoclay, nano-alumina, carbon nanotubes, carbon nanofibers, graphene, PA66 nanofibrous mats, milled glass fibers, starch, aramid powder, and carbon black are frequently used as fillers in epoxy matrixes. They improve mechanical strength after impact and reduce residual deformations and damage such as matrix cracking, delamination, fiber breakage, and fiber pullout. The interlayering nanofibrous mats can prevent the spreading of microcracks and delamination. Under breakage, they act as linkage for two adjacent layers and improve interlaminar toughness. The nanoclay improves impact strength and hardness, enhances adhesion between matrix and fiber, and reduces fiber pullout. The carbon nanotubes reduce deformation, matrix cracking, and delamination and enhance residual tensile strength.

- The addition of metal wires can reduce various damage mechanisms, such as fiber breakage, fiber pullout, and fabric splitting. The hybrid composites with tinned copper wires can withstand higher impact energies without perforation, although they still suffer damage mechanisms like matrix cracking, delamination, and fiber breakage. Adding fillers to the epoxy matrix can further improve their impact response and reduce damage level. The addition of shape memory alloy wires can reduce matrix cracking, increase flexural strength, and improve interlaminar fracture toughness. However, the addition of metal wires can increase the weight and cost of the composite. At very high impact energies, the performance of composites with metal wires may be lower than those without them.

- The natural fiber-reinforced epoxy composites exhibit significant impact resistance due to the fiber structure, which effectively blocks crack propagation and delamination. Hemp, kenaf, flax, banana, sisal, and jute are among the most commonly used natural fibers for reinforcing epoxy composites. The failure modes of natural fiber-reinforced epoxy composites typically include fiber breakage, debonding, and matrix cracking. The addition of fillers, such as rubber, can improve the resistance to damage.

- Chemical treatments can affect the impact behavior of natural fibers. For example, alkali treatment can improve the fiber-matrix interface but reduce energy absorption. Their moving in the warp and weft directions blocks the propagation of cracks and delamination due to the presence of lateral filaments within these fibers. Also, the type of natural fiber can influence both impact strength and mechanical resistance. In this case, it was found that the palm leaflet fabric exhibits higher impact strength but lower bending resistance compared to the palm leaflet tree.

- Natural fiber-reinforced epoxy composites present a good solution to replace synthetic fibers in many industrial applications due to their biodegradability, compostability, and sustainability.

Author details

Marina Bunea[1,2]

1 “Dunarea de Jos” University of Galati, Romania

2 “Bogdan Petriceicu Hasdeu” State University of Cahul, Republic of Moldova

*Address all correspondence to: marina.bunea@ugal.ro

References

[1] Skalková P, Krmelová V, Krmela J, Ondrušová D, Crkoň A, Benčiková E. Composite materials with epoxy matrix and their properties. IOP Conference Series: Materials Science and Engineering. 2021;**1199**(1):012030. DOI: 10.1088/1757-899X/1199/1/012030

[2] Zadafiya K, Bandhu D, Kumari S, Chatterjee S, Abhishek K. Recent trends in drilling of carbon fiber reinforced polymers (CFRPs): A state-of-the-art review. Journal of Manufacturing Processes. 2021;**69**:47-68. DOI: 10.1016/j.jmapro.2021.07.029

[3] Zhao X, Lu S, Li W, Zhang S, Li K, Nawaz K, et al. Epoxy as filler or matrix for polymer composites. In: Jerold Samuel Chelladurai S, Arthanari R, Meera MR, editors. Epoxy-Based Composites. London, UK: IntechOpen; 2022. DOI: 10.5772/intechopen.102448

[4] Maxineasa SG, Taranu N. Life cycle analysis of strengthening concrete beams with FRP. In: Eco-Efficient Repair and Rehabilitation of Concrete Infrastructures. Duxford: Woodhead Publishing; 2018. pp. 673-721. DOI: 10.1016/B978-0-08-102181-1.00024-1

[5] Bunea M, Bria V, Silva FS, Bîrsan IG, Buciumeanu M. Influence of fiber orientation and fillers on low velocity impact response of the fabric reinforced epoxy composites. Applied Composite Materials. 2021;**28**(4):1277-1290. DOI: 10.1007/s10443-021-09910-1

[6] Tian J, Xu T, An L, Wang S, Tan Y, Chen G. Study on behavior and mechanism of low-velocity impact and post-impact flexural properties of carbon-aramid/epoxy resin laminated composites. Composite Structures. 2022; **300**:116166. DOI: 10.1016/j.compstruct.2022.116166

[7] Loos M. Composites. In: Carbon Nanotube Reinforced Composites. Amsterdam: William Andrew; 2015. pp. 37-72. DOI: 10.1016/B978-1-4557-3195-4.00002-3

[8] Ramakrishna S, Huang ZM. Biocomposites. In: Reference Module in Materials Science and Materials Engineering. Amsterdam: Elsevier; 2016. pp. 215-296. DOI: 10.1016/B978-0-12-803581-8.00965-6

[9] Pincheira G, Canales C, Medina C, Fernández E, Flores P. Influence of aramid fibers on the mechanical behavior of a hybrid carbon–aramid–reinforced epoxy composite. Proceedings of the Institution of Mechanical Engineers, Part L: Journal of Materials: Design and Applications. 2018;**232**(1):58-66. DOI: 10.1177/1464420715612827

[10] Lei ZX, Ma J, Sun WK, Yin BB, Liew KM. Low-velocity impact and compression-after-impact behaviors of twill woven carbon fiber/glass fiber hybrid composite laminates with flame retardant epoxy resin. Composite Structures. 2023;**321**:117253. DOI: 10.1016/j.compstruct.2023.117253

[11] Polimeno U, Meo M. Detecting barely visible impact damage detection on aircraft composites structures. Composite Structures. 2009;**91**(4):398-402. DOI: 10.1016/j.compstruct.2009.04.014

[12] Wronkowicz-Katunin A, Katunin A, Dragan K. Reconstruction of barely visible impact damage in composite structures based on non-destructive evaluation results. Sensors. 2019;**19**(21):4629. DOI: 10.3390/s19214629

[13] Hosur MV, Adbullah M, Jeelani S. Studies on the low-velocity impact response of woven hybrid composites. Composite Structures. 2005;**67**(3): 253-262. DOI: 10.1016/j.compstruct.2004.07.024

[14] Bunea M, Cîrciumaru A, Buciumeanu M, Bîrsan IG, Silva FS. Low velocity impact response of fabric reinforced hybrid composites with stratified filled epoxy matrix. Composites Science and Technology. 2019;**169**:242-248. DOI: 10.1016/j.compscitech.2018.11.024

[15] David-West OS, Alexander NV, Nash DH, Banks WM. Energy absorption and bending stiffness in CFRP laminates: The effect of 45° plies. Thin-Walled Structures. 2008;**46**(7–9):860-869. DOI: 10.1016/j.tws.2008.01.024

[16] Ma B, Cao X, Feng Y, Song Y, Yang F, Li Y, et al. A comparative study on the low velocity impact behavior of UD, woven, and hybrid UD/woven FRP composite laminates. Composites Part B: Engineering. 2024;**271**:111133. DOI: 10.1016/j.compositesb.2023.111133

[17] Stefanescu V, Bunea M, Cîrciumaru A. Impact analysis of fabric reinforced plates. Materiale Plastice. 2015;**52**(2):198-203

[18] Hoseinlaghab S, Farahani M, Safarabadi M, Nikkhah M. Tension-after-impact analysis and damage mechanism evaluation in laminated composites using AE monitoring. Mechanical Systems and Signal Processing. 2023;**186**:109844. DOI: 10.1016/j.ymssp.2022.109844

[19] Shabani P, Li L, Laliberte J, Qi G. Compression after impact (CAI) failure mechanisms and damage evolution in large composite laminates: High-fidelity simulation and experimental study. Composite Structures. 2024;**339**:118143. DOI: 10.1016/j.compstruct.2024.118143

[20] Quang PX, Wicaksono S, Dirgantara T, Hadi BK. The effect of low-velocity impact on the flexural strength of E-glass/epoxy composite plates. Journal of Mechanical Science and Technology. 2020;**34**(5):1879-1886. DOI: 10.1007/s12206-020-0409-8

[21] Sunith, Babu L, Ashok Kumar K, Jaya Christiyan KG, Byary MA, Manu Puranic V, Mohammed Jawad AA, et al. Effect of laminate thickness on low-velocity impact of GFRP/epoxy composites. Materials Today: Proceedings. 2023:S2214785323045005. DOI: 10.1016/j.matpr.2023.08.229

[22] Yu T, Xue P. Utilizing plastic deformation for energy absorption. In: Introduction to Engineering Plasticity. Amsterdam: Elsevier; 2022. pp. 293-326. DOI: 10.1016/B978-0-323-98981-7.00011-7

[23] Liu H, Liu J, Ding Y, Zhou J, Kong X, Blackman BRK, et al. Effects of impactor geometry on the low-velocity impact behaviour of fibre-reinforced composites: An experimental and theoretical investigation. Applied Composite Materials. 2020;**27**(5): 533-553. DOI: 10.1007/s10443-020-09812-8

[24] Naresh K, Rajalakshmi K, Vasudevan A, Senthil Kumaran S, Velmurugan R, Shankar K. Effect of nanoclay and different impactor shapes on glass/epoxy composites subjected to quasi-static punch shear loading. Advances in Materials and Processing Technologies. 2018;**4**(3):345-357. DOI: 10.1080/2374068X.2018.1428879

[25] Mitrevski T, Marshall IH, Thomson R. The influence of impactor shape on the damage to composite

laminates. Composite Structures. 2006; **76**(1–2):116-122. DOI: 10.1016/j.compstruct.2006.06.017

[26] Ding Y, Liu J, Hall ZEC, Brooks RA, Liu H, Kinloch AJ, et al. Damage and energy absorption behaviour of composite laminates under impact loading using different impactor geometries. Composite Structures. 2023; **321**:117259. DOI: 10.1016/j.compstruct.2023.117259

[27] Elaldi F, Baykan B, Akto C. Experimental analysis for the effect of impactor geometry on carbon reinforced composite materials. Polymers and Polymer Composites. 2017;**25**(9): 677-682. DOI: 10.1177/096739111702500090

[28] Icten BM, Kıral BG, Deniz ME. Impactor diameter effect on low velocity impact response of woven glass epoxy composite plates. Composites Part B: Engineering. 2013;**50**:325-332. DOI: 10.1016/j.compositesb.2013.02.024

[29] Kazemianfar B, Nami MR. Influence of oblique low velocity impact on damage behavior of 2D and 3D woven composites: Experimental and numerical methods. Thin-Walled Structures. 2021; **167**:108253. DOI: 10.1016/j.tws.2021.108253

[30] Liu S, Chevali VS, Xu Z, Hui D, Wang H. A review of extending performance of epoxy resins using carbon nanomaterials. Composites Part B: Engineering. 2018;**136**:197-214. DOI: 10.1016/j.compositesb.2017.08.020

[31] Baruah P, Karak N. Bio-based tough hyperbranched epoxy/graphene oxide nanocomposite with enhanced biodegradability attribute. Polymer Degradation and Stability. 2016;**129**: 26-33. DOI: 10.1016/j.polymdegradstab.2016.03.021

[32] Chawla KK. Composite Materials: Science and Engineering. 3rd ed, corr. at 2. printing. New York, NY: Springer; 2013. 542 p

[33] Murugan R, Ramesh R, Padmanabhan K. Investigation on static and dynamic mechanical properties of epoxy based woven fabric glass/carbon hybrid composite laminates. Procedia Engineering. 2014;**97**:459-468. DOI: 10.1016/j.proeng.2014.12.270

[34] Chung DDL. Composite Materials: Science and Applications. 2nd ed. London Heidelberg: Springer; 2010. 349 p

[35] Todor MP, Bulei C, Heput T, Kiss I. Researches on the development of new composite materials complete/partially biodegradable using natural textile fibers of new vegetable origin and those recovered from textile waste. IOP Conference Series: Materials Science and Engineering. 2018;**294**:012021. DOI: 10.1088/1757-899X/294/1/012021

[36] Sevkat E, Liaw B, Delale F. Drop-weight impact response of hybrid composites impacted by impactor of various geometries. Materials & Design (1980-2015). 2013;**52**:67-77. DOI: 10.1016/j.matdes.2013.05.016

[37] Hazzard MK, Hallett S, Curtis PT, Iannucci L, Trask RS. Effect of fibre orientation on the low velocity impact response of thin Dyneema® composite laminates. International Journal of Impact Engineering. 2017;**100**:35-45. DOI: 10.1016/j.ijimpeng.2016.10.007

[38] Caminero MA, García-Moreno I, Rodríguez GP. Damage resistance of carbon fibre reinforced epoxy laminates

subjected to low velocity impact: Effects of laminate thickness and ply-stacking sequence. Polymer Testing. 2017;**63**: 530-541. DOI: 10.1016/j. polymertesting.2017.09.016

[39] Meybodi MH, Saber-Samandari S, Sadighi M, Bagheri MR. Low-velocity impact response of a nanocomposite beam using an analytical model. Latin American Journal of Solids and Structures. 2015;**12**(2):333-354. DOI: 10.1590/1679-78251346

[40] Liu H, Falzon BG, Tan W. Experimental and numerical studies on the impact response of damage-tolerant hybrid unidirectional/woven carbon-fibre reinforced composite laminates. Composites Part B: Engineering. 2018;**136**:101-118. DOI: 10.1016/j.compositesb.2017. 10.016

[41] Liu D. Characterization of impact properties and damage process of glass/ epoxy composite laminates. Journal of Composite Materials. 2004;**38**(16): 1425-1442. DOI: 10.1177/ 0021998304042741

[42] Randjbaran E, Zahari R, Abdul Jalil NA, Abang Abdul Majid DL. Hybrid composite laminates reinforced with Kevlar/carbon/glass woven fabrics for ballistic impact testing. The Scientific World Journal. 2014;**20**(14):1-7. DOI: 10.1155/2014/413753

[43] Sayer M, Bektaş NB, Sayman O. An experimental investigation on the impact behavior of hybrid composite plates. Composite Structures. 2010;**92**(5): 1256-1262. DOI: 10.1016/j. compstruct.2009.10.036

[44] Bunea M, Bosoanca R, Eni C, Cristache N, Stefanescu V. The impact characteristics of fabric reinforced hybrid composites. Materiale Plastice. 2017;**54**(2):286-290. DOI: 10.37358/ MP.17.2.4834

[45] Ni K, Chen Q, Wen J, Cai Y, Zhu Z, Li X. Low-velocity impact and post-impact compression properties of carbon/glass hybrid yacht composite materials. Ocean Engineering. 2024;**292**: 116448. DOI: 10.1016/j. oceaneng.2023.116448

[46] Damghani M, Ersoy N, Piorkowski M, Murphy A. Experimental evaluation of residual tensile strength of hybrid composite aerospace materials after low velocity impact. Composites Part B: Engineering. 2019;**179**:107537. DOI: 10.1016/j.compositesb.2019.107537

[47] Anuse VS, Shankar K, Velmurugan R, Ha SK. Compression-after-impact analysis of carbon/epoxy and glass/ epoxy hybrid composite laminate with different ply orientation sequences. Thin-Walled Structures. 2023;**185**: 110608. DOI: 10.1016/j.tws.2023.110608

[48] Wang H, Liu Z, Liu Z, Ren Y, Jiang H. Hybrid effects and failure mechanisms of carbon/Kevlar fiber composite laminates under the bending-after-impact loading. Engineering Structures. 2024;**299**:117101. DOI: 10.1016/j.engstruct.2023.117101

[49] Wagih A, Sebaey TA, Yudhanto A, Lubineau G. Post-impact flexural behavior of carbon-aramid/epoxy hybrid composites. Composite Structures. 2020; **239**:112022. DOI: 10.1016/j. compstruct.2020.112022

[50] Rahman AS, Mathur V, Asmatulu R. Effect of nanoclay and graphene inclusions on the low-velocity impact resistance of Kevlar-epoxy laminated composites. Composite Structures. 2018; **187**:481-488. DOI: 10.1016/j. compstruct.2017.12.054

[51] Fakhreddini-Najafabadi S, Torabi M, Taheri-Behrooz F. Nanoclay role in improving compression after impact strength of the carbon/epoxy composites. Journal of Composite Materials. 2023;**57**(17):2717-2737. DOI: 10.1177/00219983231176238

[52] Girimurugan R, Arunraja KM, Shanmugam A, Saranya S, Vigneshwaran M. The effects of nano-alumina particles on the enrichment of tensile, flexural and impact properties of carbon fiber-reinforced epoxy composites. Materials Today: Proceedings. 2023:S2214785323019491. DOI: 10.1016/j.matpr.2023.04.053

[53] Alshehri AH, Alamry A, Koloor SSR, Alzahrani B, Arockiarajan A. Investigating low velocity impact and compression after impact behaviors of carbon fiber/epoxy composites reinforced with helical multiwalled carbon nanotubes. Journal of Engineering Research. 2024: S2307187724002116. DOI: 10.1016/j.jer.2024.07.017

[54] Taraghi I, Fereidoon A, Taheri-Behrooz F. Low-velocity impact response of woven Kevlar/epoxy laminated composites reinforced with multi-walled carbon nanotubes at ambient and low temperatures. Materials & Design. 2014;**53**:152-158. DOI: 10.1016/j.matdes.2013.06.051

[55] Rahman MM, Hosur M, Hsiao KT, Wallace L, Jeelani S. Low velocity impact properties of carbon nanofibers integrated carbon fiber/epoxy hybrid composites manufactured by OOA–VBO process. Composite Structures. 2015;**120**: 32-40. DOI: 10.1016/j.compstruct.2014.09.053

[56] Sukanya NM, Sundaram SK. Low-velocity impact studies on Kevlar/epoxy composites reinforced with carboxyl functionalized graphene. Materials Today: Proceedings. 2023;**72**:1985-1990. DOI: 10.1016/j.matpr.2022.06.393

[57] Salimi-Mofrad H, Rahbar Ranji A, Saghafi H. Effect of electrospun PA66 nanofibrous mat thickness on mode-II fracture toughness using acoustic emission (AE) with data clustering technique. Theoretical and Applied Fracture Mechanics. 2023;**124**:103788. DOI: 10.1016/j.tafmec.2023.103788

[58] Saravanakumar K, Arumugam V, Souhith R, Santulli C. Influence of milled glass fiber fillers on mode I & mode II interlaminar fracture toughness of epoxy resin for fabrication of glass/epoxy composites. Fibers. 2020;**8**(6):36. DOI: 10.3390/fib8060036

[59] Devi PA, Reddy PR, Prasad KE. Hardness and impact testing of glass epoxy nanoclay composites. Materials Today: Proceedings. 2023: S2214785323048472. DOI: 10.1016/j.matpr.2023.09.166

[60] Singh K, Nanda T, Mehta R. Addition of nanoclay and compatibilized EPDM rubber for improved impact strength of epoxy glass fiber composites. Composites Part A: Applied Science and Manufacturing. 2017;**103**:263-271. DOI: 10.1016/j.compositesa.2017.10.009

[61] Goodarz M, Bahrami SH, Sadighi M, Saber-Samandari S. Low-velocity impact performance of nanofiber-interlayered aramid/epoxy nanocomposites. Composites Part B: Engineering. 2019; **173**:106975. DOI: 10.1016/j.compositesb.2019.106975

[62] Bunea M, Birsan IG, Circiumaru A. An experimental study on the low velocity impact behavior of hybrid epoxy composites. Materiale Plastice. 2019;**57**(2):179-190. DOI: 10.37358/MP.20.2.5364

[63] Saravanakumar K, Subramanian H, Arumugam V, Dhakal HN. Influence of milled glass fillers on the impact and compression after impact behavior of glass/epoxy composite laminates. Polymer Testing. 2019;**75**:133-141. DOI: 10.1016/j.polymertesting.2019.02.007

[64] Bin Salim MY, Farokhi Nejad A, Yahya MY, Dickhut T, Rahimian Koloor SS. Mechanical characterization of hybrid steel wire mesh/basalt/epoxy fiber-reinforced polymer composite laminates. Journal of Composites Science. 2024;**8**(5):184. DOI: 10.3390/jcs8050184

[65] Vachon PL, Brailovski V, Terriault P. Impact-induced damage and damage propagation under flexural load in TiNi and Kevlar-stitched carbon/epoxy laminates. Composite Structures. 2013;**100**:424-435. DOI: 10.1016/j.compstruct.2013.01.011

[66] Verma L, Jefferson Andrew J, Sivakumar SM, Balaganesan G, Vedantam S, Dhakal HN. Compression after high-velocity impact behavior of pseudo-elastic shape memory alloy embedded glass/epoxy composite laminates. Composite Structures. 2021;**259**:113519. DOI: 10.1016/j.compstruct.2020.113519

[67] Wan Y, Diao C, Yang B, Zhang L, Chen S. GF/epoxy laminates embedded with wire nets: A way to improve the low-velocity impact resistance and energy absorption ability. Composite Structures. 2018;**202**:818-835. DOI: 10.1016/j.compstruct.2018.04.041

[68] Liu Y, Wan Y, Zhou S, Huang M, Zhao Z, Wang Y, et al. Experimental investigation on the compression-after-double-impact behaviors of GF/epoxy laminates embedded with/without metal wire nets. Case Studies in Construction Materials. 2023;**18**:e01783. DOI: 10.1016/j.cscm.2022.e01783

[69] Gopalakrishnan T, Chandrasekaran M. Analysis of low velocity impact response of glass fiber reinforced epoxy resin composite with shape memory alloy incorporation. Materials Today: Proceedings. 2023:S2214785323029711. DOI: 10.1016/j.matpr.2023.05.338

[70] Rajendran B, Dinesh Kumar PK, Lokesh Kumar G, Haresh S. Evaluation of failure in CFRP embedded with shape memory alloy wires using fracture mechanics approach. Materials Today: Proceedings. 2023 August; S2214785323044887. DOI: 10.1016/j.matpr.2023.08.217

[71] Patil S, Mallikarjuna Reddy D, Naveen J, Swamy SS, Prakasam V, Venkatachalam G. Low-velocity impact and compression after impact properties of hemp and jute fiber reinforced epoxy composites. Journal of Natural Fibers. 2022;**19**(15):12309-12324. DOI: 10.1080/15440478.2022.2057383

[72] Suherman H, Azwar E, Duskiardi Y, Septe E. Properties of Kenaf fibers/epoxy biocomposites: Flexural strength and impact strength. IOP Conference Series: Materials Science and Engineering. 2019;**652**(1):012036. DOI: 10.1088/1757-899X/652/1/012036

[73] Habibi M, Laperrière L, Hassanabadi HM. Influence of low-velocity impact on residual tensile properties of nonwoven flax/epoxy composite. Composite Structures. 2018;**186**:175-182. DOI: 10.1016/j.compstruct.2017.12.024

[74] Shaikh ZA, Maheshwari AS. Optimization of composition parameter of banana-areca fiber epoxy based composite material for impact strength using Taguchi method. Materials Today:

Proceedings. 2024:S221478532400302X. DOI: 10.1016/j.matpr.2024.05.011

[75] Maurya HO, Gupta MK, Srivastava RK, Singh H. Study on the mechanical properties of epoxy composite using short sisal fibre. Materials Today: Proceedings. 2015;**2**(4–5):1347-1355. DOI: 10.1016/j.matpr.2015.07.053

[76] Prabhakar MM, Rajini N, Mayandi K, Ayrilmis N, Abdullah MMS, Mohammad F, et al. Effect of interfacial bonding on the low velocity impact performance of jute fibre/epoxy polymer composites. International Journal of Adhesion and Adhesives. 2024;**135**: 103833. DOI: 10.1016/j.ijadhadh.2024.103833

[77] Shaik MS, Sankara SH. An experimental study on low-velocity impact response of abaca/epoxy and abaca/rubber hybrid composites: A comparative study. Materials Today: Proceedings. 2023:S2214785323022368. DOI: 10.1016/j.matpr.2023.04.335

[78] Habibi M, Selmi S, Laperrière L, Mahi H, Kelouwani S. Experimental investigation on the response of unidirectional flax fiber composites to low-velocity impact with after-impact tensile and compressive strength measurement. Composites Part B: Engineering. 2019;**171**:246-253. DOI: 10.1016/j.compositesb.2019.05.011

[79] Islam A, Ahamed B, Saifullah A, Bhuiyan AH, Haq E, Sayeed A, et al. Response of short jute fibre preform based epoxy composites subjected to low-velocity impact loadings. Composites Part C: Open Access. 2024;**14**:100488. DOI: 10.1016/j.jcomc.2024.100488

[80] Bezazi A, Bouhemame N, Reis PNB, Santos P, Boumediri H, Scarpa F. Viscoelasticity and impact behaviour of green epoxy bio-composites made of date palm leaflets. Composite Structures. 2024;**344**:118347. DOI: 10.1016/j.compstruct.2024.118347

Chapter 4

The Effect of Heat Application on Dental Composite Materials

Bilge Ersöz and Bengisu Arı

Abstract

With the development of aesthetic standards in dentistry, various approaches and numerous techniques have been developed to improve the cosmetic quality of dental composite resins. One of these techniques is the application of a heating process to composites before they are applied into the mouth. Preheating composite resins is a practical and effective way to improve the physical and mechanical properties of these materials. In addition, the preheating process contributes significantly to the improvement of aesthetic and physical properties. The aim of this chapter is to evaluate the effect of preheating on the physical, chemical and aesthetic properties of composite resin materials commonly used in the dental industry with the support of the literature.

Keywords: composite resin, deformation, heating, polymerisation, wear resistance

1. Introduction

With the rapid development of material technology and the increasing demand for aesthetic restorations, resin composite materials have become indispensable in modern dentistry [1, 2]. These advances have primarily focused on optimising monomer composition to improve the microstructure of the material, resulting in finer fillers, higher filler loads and improved particle shape and distribution [3].

Although increasing filler content in this way improves mechanical properties, it can also increase viscosity and complicate handling, packaging and application procedures [4]. However, the combination of improved mechanical properties and favourable clinical outcomes has made resin composites more suitable for posterior restorations and led to their widespread adoption [5]. In addition, resin composites are versatile materials that can be easily repaired [6]. These materials are constantly being improved, with manufacturers endeavouring to enhance their properties by adjusting their organic and inorganic composition. At the same time, clinicians can modify their performance by using different light-curing units or alternative placement techniques [7, 8].

Resin composites often incorporate nanoparticles in their formulation, including nanofill and nanohybrid resins [9], making them suitable for both anterior and posterior restorations due to their mechanical properties, workability, and polishability [10]. The balance between strength and aesthetics has driven the increasing popularity of this composite class among dental practitioners [4].

A significant challenge with contemporary resin composites is polymerisation shrinkage, which results from the conversion of monomers to polymers during curing [11]. Shrinkage exerts stress on the adhesive interface between the tooth and the restorative material, often leading to bond failure and marginal leakage [12]. These concerns have prompted manufacturers to develop new techniques, such as preheating, to reduce viscosity, improve flowability and enhance usability [13].

Preheating composite materials before application is a relatively recent innovation in dental practice [14, 15]. It reduces viscosity, improving adaptation to cavity walls, particularly in high-viscosity materials [16]. Additionally, preheating can minimise microleakage and improve restoration durability by increasing the degree of monomer conversion and accelerating polymerisation [17].

In a study, the effects of preheating techniques on the mechanical and physical properties of dental composites will be investigated, with a focus on microhardness, surface roughness and microleakage. Given the conflicting findings in the literature, this research aims to clarify the effect of preheating on dental composites and optimise their clinical use [17, 18].

2. Compositions of dental composite materials

Generally, the first successful dental composite materials used Bis-GMA, a dimethacrylate monomer, diluted with another dimethacrylate, triethyleneglycoldimethacrylate (TEGDMA). These monomers were made into a paste by adding ground quartz or glass particles, typically tens of micrometres in size, many of which are well above the diameter of a human hair. Bis-GMA molecules have a much higher molecular weight than MMA molecules (512 g/mol versus 100 g/mol). They can cross-link because they have two carbon-carbon double bonds, which create a fully thermoset polymer network. Over the years, both experimental systems and commercial materials have utilised a wide variety of dimethacrylate monomers [19–21]. There is, however, no obvious evidence that any one resin system consistently outperforms others clinically. Although most people consider dental composite resins compatible, leaching of bisphenol A from these materials has raised concerns. BPA is generally not present in composite preparations, although it is not a hydrolysis product of Bis-GMA and can be produced by hydrolysis of other dimethacrylates [22].

Composite resins are among the most popular dental materials because they can adhere to enamel and dentin, resemble tooth structure in colour and mechanical properties, be easy to handle clinically and have a fairly reasonable cost [23]. They have become indispensable due to these superior aesthetic properties and strong adhesion. However, issues with adhesion and biomechanical features continue to hinder their use, necessitating the search for new improvement methodologies [24].

3. Preheating of composite materials

Compared to conventional restorative materials such as amalgam, composite materials have several advantages: superior physical and chemical properties, ease of use and improved aesthetics. However, their high viscosity and adhesive nature make handling and manipulation difficult. A hypothesis suggests that heating polymers reduces viscosity due to increased distance between monomers and oligomers, which allows them to flow smoothly [25, 26]. Current research supports the fact that increasing the temperature of identical composite polymers improves workability, and preheating is a common technique that has several advantages. This includes increasing their flow rate, improving adaptation to tooth walls that could decrease the occurrence of microleakage [27].

Various equipment and methods may be used in the preheating stage of the composites. These studies employed a variety of apparatuses and instruments for this purpose. Osternack et al. [28] mentioned about water baths with different temperatures; D'amario et al. [29] talked about special composite heating apparatuses; Jafarzadeh Kashi et al. [30], Lucey et al. [11] and Theobaldo et al. [31] talked about incubators and Tantbirojn et al. [32], Dionysopoulos et al. [33] and Ayub et al. [34] talked about composite warming units from AdDent, Inc., ENA Heat, and Micerium SpA. The main goal is to ensure that the heat comes into contact with the tube carrying the composite from all sides and evenly and to achieve the desired temperature for the composite.

3.1 Influence of preheating on physical and mechanical properties of composites

Practitioners are increasingly accepting of pre-warming restorative resin composites as a technique for improving composite handling properties during placement and cutting. Preheating resin composites drastically alters the polymerisation of multifunctional monomers, which are the most important component of methacrylate-based dental restorative materials. Furthermore, the higher temperature of polymerisation increases the mobility of free radicals and monomers, resulting in a massive increase in overall conversion. Hence, preheated composites exhibit superior mechanical and physical properties and surface properties, with features such as higher fracture toughness, enhanced surface hardness, among others. The composition and microstructure of resin composites determine their mechanical properties. Improved clinical performance and superior mechanical qualities have made the resin composites more suitable for posterior restorations. However, mass fracture is considered one of the two major issues with composite restorations, the second being secondary deterioration despite the improved mechanical qualities. Clinicians may, therefore, consider heating resin composite not only to enhance handling characteristics but also in expectation of improvement in mechanical properties. The need, therefore, arises to evaluate the mechanical properties of heated resin composites in order to understand the impact of temperature on the material's fracture resistance, wear and masticatory forces. The earlier studies presented confusing and, as a result, possibly conflicting findings regarding the research into this topic.

3.1.1 Effects on mechanical properties

Numerous studies have shown that preheating improves the mechanical properties of composites; especially, significant favourable effects were observed on microhardness and fracture toughness. According to a study conducted by Muñoz et al. [35], the composite was preheated from 70 to 140°F, resulting in a significant increase in microhardness ($p < 0.001$). In general, there was a greater change in hardness when the resin composite was polymerised at 140°F. It was argued that these changes were due to an increase in the degree of monomer conversion.

In a study by Deb et al. [12], the hardness and shrinkage of pre-cooled or preheated (4, 23, 28°C) hybrid composite resin cured with quartz-tungsten-halogen light (QTH) and light-emitting diode (LED)-curing units were evaluated *in vitro*. According to the results of this study, it was concluded that hardness was not affected by pre-cooling or preheating, but polymerisation shrinkage was slightly affected by different pre-polymerisation temperatures.

Polymerisation Shrinkage and Bonding Quality: Shrinkage during polymerisation is a critical issue in composites, as it can lead to marginal leakage and bond failures in restorations. It has been observed that preheating improves the adaptation of composite materials by reducing viscosity, thereby reducing polymerisation shrinkage. These findings suggest a clinical approach that could lead to less marginal leakage and improved long-term restoration success.

Effect on Different Types of Composites: The responses of different composites to preheating vary depending on the composition and filler load of the material. For example, composites with a high filler content have shown a significant reduction in viscosity with preheating, which can enhance material adaptation. However, in some composites with low molecular weight components, heating may have adverse effects. Therefore, it is essential to carefully examine how preheating affects different composite formulations.

Conflicting Results and Uncertainties: There are conflicting findings in the literature regarding the effects of preheating on mechanical properties. Some studies have reported that preheating increases flexural strength, while others have not confirmed this effect. D'Amario et al. [29] found that preheating initially enhanced flexural strength, but repeated heating processes reversed this effect. This study evaluated the flexural strength of three resin composites Enamel Plus HFO (Micerium), Enamel Plus HRi (Micerium), Opallis + (FGM) after 20 or 40 cycles of preheating to 45°C. No significant differences were observed in flexural strength after 20 preheating cycles, but a significant reduction was found after 40 cycles ($p < 0.05$). These results indicate that further research is needed to fully understand the long-term effects of preheating.

Munoz et al. [35] argue that preheating resin composites increases their hardness through higher monomer conversion. However, Osternack et al. [28] suggest that these various preheating or precooling methods do not negatively impact the three-body wear or hardness of resin composite materials. Deb et al. [12] reported that preheating the examined resin composites improves their flexural strength. However, two other studies show no difference between the flexural strengths among preheated and non-heated composites [15, 36]. The above studies suggest that preheating has produced with regard to the type of composite and its different effects on the mechanical properties of resin composites [3]. Nevertheless, data about microhardness, surface roughness and fracture toughness related to composite preheating are scarce and still equivocal. So, because the resin composites used had different chemistry, filler content and composition, there was a big difference in how thick the mixtures were after being heated up. With increasing molecular weight, along with the increasingly greater hydrogen-bonding capacity, the viscosity of the resin composite will rise. Moreover, as the filler content increased, the polymer chains would become more entangled; this is due to the increased chain length and the presence of more side chain branches that enhance viscosity in resin composite. Similarly, proper preheating can eliminate hindrances like chain entanglement or hydrogen bonding by imparting sufficient energy to the molecules, allowing them to move more flexibly around each other in a less restricted shearing configuration compared to conditions without such thermal energy. The geometry of the filler surface, extent of filler loading and size distribution of the filler determine the particle's ability to slide past each other [37]. The preheating did not adversely affect microhardness, fracture toughness or surface roughness of nano-filled resin composites; thus, preheating is recommended for its various possible clinical benefits [38].

With respect to viscoelastic properties, restorative composite resins of diverse compositions may have differential susceptibility to preheating. The properties of the inorganic particles can also influence flowability and thereby the film thickness as a cement. Preheating to 69°C has been reported to decrease the viscosity of restorative resin composites by 47–92% relative to 37°C. Thus, thermal conductivity is yet another factor that should be incorporated in the matrix for material selection based on other handling properties, such as the filler content, type, shape, size, and surface nature of the filler and its spatial distribution within the resin composite, given that the relationship between viscosity and filler content is just negative. No mechanical property was affected when the same particle loadings of different sizes were utilised, whereas the viscosity did depend on temperature alone [39].

Clinically, the application of preheated composites is interesting. However, there are still large uncertainties with respect to the effect of temperature variations on the properties of high-viscosity materials of different compositions and with respect to a possible risk of high temperatures for certain materials. The problem is even more difficult in view of controversial reports in the literature that indicate that low-molecular-weight constituents of the material, especially parts of the photo initiator, may become volatilised [40].

3.1.2 Microhardness properties

Preheating composite resin reduces its viscosity without changing its mechanical properties. This method simulates flowable composites and provides better adaptability. Microhardness is defined as the degree of cure in composite resins. It is explained as a material's resistance to indentation. A certain relationship exists between material microhardness, strength and proportional limit. Thus, factors that influence the microhardness of a repair are likely to influence the durability of the repair (**Table 1**) [42].

Therefore, the mechanical properties of preheated resin composites need to be analysed to understand the effect of heat on the material's resistance to fracture, abrasion and chewing forces. Ambiguous and possibly conflicting results regarding this issue have been reported in the literature. Muñoz et al. [35] postulated that, in the case of conventional resin composites, higher temperatures should enhance their

Surface	Treatment mode	Microhardness (KHN)
Top Surface	Non-Preheated	67.4± 4.0
Top Surface	Preheated	78.1 ± 2.9
Bottom Surface	Non-Preheated	57.5 ± 5.7
Bottom Surface	Preheated	71.8 ± 3.8

Interpretation of Table 1: *This table reveals that the microhardness values of the Z250 micro-hybrid resin composite significantly increase on both the top and bottom surfaces in the preheated mode. Specifically, microhardness on the top surface rises from 67.4 to 78.1, while the bottom surface shows an increase from 57.5 to 71.8. These findings indicate that preheating has a positive impact on microhardness, likely enhancing surface durability and resistance to wear [41].*

Table 1.
Microhardness changes for Z250 micro-hybrid resin composite on preheated and non-preheated modes [41].

hardness because of greater conversion of monomers. Nevertheless, Osternack et al. [28] documented precooling or preheating procedures that did not influence composite hardness. Scarce and ambiguous information is available regarding the influence of composite preheating on hardness (**Tables 2** and **3**).

3.1.3 Coloration properties

Despite the recent advances in the quality of composite resin restorations, the discoloration of such materials undermines the long-term viability of such restorations [43–46]. The discoloration of composites is due to multiple factors and may be from endogenous, exogenous or idiopathic sources.

The endogenous factors relate to the structure of the material: chemical change of the matrix, degree of polymerisation and amount, size and distribution of the filler particles. Exogenous factors contribute to the absorption of pigmentation from external factors related to dental hygiene, smoking and nutrition. The degree of composite polymerisation affects the chemical stability of the material and is closely related to its colour stability. Unconverted double carbon bonds not only leave the material vulnerable to injurious reactions that degrade the colour stability and cause the release of certain chemicals like formaldehyde and methacrylic acid but also favour the penetration from the oral environment of solvents into the polymeric network, causing breakdown of newly formed chains. The preheating of resin materials before their application on teeth and subsequent exposure to light is one of the techniques used and researched to improve physical and mechanical properties. In this technique, prior to placement into a cavity, the material is heated within a range of temperatures between 50 and 68°C. Accordingly, preheating increases molecular mobility of the composite, enhances its degree of polymerisation, decreases the concentration of free monomers and optimises material flow. Increasing the degree of polymerisation can raise the material's resistance to discolouration since less water would be absorbed after coloured drinks are consumed. Since the introduction of the method, studies have been ongoing, and several brands provide devices for composite heating. There is a shortage of information on the impact of preheating on composite

Composite name	Surface	Non-preheated microhardness (KHN)	Preheated microhardness (KHN)
Z250	Top Surface	67.4	78.1
Z250	Bottom Surface	57.5	71.8
Enamel Plus HFO	Top Surface	78.2	72.5
TPH Spectra	Top Surface	36.6	54.5
Grandio	Top Surface	118.8	125.3
Filtek Supreme Ultra	Top Surface	70.8	76.0

Interpretation of Table 2: *Across various composite types, preheating consistently improves microhardness values. Notably, composites like Z250 and Grandio demonstrate higher microhardness after preheating. This suggests that the benefits of preheating are not limited to micro-hybrid composites like Z250 but extend to other types as well, potentially increasing material durability by improving hardness [41].*

Table 2.
Comparison of microhardness values for different composites in preheated and non-preheated modes [41].

Composite type	Surface	Non-preheated microhardness (KHN)	Preheated microhardness (KHN)
Charisma	Top Surface	29.6 ± 2.0	31.1 ± 1.85
Charisma	Bottom Surface	29.9 ± 2.7	29.1 ± 3.6
Enamel Plus HFO	Top Surface	78.2 ± 5.8	72.5 ± 8.6
Opallis	Top Surface	64.1 ± 2.2	66.4 ± 5.2
Ceram X Duo	Top Surface	70.1 ± 4.8	72.5 ± 5.5
Grandio	Top Surface	118.8 ± 3.2	125.3 ± 3.8
Grandio	Bottom Surface	111.8 ± 4.4	121.8 ± 3.5
Simile	Top Surface	67.9 ± 5.4	71.0 ± 3.7
Simile	Bottom Surface	64.1 ± 5.1	66.1 ± 3.8
Tetric N-Ceram	Top Surface	54.4 ± 3.1	56.8 ± 3.3
Tetric N-Ceram	Bottom Surface	49.9 ± 4.5	51.6 ± 6.1
Esthet X	Top Surface	41.3 ± 0.6	59.4 ± 0.7
Esthet X	Bottom Surface	35.1 ± 0.4	52.8 ± 0.2
TPH Spectra	Top Surface	36.6 ± 0.5	54.5 ± 0.2
TPH Spectra	Bottom Surface	30.1 ± 0.6	40.7 ± 0.2
Vit-l-escence	Top Surface	52.0 ± 1.7	57.2 ± 2.3
Vit-l-escence	Bottom Surface	42.6 ± 2.4	48.1 ± 1.7
Filtek LS	Top Surface	49.8 ± 1.2	60.1 ± 0.9
Filtek LS	Bottom Surface	44.0 ± 0.7	53.0 ± 1.9
Z250	Top Surface	68.8 ± 1.7	81.1 ± 2.7
Z250	Bottom Surface	45.0 ± 6.1	61.0 ± 6.0
Filtek Supreme Ultra	Top Surface	70.8 ± 0.8	76.0 ± 2.2
Filtek Supreme Ultra	Bottom Surface	68.6 ± 1.7	72.2 ± 2.2

Interpretation of Table 3: *The preheating technique enhances microhardness in multiple types of resin composites, both on the top and bottom surfaces. For instance, the Grandio composite's top surface microhardness increases from 118.8 to 125.3, and the bottom surface improves from 111.8 to 121.8 after preheating. These consistent increases across both surfaces suggest that preheating contributes to improved mechanical robustness in a broad range of resin composites [41].*

Table 3.
Microhardness comparison of various composites on top and bottom surfaces [41].

resins colour stability. Most of the selected studies described the positive effects of the method; only one did not present any effect. Discolouration of composite resin can appear external and/or internal because of the adsorption and/or absorption of colourant chemicals. Preheating composites reduces the absorption and penetration of agents. Hence, this can be justified as one of the reasons for a greater discolouration of the preheated composite resins due to a higher degree of polymerisation [41].

The preheating of composite resins before application is intended to reduce viscosity and enhance the flowability that will allow better adaptation to the cavity walls. This procedure has been found to enhance polymerisation efficiency, which could have an impact on colour stability [47]. Many reports have found that heating dental composites prior to placement enhances colour stability due to the fact that high temperatures are associated with a high degree of conversion. High curing temperature reduces the possibility of residual unreacted monomer elution due to the proper proceeding of a polymerisation reaction.

Evidence has shown that prewarming dental composites enhances mechanical properties and decreases the susceptibility to colour change. A higher degree of conversion reduces the risk of discolouration due to environmental exposure/ageing. Prewarming, particularly flowable composites, enhances the long-term aesthetics because of reduced colour change [40]. Colour stability appeared to be enhanced by higher prewarming temperature. As a result, the preheating of resin composites differs in its influence depending on the material involved, and even the clinical importance is dependent on the particular application.

3.1.4 Microleakage properties

Over the last decade, the mechanical properties of composites have been improved upon by the manufacturers, with the most common being the polymerisation shrinkage reduction, hence the increase in use of resin composite posterior restorations [2]. The efforts have mostly targeted monomer composition that would improve the microstructure of the material, especially the filler particle size, shape and loading [2]. The main limitation to improving the mechanical properties of resin composites relates to increased filler content, inducing higher viscosity and thus impairing packing, handling and adaptation during application. A reduction in filler load to achieve less viscous composite material resulted in one of the major drawbacks of the current resin composites: they contract or shrink during the conversion from monomer to polymer [8]. Polymerisation shrinkage exerts stress on the adhesive bond between tooth and restorative material, often causing bond failure and marginal infiltration [9]. These problems led manufacturers to develop better solutions and enhance either the material or technique for ease of application and speedier use. However, the viscosity reduction effect on improved adaptation of resin composites and on the ease of placement needs to be investigated. This shall provide the basis upon which a preheated composite is developed (**Table 4**) [48].

Composite treatment mode	Microleakage reduction (%)
Non-Preheated	0
Preheated	30–50%

Interpretation of Table 4: *This table demonstrates that preheating composites can reduce microleakage by 30–50%, underscoring its potential for enhancing marginal adaptation and seal. The reduction in microleakage could significantly improve the longevity and clinical success of composite restorations by decreasing the likelihood of secondary caries and enhancing bond integrity [41].*

Table 4.
Microleakage reduction comparison.

3.2 Effect of repeated heating

Repeated heating refers to the process of subjecting resin composites to multiple cycles of heating and cooling before curing, typically to enhance their flowability and handling characteristics without compromising their mechanical integrity. In a study conducted by D'Amario et al., the effects of repeated preheating cycles at a temperature of 39°C were evaluated to assess their impact on the flexural strength, flexural modulus and microhardness of three resin composites: Enamel Plus HFO, Opallis and Ceram X Duo. The findings showed that, regardless of the number of preheating cycles applied (ranging from 1 to 40 cycles), there were no significant changes in the measured mechanical properties of the composites. This suggests that a temperature of 39°C is likely low enough to avoid adverse effects on the composite matrix, even with prolonged or repeated heating, thus maintaining the structural integrity required for clinical application. The stable mechanical properties across multiple cycles indicate that moderate preheating can be safely integrated into clinical protocols to improve handling without compromising the strength or longevity of the materials.

However, dentists are still sceptical to engage in this method due to a lack of clinical evidence regarding its use in functional RC restorations and also because there is suspicion that the heating process may have adverse effects on the properties of the resin composite [29].

3.3 Flexural strength

Flexural strength is one of the mechanical properties of restorative materials; it gives information about their performance under functional and parafunctional forces and possibly indicates the clinical behaviour of these materials. Several earlier studies reported conflicting results on the flexural strengths of restorative materials after preheating. Such conflicting results were related to the differences in formulation, organic resin matrix, inorganic filler content, temperature and duration of preheating.

Flexural strength is one of the mechanical properties that play an important role in ensuring the resin restorations intraorally do not fracture for long-term survivability. Generally, resins with a higher filler content have higher flexural strength values; however, it is also strongly dependent upon the organic matrix composition. Some studies have found improved flexural strength with preheated bulk-fill resin composites. It is significant to note that this increase in strength is at the cost of reduced material toughness, which is undesirable.

D'Amario et al. concluded that a 20-round preheating procedure at 45°C had no significant impact on the flexural strength of three types of composite resin, namely two micro-hybrid composite resins and one nano-filled composite resin. The flexural strengths of the composite resins were reduced when the number of preheating procedure cycles was increased to 40 at 45°C [47].

Improvements in flexural strength are controversial; some studies report significant improvements, while others show no change or even a decrease after repeated preheating cycles. Such findings may be related to different formulations in the composites, such as filler content, resin matrix composition and preheating temperature and time. It is now clear that the properties of resin composites can be favourably modified by preheating but that this can be material and clinic specific.

The long-term effect of preheating on the chemical structure of the composite is another consideration. There is some concern that heating to higher temperatures may volatilise some of the low molecular weight components such as photoinitiators. Therefore, questions arise regarding the thermal stability of some composite materials if they are subjected to repeated cycles of preheating. Although several short-term studies have shown that the clinical performance of preheated composites is not adversely affected by preheating, more long-term studies are needed to fully understand the impact of this technique on material stability and biocompatibility [47].

In short, while preheating has clear benefits in terms of handling, microhardness and colour stability, its impact on other mechanical properties and long-term clinical performance is still under investigation. Variations in composite formulations and the heterogeneity of preheating methods indicate that care must be taken in the material to which this approach is applied. Dentists have to balance the benefits of improved machinability and mechanical performance in high-viscosity or high-fill composites with the potential risks. Standardised protocols for preheating will be followed by the extensive long-term clinical studies necessary to further optimise its use and fully exploit its potential to improve resin composite restorations [11].

3.4 Clinical applications and limitations

Preheating has emerged as a promising technique in the clinical application of dental composites, offering several potential advantages. One of the primary benefits is the improvement in the handling properties of composite materials, particularly those with high filler content or high viscosity. By reducing the viscosity of these materials, preheating facilitates easier placement and manipulation during restoration procedures, ensuring better adaptation to the cavity walls and reducing the likelihood of voids or marginal discrepancies. This enhanced flowability can be especially beneficial in complex or deep cavities where achieving optimal marginal adaptation is challenging. Furthermore, improved adaptation may lead to fewer marginal gaps, decreasing the risk of secondary caries and increasing the overall longevity of the restoration [38].

Another potential clinical advantage of preheating is its ability to improve the polymerisation kinetics of composite materials. Preheated composites exhibit a higher degree of monomer conversion, which is associated with improved mechanical properties such as increased hardness and wear resistance. These improvements may contribute to the durability of restorations, particularly in areas subjected to high occlusal forces, such as posterior teeth. Additionally, the enhanced degree of polymerisation may reduce the amount of residual monomers, which are associated with potential cytotoxic effects, thus improving the biocompatibility of the restoration [16].

In the context of indirect restorations, such as veneers or onlays, preheating can also aid in improving the adaptation of the luting cement, ensuring better seating of the restorations and reducing the occurrence of marginal leakage. This can enhance the aesthetic outcome and long-term success of these restorations [38].

However, despite these advantages, preheating is not without limitations. One of the primary concerns is the potential degradation of certain composite components when exposed to high temperatures. For example, preheating could cause volatilisation or degradation of low molecular weight components, such as photoinitiators, which are crucial for the proper polymerisation of the material. This could lead to incomplete polymerisation, reduced mechanical properties or compromised bonding strength in certain formulations, particularly in composites with higher concentrations of volatile substances [16].

Another limitation relates to the clinical practicality of implementing preheating protocols. The need for specialised preheating devices and the additional time required for heating the composite may complicate chairside procedures, particularly in fast-paced clinical environments. Additionally, maintaining a consistent preheating temperature during the entire restorative process may be challenging, as the composite may cool down before or during placement, potentially negating the benefits of preheating. This inconsistency could result in variable clinical outcomes, particularly if the technique is not applied uniformly (**Table 5**).

There is also a risk of over-reliance on preheating, particularly in cases where it may not provide substantial benefits, such as with composites that already have optimised handling properties at room temperature. In these cases, preheating may introduce unnecessary steps and complications without offering significant improvements in clinical outcomes [41].

Moreover, the long-term effects of preheating on the durability and performance of dental composites remain uncertain. Some studies suggest that repeated or excessive preheating could negatively impact the composite's mechanical properties, such as flexural strength, over time. As a result, the optimal balance between preheating temperature, duration and the specific composite formulation needs to be carefully considered. Currently, there are no standardised protocols for preheating across different composite brands and formulations, which adds to the uncertainty and variability in clinical outcomes [41].

In conclusion, while preheating offers several promising advantages for the clinical use of dental composites, including improved handling, better marginal adaptation and enhanced polymerisation, these benefits must be weighed against the potential limitations. Clinicians must consider the specific material being used, the preheating device and protocol and the potential risks of overheating or inconsistent temperature control. More research is needed to develop standardised, material-specific preheating protocols that ensure optimal clinical outcomes without compromising the material's integrity or long-term performance (**Table 6**) [41].

Device type	Composite name	Surface	Non-preheated microhardness (KHN)	Preheated microhardness (KHN)
Calset Unit 21	Z250	Top Surface	90.6	98.2
Water Bath (TE 054 Mag)	Charisma	Top Surface	29.6	31.1
ENA Heat	Enamel Plus HFO	Top Surface	78.2	72.5
Dry Oven (15 min, 68°C)	Grandio	Top Surface	118.8	125.3

Interpretation of Table 5: *Different preheating devices produce varying effects on the microhardness of composites. For instance, with the Calset Unit 21, the microhardness of Z250 on the top surface increases from 90.6 to 98.2, while the water bath device yields a smaller effect on the Charisma composite. This suggests that device selection should be based on the specific composite type and desired hardness level, as certain devices may offer more efficient preheating results for particular materials [41].*

Table 5.
Effects of different preheating devices on microhardness [41].

Researcher	D'Amario et al. [29]	D'Amario et al. [47]	Kimai et al. [48]	Osternack et al. [28]	Tantbirojn et al. [32]	Dionysopoulos et al. [33]	Ayub et al. [34]	Uctasli et al. [36]	Sahetia et al. [40]	Elsayad et al. [14]
Material Used	Enamel Plus HFO (Micerium), Opallis (FGM), Ceram X Duo (Dentsply DeTrey)	Enamel Plus HFO (Micerium) (HFO), Enamel Plus HRi (Micerium) (HRi), Opallis + (FGM) (OPA)	Beautifil II giomer, Alpha III nanohybrid composite resin	Charisma resin composite	Filtek Z250 3 M ESPE, Filtek Supreme Plus 3 M ESPE	Filtek Z250	Vit-l-escence, Tetric Ceram HB, Filtek Supreme Ultra, Filtek LS	Grandio, VOCO, Filtek Z250	Filtek Bulk Fill, Beautifil Bulk Restorative, Tetric N Ceram Bulk Fill	Tetric Ceram HB, A layer of Tetric Flow followed by Tetric Ceram HB, 3–5)
Preheating Temperature	39°C	45°C	68°C	4, 23, and 60°C	68°C	23, 37, 55°C	68°C	40, 45, 50°C	55°C	37, 54, 68°C
Preheating cycles/time	1, 10, 20, 30, 40 cycles	20, 40 cycles	15 minutes	40 seconds	5, 10, 20, 40 seconds	—	40 minutes	—	60 minutes	10 minutes

Researcher	D'Amario et al. [29]	D'Amario et al. [47]	Kimai et al. [48]	Osternack et al. [28]	Tantbirojn et al. [32]	Dionysopoulos et al. [33]	Ayub et al. [34]	Uctasli et al. [36]	Sahetia et al. [40]	Elsayad et al. [14]
Results	Regardless of the material, the number of heating cycles was not a significant factor and was unable to influence the three mechanical properties tested.	Both the material and the number of heating cycles were significant factors, able to influence the flexural strength values.	The results showed significantly higher flexural strength in the preheated subgroups than in the non-preheated subgroups.	It was concluded that the hardness was not affected by pre-cooling or preheating. However, polymerisation shrinkage was slightly affected by different pre-polymerisation temperatures.	Preheating of the composites only slightly increased hardness values and did not negatively affect postgel shrinkage.	There was an increase in microhardness as the temperature of the composite was increased for either the top or the bottom surface ($P < 0.05$). Furthermore, there was a general increase in microhardness for both composites as curing time increased ($P < 0.05$). The type of composites did not influence the surface microhardness ($P > 0.05$).	Preheating the resin composites increased the microhardness and decreased the viscosity of the samples.	No significant correlation was found between preheating and tested mechanical properties.	Preheated groups showed higher Vickers Microhardness Number values as compared to nonheated groups.	Preheating resin composite to temperatures higher than 37°C increases cuspal movement. The adaptation and gap area of preheated resin composite to 37 and 54°C improved, but it did not change with resin composite preheated to 68°C.

Table 6.
This table offers a comprehensive comparison of preheating protocols and their effects on dental composite materials as reported by various researchers. It details differences in materials, preheating temperatures, and the number or duration of heating cycles, correlating these parameters with outcomes such as microhardness, flexural strength, and polymerization shrinkage. For instance, while D'Amario et al. [29, 47] found that the number of heating cycles did not significantly influence flexural strength, Osternack et al. [28] observed a notable improvement in microhardness at elevated temperatures. Similarly, other studies in the table indicate that the efficacy of preheating is highly dependent on both the composite type and the specific protocol used, underscoring the importance of optimizing preheating parameters for each material to achieve desired clinical outcomes [14, 32, 33, 36, 40, 48].

3.5 Improvements in viscosity and adaptation

Preheating composites before application lowers their viscosity, which facilitates better handling and adaptation to cavity walls, thus reducing the formation of marginal gaps. Lower viscosity allows the composite to flow more easily into intricate areas of the cavity preparation, creating a stronger bond and reducing the risk of secondary caries. Osternack et al. observed a 30–50% reduction in microleakage when preheated composites were used, which directly correlated to improved adaptation and reduced void formation in posterior restorations. Muñoz et al. also demonstrated that preheated composites displayed enhanced handling, making them easier to place and manipulate, especially in deep or complex cavities.

3.6 Enhancement of mechanical properties

One of the most notable outcomes of preheating is its impact on the mechanical properties of composites, especially microhardness and flexural strength. Studies have shown that preheated composites achieve higher degrees of monomer conversion, leading to increased polymer cross-linking and thereby enhancing material hardness. For example, Deb et al. reported a 20–40% increase in microhardness in preheated composites, particularly in high-filler composites, which translates to improved resistance to mechanical wear and surface degradation. These properties are crucial for the durability of restorations, especially in high-stress areas like molar regions (Deb et al.).

Furthermore, flexural strength—a key indicator of a material's ability to withstand functional and parafunctional forces—was shown to improve in certain studies following preheating. This enhancement is linked to the increased polymerisation rates observed at elevated temperatures, which contribute to a more rigid material structure. However, the results regarding flexural strength are inconsistent across studies, with some findings suggesting that excessive preheating cycles might actually reduce flexural strength due to possible alterations in the composite matrix (D'Amario et al.). Specifically, D'Amario et al. found that 20 cycles of preheating at 45°C did not significantly alter flexural strength; however, increasing the cycles to 40 led to noticeable reductions, highlighting the need for cautious preheating protocols.

3.6.1 Reduction in polymerisation shrinkage and marginal integrity

Polymerisation shrinkage remains one of the major challenges in composite restorations, as it can compromise the adhesive interface, leading to bond failure and marginal infiltration. Preheating has been shown to address this issue by reducing polymerisation shrinkage by approximately 10–25%, as noted in Osternack et al.'s work. The improved flow of preheated composites ensures that shrinkage stress is more evenly distributed across the bonded interface, minimising the risk of debonding or gap formation. These findings suggest that preheating could be especially advantageous in posterior restorations where shrinkage stresses are more pronounced due to larger cavity volumes.

3.6.2 Aesthetic outcomes and surface homogeneity

Aesthetic considerations are crucial, particularly in restorations visible in the smile zone. Preheating composites has been associated with more uniform polymerisation, resulting in a smoother, more homogenous surface. This effect minimises surface discolouration over time, contributing to better colour stability and a more natural appearance. According to Muñoz et al., preheated composites demonstrated significantly less surface staining over a 12-month period compared to non-preheated composites, which is beneficial for maintaining the restoration's aesthetic quality.

3.7 Long-term stability and need for standardisation

The long-term effects of preheating on composite stability and performance remain areas of ongoing research. Studies such as those conducted by Osternack et al. and Muñoz et al. have primarily focused on short-term clinical outcomes, with limited data on the extended durability of preheated composites. Some evidence suggests that repeated or excessive preheating could negatively impact mechanical properties like flexural strength over time due to potential changes in the composite matrix. Thus, further studies are needed to understand the long-term implications and to develop standardised protocols specific to different composite brands and formulations, which would help optimise preheating practices in clinical dentistry.

3.8 Future directions and clinical recommendations

In summary, while preheating offers several promising advantages—such as improved handling, better marginal adaptation, enhanced polymerisation and increased microhardness—these benefits must be balanced with its limitations. Dentists are encouraged to consider the specific composite material being used, the preheating protocol, and the potential risks associated with component degradation. As future research refines preheating methodologies and material formulations, this technique will likely become more widely integrated into clinical practice.

Ultimately, incorporating preheating into routine practice has the potential to enhance both the functional and aesthetic outcomes of dental composite restorations, provided that standardised protocols are established. Further long-term studies will be essential to fully understand the clinical implications of preheating, ensuring that it contributes positively to the durability and performance of composite restorations. With these considerations in mind, preheating remains a valuable, yet evolving, technique within restorative dentistry that holds promise for optimising patient outcomes.

Preheating emerges as a promising approach in restorative dentistry, offering essential advancements in the handling, adaptation and durability of composite restorations. Preheating addresses several core challenges in composite application by effectively reducing viscosity, enhancing microhardness and improving both aesthetic stability and polymerisation integrity. Although considerations around equipment needs and material stability with repeated heating cycles may present limitations, the advantages of controlled preheating appear substantial. With continued research and the adoption of standardised protocols, preheating has the potential to become a cornerstone practice, enabling clinicians to achieve superior restorative outcomes while meeting both functional and aesthetic demands in patient care.

4. Conclusions

Numerous techniques have been developed to improve the mechanical properties of dental composite materials. The current one of these techniques is preheating.

This technique shows promise in improving the mechanical, physical and aesthetic properties of restorative materials. Given the problems associated with composite applications, especially in posterior restorations, the improvements attributed to preheating have been beneficial in achieving long-term clinical success. Based on the data obtained in this chapter, it can be concluded that preheating significantly improves the physical, mechanical and aesthetic properties of composite materials. However, *in vivo* and *in vitro* studies are needed in this regard.

4.1 Significance and applicability of preheating

Preheating represents a valuable advancement in restorative dentistry, addressing challenges such as the high viscosity of composite materials and ensuring better clinical outcomes. By reducing viscosity, preheating enhances the handling and adaptability of composites, particularly in complex restorations, leading to improved marginal adaptation and reduced voids.

In addition, preheating increases monomer conversion rates, resulting in enhanced mechanical properties, such as hardness and wear resistance, while minimising residual monomer content, thus improving biocompatibility. These benefits extend to both direct restorations, like fillings, and indirect restorations, such as veneers and onlays, where preheating improves material adaptation and reduces marginal leakage.

Despite its advantages, the lack of standardised preheating protocols limits its routine clinical use. Future research should focus on developing evidence-based guidelines to maximise the potential of preheating and integrate it seamlessly into clinical practice, ensuring reliable and durable restorative outcomes.

Conflict of interest

The authors declare no conflict of interest.

Author details

Bilge Ersöz* and Bengisu Arı
Health Science University Gulhane Faculty of Dentistry Department of Restorative Dentistry, Ankara, Turkiye

*Address all correspondence to: bilgecaliskan9@gmail.com

References

[1] Ilie N, Jelen E, Clementino-Luedemann T, Hickel R. Low-shrinkage composite for dental application. Dental Materials Journal. 2007;**26**:149-155. DOI: 10.4012/dmj.26.149

[2] Ferracane JL. Resin composite--state of the art. Dental Materials. 2011;**27**: 29-38. DOI: 10.1016/j.dental.2010.10.020

[3] Mohammadi N, Jafari-Navimipour E, Kimyai S, Ajami AA, Bahari M, Ansarin M, et al. Effect of pre-heating on the mechanical properties of silorane-based and methacrylate-based composites. Journal of Clinical and Experimental Dentistry. 2016;**8**(4):e373-e378. DOI: 10.4317/jced.52807

[4] Christensen GJ. Should resin-based composite dominate restorative dentistry today? Journal of the American Dental Association (1939). 2010;**141**:1490-1493. DOI: 10.14219/jada.archive.2010.0112

[5] Al-Sharaa KA, Watts DC. Stickiness prior to setting of some light cured resin-composites. Dental Materials. 2003;**19**:182-187. DOI: 10.1016/S0109-5641(02)00097-0

[6] Manhart J, Kunzelmann KH, Chen H, Hickel R. Mechanical properties and wear behavior of light-cured packable composite resins. Dental Materials. 2000;**16**:33-40. DOI: 10.1016/S0109-5641(99)00082-2

[7] Cramer NB, Stansbury JW, Bowman CN. Recent advances and developments in composite dental restorative materials. Journal of Dental Research. 2011;**90**:402-416. DOI: 10.1177/0022034510381263

[8] Van Meerbeek B, Perdigão J, Lambrechts P, Vanherle G. The clinical performance of adhesives. Journal of Dentistry. 1998;**26**:1-20. DOI: 10.1016/S0300-5712(96)00070-x

[9] Browne RM, Tobias RS. Microbial microleakage and pulpal inflammation: A review. Endodontics and Dental Traumatology. 1986;**2**:177-183. DOI: 10.1111/j.1600-9657.1986.tb00141.x

[10] Clelland NL, Pagnotto MP, Kerby RE, Seghi RR. Relative wear of flowable and highly filled composite. The Journal of Prosthetic Dentistry. 2005;**93**:153-157. DOI: 10.1016/j.prosdent.2004.11.006

[11] Lucey S, Lynch CD, Ray NJ, Burke FM, Hannigan A. Effect of pre-heating on the viscosity and microhardness of a resin composite. Journal of Oral Rehabilitation. 2010;**37**:278-282. DOI: 10.1111/j.1365-2842.2009.02045.x

[12] Deb S, Di Silvio L, Mackler HE, Millar BJ. Pre-warming of dental composites. Dental Materials: Official Publication of the Academy of Dental Materials. 2011;**27**(4):e51-e59. DOI: 10.1016/j.dental.2010.11.009

[13] Wagner WC, Aksu MN, Neme AM, Linger JB, Pink FE, Walker S. Effect of pre-heating resin composite on restoration microleakage. Operative Dentistry. 2008;**33**:72-78. DOI: 10.2341/07-41

[14] Elsayad I. Cuspal movement and gap formation in premolars restored with preheated resin composite. Operative Dentistry. 2009;**34**:725-731. DOI: 10.2341/09-012-L

[15] Fróes-Salgado NR, Silva LM, Kawano Y, Francci C, Reis A, Loguercio AD. Composite pre-heating:

Effects on marginal adaptation, degree of conversion and mechanical properties. Dental Materials. 2010;**26**:908-914. DOI: 10.1016/j.dental.2010.03.023

[16] Daronch M, Rueggeberg FA, De Goes MF. Monomer conversion of pre-heated composite. Journal of Dental Research. 2005;**84**:663-667. DOI: 10.1177/154405910508400716

[17] El-Korashy DI. Post-gel shrinkage strain and degree of conversion of preheated resin composite cured using different regimens. Operative Dentistry. 2010;**35**:172-179. DOI: 10.2341/09-072-L

[18] Sarrett DC. Clinical challenges and the relevance of materials testing for posterior composite restorations. Dental Materials. 2005;**21**:9-20. DOI: 10.1016/j.dental.2004.10.001

[19] Fugolin APP, Pfeifer CS. New resins for dental composites. Journal of Dental Research. 2017;**96**:1085-1091. DOI: 10.1177/0022034517720658

[20] He J, Lassila L, Garoushi S, Vallittu P. Tailoring the monomers to overcome the shortcomings of current dental resin composites. Biomaterial Investigations in Dentistry. 2023;**10**:2191621. DOI: 10.1080/26415275.2023.2191621

[21] Peutzfeldt A. Resin composites in dentistry: The monomer systems. European Journal of Oral Sciences. 1997;**105**:97-116. DOI: 10.1111/j.1600-0722.1997.tb00188.x

[22] Schmalz G, Widbiller M. Biocompatibility of amalgam vs composite: A review. Oral Health and Preventive Dentistry. 2022;**20**:149-155. DOI: 10.3290/j.ohpd.b2831749

[23] Stoleriu S, Iovan V, Pancu G, Sandu V, Taraboanta I, et al. Comparative study of color stability of three composite materials treated by finishing and coated sealing, after immersion in different solutions. Revista De Chimie. 2019;**70**:1681-1684. DOI: 10.37358/Rev.Chim.1949

[24] Saini RS, Binduhayyim RIH, Gurumurthy V, Alshadidi AAF, Aldosari LIN, Okshah A, et al. Dental biomaterials redefined: Molecular docking and dynamics-driven dental resin composite optimization. BMC Oral Health. 2024;**24**(1):557. DOI: 10.1186/s12903-024-04343-1

[25] Tsujimoto A, Barkmeier WW, Fischer NG, et al. Wear of resin composites: Current insights into underlying mechanisms, evaluation methods, and influential factors. Japanese Dental Science Review. 2018;**54**:76-87. DOI: 10.1016/j.jdsr.2017.11.002

[26] Baroudi K, Rodrigues JC. Flowable resin composites: A systematic review and clinical considerations. Journal of Clinical and Diagnostic Research. 2015;**9**:0-24. DOI: 10.7860/JCDR/2015/12294.6129

[27] Nikolaos-Stefanos K. Resin composite pre-heating - a literature review of the laboratory results. International Journal of Oral and Dental Health. 2018;**4**:74. DOI: 10.23937/2469-5734/1510074

[28] Osternack FH, Caldas DB, Almeida JB, Souza EM, Mazur RF. Effects of preheating and precooling on the hardness and shrinkage of a composite resin cured with QTH and LED. Operative Dentistry. 2013;**38**:e8. DOI: 10.2341/11-411-L

[29] D'Amario M, De Angelis F, Vadini M, Marchili N, Mummolo S, D'Arcangelo C. Influence of a repeated preheating procedure on mechanical

properties of three resin composites. Operative Dentistry. 2015;**40**:181-189. DOI: 10.2341/13-238-L

[30] Jafarzadeh Kashi TS, Fereidouni F, Khoshroo K, Heidari S, Masaeli R, Vammadian M. Effect of preheating on the microhardness of nanohybrid resin-based composites. Frontiers in Biomedical Technologies. 2015;**2**:15-22. DOI: 10.32598/fbtj.2.1.15

[31] Theobaldo JD, Aguiar FH, Pini NI, Lima DA, Liporoni PC, Catelan A. Effect of preheating and light-curing unit on physicochemical properties of a bulk fill composite. Clinical, Cosmetic and Investigational Dentistry. 2017;**9**:39-43. DOI: 10.2147/CCIDE.S130803

[32] Tantbirojn D, Chongvisal S, Augustson DG, Versluis A. Hardness and postgel shrinkage of preheated composites. Quintessence International. 2011;**42**:e59

[33] Dionysopoulos D, Papadopoulos C, Koliniotou-Koumpia E. Effect of temperature, curing time, and filler composition on surface microhardness of composite resins. Journal of Conservative Dentistry. 2015;**18**:114-118. DOI: 10.4103/0972-0707.153071

[34] Ayub KV, Santos GC Jr, Rizkalla AS, Bohay R, Pegoraro LF, Rubo JH, et al. Effect of preheating on microhardness and viscosity of 4 resin composites. Journal of the Canadian Dental Association. 2014;**80**:e12

[35] Muñoz CA, Bond PR, Sy-Muñoz J, Tan D, Peterson J. Effect of pre-heating on depth of cure and surface hardness of light-polymerized resin composites. American Journal of Dentistry. 2008;**21**:215-222

[36] Uctasli MB, Arisu HD, Lassila LV, Vallittu PK. Effect of preheating on the mechanical properties of resin composites. European Journal of Dentistry. 2008;**2**:263-268

[37] Elkaffass AA, Eltoukhy RI, Elnegoly SA, Mahmoud SH. Influence of preheating on mechanical and surface properties of nanofilled resin composites. Journal of Clinical and Experimental Dentistry. 2020;**12**(5):e494-e500. DOI: 10.4317/jced.56469

[38] Poubel DLDN, da Silva RC, Ribeiro APD, Garcia FCP. Effect of preheating on the viscosity of composite resins. Journal of Conservative Dentistry and Endodontics. 2024;**27**(4):360-365. DOI: 10.4103/JCDE.JCDE_75_23

[39] Bueno T, Masoud N, Akkus A, Silva I, McPherson K, Furuse A, et al. Effects of pre-heating on physical-mechanical-chemical properties of contemporary resin composites. Odontology. 2024. DOI: 10.1007/s10266-024-00953-x

[40] Sahetia AS, Sirsat PP, Gulve MN, Kolhe SJ, Patel SP. Preheating effect on microhardness and depth of cure of three bulk-fill composite resins: An in vitro study. Endodontology. 2024;**36**(2):125-130. DOI: 10.4103/endo.endo_56_23

[41] Elkaffass AA, Eltoukhy RI, Elnegoly SA, Mahmoud SH. The effect of preheating resin composites on surface hardness: A systematic review and meta-analysis. Restorative Dentistry and Endodontics. 2019;**44**(4). DOI: 10.5395/rde.2019.44.e41

[42] Burke FJ, Lucarotti PS. How long do direct restorations placed within the general dental services in England and Wales survive? British Dental Journal. 2009;**206**(1):E2-27. DOI: 10.1038/sj.bdj.2008.1042

[43] Ahmadizenouz G, Esmaeili B, Ahangari Z, Khafri S, Rahmani A.

Effect of energy drinks on discoloration of silorane and dimethacrylate-based composite resins. Journal of dentistry (Tehran). 2016;**13**:261-270

[44] Dietschi D, Shahidi C, Krejci I. Clinical performance of direct anterior composite restorations: A systematic literature review and critical appraisal. International Journal of Esthetic Dentistry. 2019;**14**:252-270

[45] Ren YF, Feng L, Serban D, Malmstrom HS. Effects of common beverage colorants on color stability of dental composite resins: The utility of a thermocycling stain challenge model in vitro. Journal of Dentistry. 2012;**40**(Suppl 1):e48-e56. DOI: 10.1016/j.jdent.2012.04.017

[46] Darabi F, Seyed-Monir A, Mihandoust S, Maleki D. The effect of preheating of composite resin on its color stability after immersion in tea and coffee solutions: An in-vitro study. Journal of Clinical and Experimental Dentistry. 2019;**11**(12):e1151-e1156. DOI: 10.4317/jced.56438

[47] D'Amario M, Pacioni S, Capogreco M, Gatto R, Baldi M. Effect of repeated preheating cycles on flexural strength of resin composites. Operative Dentistry. 2013;**38**(1):33-38. DOI: 10.2341/11-476-L

[48] Kimyai S, Mashayekhi Z, Mohammadi N, Bahari M, Abed Kahnamouei M, Ebrahimi Chaharom ME. Comparison of the effect of preheating on the flexural strength of giomer and nanohybrid composite resin. Journal of Dental Research, Dental Clinics, Dental Prospects. 2022;**16**(3):159-163. DOI: 10.34172/joddd.2022.027

Chapter 5

The Utilization of Composite Bearings in Heavy Agricultural, Construction, Forestry, and Mining Equipment Design Applications

Thomas Barnes, Brecken Beyer, Wyatt Griffey, Caleb Huffmeyer, Clayton Ness, Owen Nifong, Tyler J. McPheron and Robert M. Stwalley III

Abstract

This design overview will describe composite material bearings and their utilization in agricultural, construction, forestry, and mining heavy equipment. Currently, metallic bearings are used in heavy machinery and implement applications. Most metallic bearings use grease or oil for lubrication. However, metallic bearings generally require periodic maintenance after repetitive use, and they are prone to corrosion. Maintenance intervals are frequent and increase operational down-time for equipment operators and technicians, further increasing operating costs and reducing productivity. Composite bearings are manufactured from fiber-reinforced materials and are produced using a variety of resin formulas to achieve the desired composite cross section. Potential design applications of self-lubricated composite bearings using polytetrafluoroethylene (PTFE) or other embedded lubrication materials to minimize friction coefficients are presented. Composite bearings are advantageous in heavy equipment applications, due to their greaseless design, corrosion resistance, high-bearing load capacities, and resistance to fatigue loads and impact. Off-road equipment operates in harsh environments and is designed with heavy-duty components that resist these conditions. An understanding of composite bearing characteristics will allow engineers to more frequently incorporate them into their designs, reducing maintenance costs, improving machine productivity, and accommodating adverse environmental conditions, all while minimizing bearing wear and friction created by mechanical motion.

Keywords: composite bearings, durability, friction, design engineering, agriculture, construction, forestry, mining, equipment manufacturing

1. Introduction

Bearings are utilized in a variety of mechanical designs and machinery applications to minimize friction coefficients and produce smooth, fluid-like mechanical motion between two working components having a single degree of articulated or rotary motion. Although bearings are commonly used in many different industries, the agricultural, construction, forestry, and mining equipment industries incorporate an abundance of heavy-duty bearings in their vehicle designs. Compared to on-road vehicles, which have bearings in suspension components, engines, transmissions, drive trains, and steering components, heavy off-road equipment typically features auxiliary motion functions that are unique to the specific piece of equipment, in addition to the same general parts that compose an on-road vehicle. For example, an agricultural combine harvester has many mechanical functions within its grain threshing system, and construction backhoe loaders have mechanical and hydraulic front loader buckets and backhoes, all of which require an array of bearings for frictionless, mechanical motion and smooth operation.

Many different types of bearings will be examined in this overview, but there are two general categories of traditional metallic bearings incorporated in most mobile equipment: plain bearings and antifriction bearings. Plain bearings refer to sliding or linear movement between the bearing surface and the contact part. Antifriction bearings have an internal 'idler gear' to facilitate rotational movement, such as a rotating power shaft or the yaw articulation of a backhoe boom [1]. Many bearings types feature mechanical designs to reduce friction coefficients, such as roller bearings, which use small cylinders inside of a steel cage to allow low friction rotation. Almost all traditional metallic bearings use a form of lubricant as their main friction-minimizing substance. This is typically a high viscosity grease or a lower viscosity film of oil between the bearing surface and the contact material. **Figure 1** shows an example of a plain journal bearing used for shaft rotation [2], while **Figure 2** shows the differences between a roller antifriction bearing and a ball antifriction bearing [3].

Most bearings that are currently used in heavy equipment designs are made of metallic materials and require regular maintenance to maintain the performance and integrity of the bearing. This introduces a variety of issues for design engineers, equipment operators, and service technicians. Daily service requirements are detrimental to equipment productivity and operating costs, due to the cost of grease and

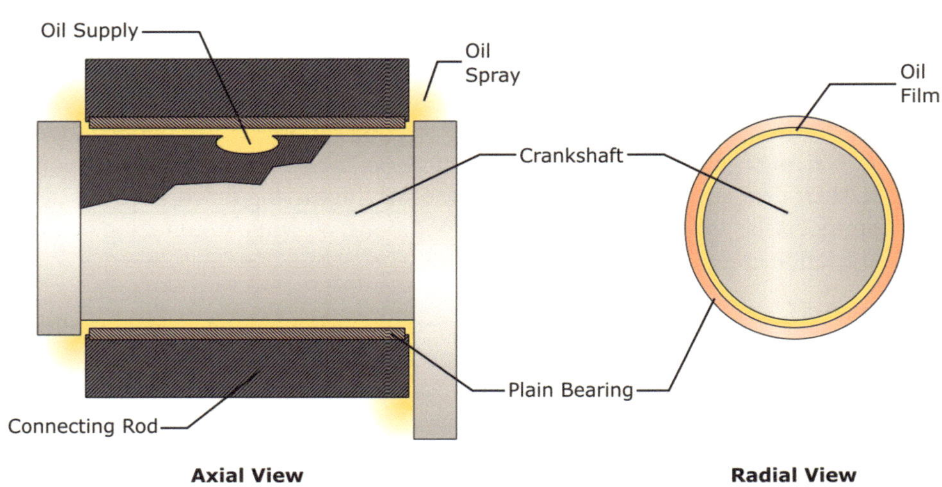

Figure 1.
Plain bearing diagram from side and in section [2].

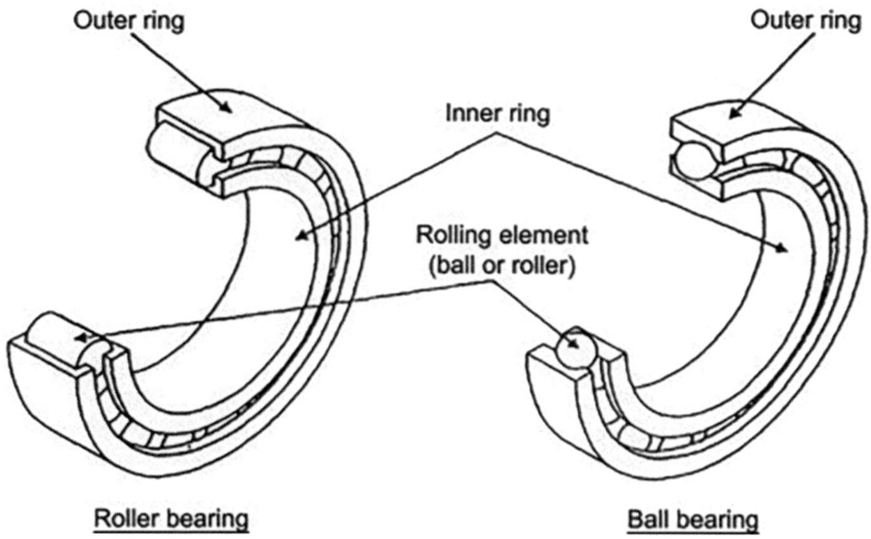

Figure 2.
Cross-section of antifriction bearings with rollers (left) and balls (right) [3].

equipment downtime. Maintenance neglect can result in catastrophic failures, fires, and expensive repair costs. To mitigate this issue, equipment longevity research indicates that composite material bearings could be a potential solution. These bearings require minimal long-term maintenance, when compared with traditional metallic bearings in heavy equipment applications.

Composite bearings utilize self-lubricating mechanisms that are embedded within the surface layer of the bearing's material during the manufacturing process. The embedded lubrication is released from the surface layer, which is typically a perforated or porous layer, when the shaft surface moves relative to the bearing surface layer. The creation of friction at the motion points causes the embedded oil to be released at the contact interface and provide a layer of lubrication between the material surface and the bearing surface. This is highly beneficial, because constant lubrication can be achieved independently of the speed of the rotating or linearly moving part [4]. Any grease or lubricant not used is reabsorbed back into the porous bearing surface, when the component is at rest. Additionally, self-lubricating bearings do not lose or deteriorate their grease over time, and they do not require regular grease or oil service intervals to maintain their functional performance. This maximizes equipment uptime and productivity and minimizes maintenance costs.

2. Literature review

Peer-reviewed journal publications discussing design recommendations and technical information for the application of composite bearings are available, but seem limited [5–7]. Journals tend to primarily concentrate on composite bearing material properties and the development of composite materials [8–13]. However, many composite bearing manufacturers provide a wealth of technical information for their products, and even include detailed design guides on their websites [14–17].

The specifications and critical details from these component design guides were heavily used to summarize and provide design recommendations in this work, as well as more thoroughly describing the general characteristics of composite bearings. Corporate-provided information from leading composite bearing manufacturers, such as Igus Motion Plastics, GGB by Timken, Polygon Composites Technology, was used to describe different mechanical properties, thermal properties, and the general bearing geometries of typical composite bearing assemblies.

Polygon Composites Technology has been manufacturing composite material products for nearly 70 years. Polygon produces composite tubing and self-lubricating composite bearings, as well as other custom composite products for a variety of industries, such as agriculture, construction, medical, oil and gas, and on-road trucks. Polygon has documented several successful applications of composite bearings, where other bearing types were inferior or where the only solution was a custom designed bearing [18]. Polygon describes design applications in the agricultural and construction sectors that were summarized for this heavy equipment-specific research, providing real-world examples of composite bearings replacing steel bearings, which optimized the overall design and decreased projected maintenance time. Another manufacturer, Igus Motion Plastics, manufactures chain and cable carriers, composite bearings, gears, and robotic equipment parts, and they validate their composite products in specific applications to ensure product quality and meet user quality specifications [19]. They have dedicated labs for testing composite bearing products under a variety of different wear conditions and structural loads. Igus makes some of their testing data available to engineers and consumers to accurately compare products to determine the most ideal product for the intended application. Finally, GGB by Timken began making composite material bearings in the late 1970s. GGB produces a hybrid style sleeve bearing, which utilizes self-lubricating, composite technology on the inner layer, with a reinforced exterior steel layer for heavy-duty applications. The GGB composite bearing assemblies have been designed specifically to support vehicular applications [20].

Bearings design metrics covering all primary bearing types, feature wear estimation calculations, and bearing load calculations for a variety of mechanical applications remain valid for composite bearings [1]. A thorough understanding of bearing lubrication, bearing failures, and bearing size selection is assumed to be within the reader's background. The current overview has utilized product literature, composite material testing, and design guides produced by composite bearing manufacturers to provide a practical overview and design guide for engineers to apply composite bearing technology in the heavy-duty agricultural, construction, forestry, and mining equipment industries. Since the composite bearing market is a relatively small part of the overall bearing industry, this overview focuses leading manufacturers in the market. The technological advancements of their products are summarized and presented through their published technical literature.

3. Traditional metallic bearings

There are two general types of bearings that provide axial support for a rotating shaft: plain bearings and antifriction bearings. In traditional designs, the selection of the bearing type would typically depend on the categorization of the motion, the type and direction of the loads, the application, and the environment. Regardless of application and type, traditional bearings can be selected to fit most reasonable design constraints. A brief analysis of traditional metallic bearings and their associated shortcomings for off-road vehicle and heavy equipment applications is presented in this section.

3.1 Plain bearings

Traditional plain bearings, also known as journal bearings, like the one shown in **Figure 1**, are made of many metallic materials, but typically bronze or brass. There are a wide variety of plain bearings, but all consist of a lubricated metal sleeve designed to reduce friction between two sliding or rotating surfaces. This design, when operated properly, results in complete hydrodynamic contact between the shaft and bearing surfaces. These bearings are capable of a high load capacity and can operate at low speeds. Due to having fewer moving parts, these bearings are cheaper to manufacture and have low operating noise. Their simplistic design allows for a variety of bearing housing sizes and a relative ease in maintenance or replacement.

The main disadvantages of the plain bearing design are its high-speed limitations and its continual high lubrication requirement, generally requiring a dedicated external lubrication system. The high contact area improves the bearing's ability to sustain high loads, but it also decreases the ability to operate at high speeds, limiting potential applications. Failure to apply the proper amount of lubrication can increase friction between the moving parts, and this can lead to significant wear, especially during start-up [2]. Finally, the lubrication and maintenance requirements of a plain journal bearing system are major disadvantages, due to the cost in time, additional system overhead, and machine service down-times. While traditional metallic plain bearings have served the off-road industry for years, the disadvantages associated with them have also caused engineers to continue to look for improvements upon this traditional bearing design.

3.2 Antifriction bearings

To fight frictional loss and increase the wear life of bearings in adverse environments, antifriction bearings make use of small rollers inside of a steel cage to act as idlers between the rotating and fixed surfaces. The basic components of a roller and a ball bearing were shown in **Figure 2**. Antifriction bearings perform well in high-load, high-speed environments. Other advantages of this design include accurate shaft alignment, low starting friction, reduced rolling torque, and a simple process for lubrication and bearing replacement [1].

3.2.1 Ball bearings

Ball bearings are the main form of antifriction bearings. Ball bearings utilize spherical rollers to decrease the friction between rotating inner races and stationary outer races. The outer race is generally held stationary by a bearing holder, and the inner race holds the shaft or pin. This design allows for ball bearings to operate at higher speeds than traditional plain bearings. Ball bearings can operate under radial and thrust loads, but they cannot provide significant axial load, unless used in a tapered configuration, shown in **Figure 3**. The addition of the conical bearing area to the outer race provides axial load support into the cone. The doubling or tripling of ball rows, as shown in **Figure 4**, can also increase the load support capability.

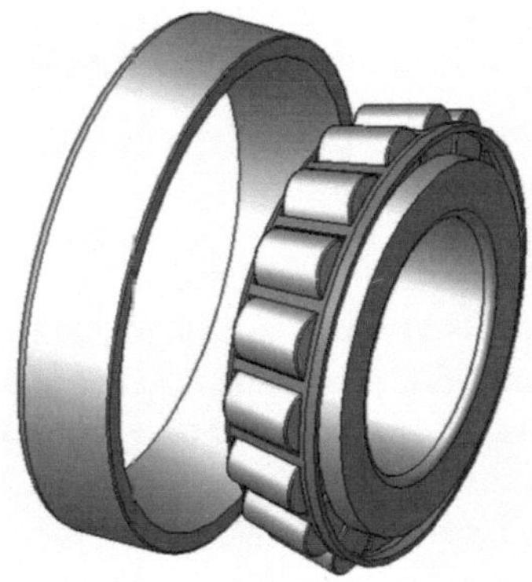

Figure 3.
Exploded view of tapered roller bearing assembly, showing the conical section to carry axial load.

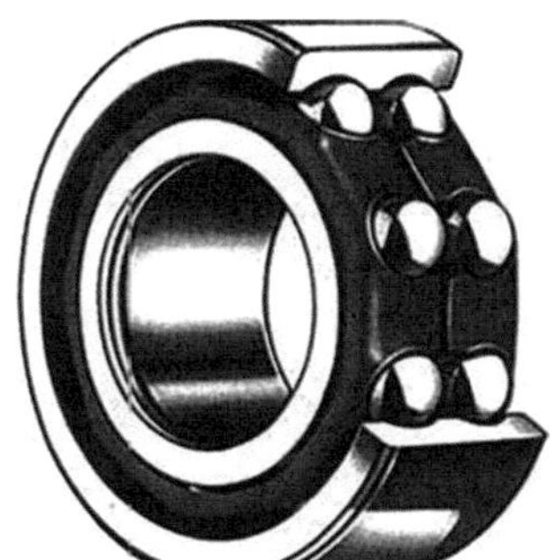

Figure 4.
Cut-away view of a double row of ball rollers in a bearing assembly.

The main disadvantages of ball and roller bearings are the lubrication requirement and reduced load carrying capacity in relation to plain bearings. The reduced load capacity comes from the smaller contact area between the balls or rollers and the races in the bearings, compared to plain bearings, which carry load across their entire engagement interface. Lubrication concerns are a major limitation for ball bearings in off-road equipment. These limitations are within the design of these bearings and cannot be overcome.

3.2.2 Roller bearings

Roller bearings provide several advantages over both plain bearings and ball bearings, particularly in axial loading applications. The cylindrical rollers allow for more contact area than balls provide, meaning a larger load can be supported. Taper allows the bearing to support a load in the radial and axial directions, and this makes tapered roller bearings a common choice for wheels in the hubs and tracks on heavy equipment. The disadvantages of roller bearings include abnormal stresses resulting from the tapered design and the precision machining necessary to manufacture them. Adequate lubrication for roller bearing performance is vital, since improper greasing risks seal failure and ultimately causes premature bearing failure. Although the required maintenance is relatively simple, it does create machine down-time, additional labor costs, and lubricant expenses. These maintenance needs can represent significant losses in the field capacity for off-road vehicles. Productive field capacity is lost when the equipment is not in operation, due to maintenance requirements. Improper greasing by applying too little grease can lead to significant metal-to-metal contact, grinding between components, and overheating, leading to permanent damage of the bearing. Over-greasing of a bearing can led to damage in the bearing seals, creating excess drag through fluid friction, and other potential environmental concerns, due to excessive lubricant being ejected from the bearing [21]. Lastly, an ineffective seal will lead to bearing contamination from solid particles or water, and this circumstance significantly reduces bearing life and will eventually cause bearing failure. Plain and antifriction bearings are useful, but their shortcomings have caused engineers to look for methods to improve upon the limitations of traditional metallic bearings.

4. Composite bearings

In general, a composite bearing looks a great deal like a plain journal bearing, with perhaps a smaller radial clearance with the shaft. Instead of needing an external lubrication source, composite bearings have an 'internal' source, based upon the bearing material containing a lubricant within its material matrix and through the base material being slippery itself. This section will examine useful composite materials, lubricants, load carrying ranges, bearing thermal properties, and manufacturing tolerances.

4.1 Materials

Composite bearings resemble journal bearings. Like plain bearings, composite bearings utilize a lubrication film to minimize friction. However, unlike traditional bearings, composite bearings do not require an external lubricant source to provide the slippery film that coats the contacting surfaces. Composite bearings often utilize self-lubricating mechanisms that are embedded in the surface layer of the bearing during the manufacturing process. The surface layer is the main sliding contact interface between the two independent parts. The embedded lubrication is released from the sliding layer, which is typically a perforated or porous layer, when the material surface moves against the bearing surface layer. The creation of friction causes the embedded oil to coat the surface contact material and provide a layer of lubrication between the material surface and the bearing surface. Some common self-lubricating materials found in composite bearings are provided in **Table 1** [22].

The sliding contact layer is typically composed of a tightly-wound polytetrafluoroethylene (PTFE) plastic and a high-strength fiber layer that is encapsulated in a self-lubricating, high-temperature epoxy resin. The type of PTFE, the fiber material, and the lubrication-filled epoxy resin utilized vary by application and bearing

Rulon	Delrin AF
Fluorosint	TriSteel PT/PI
PTFE Blends	Ultracomp
Graphite PI	Nylon 6/6, 6/12
Ertalyte Tx	UHMW

Table 1.
Common materials used in manufacturing composite bearings [22].

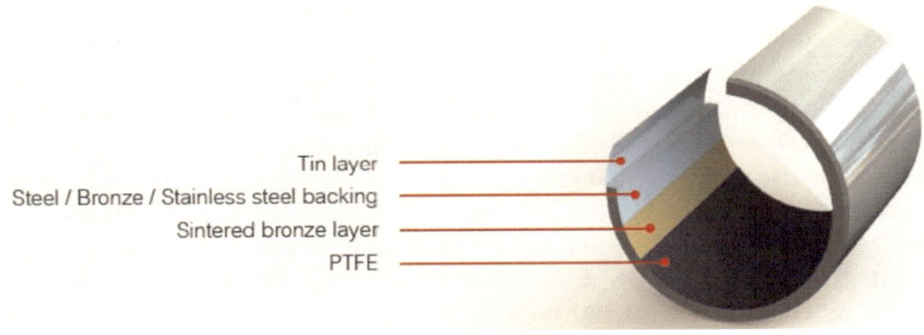

Figure 5.
Antifriction bearing showing sliding layer, surface layer, and backing layers [23].

manufacturer. To provide additional physical support to the bearing, a stiff backing layer will typically be meshed with the inside sliding contact layer. This backing layer is generally composed of continuously-wound fiberglass, coated with a high-temperature, high-stress epoxy. The ratio between sliding and backing layer that comprises the bearing varies by application and the composition of the sliding layer. A typical example of the surface and sliding layers is shown in **Figure 5** [23].

4.2 Lubrication and friction

Most composite bearings do not need lubrication. PTFE bearings fall into this category. This is due to the inherent dry sliding characteristics of the materials used in manufacturing composite bearings. Some component assemblies do require an initial 'charge' of lubrication at installation. Polyoxymethylene (POM) composite bearings

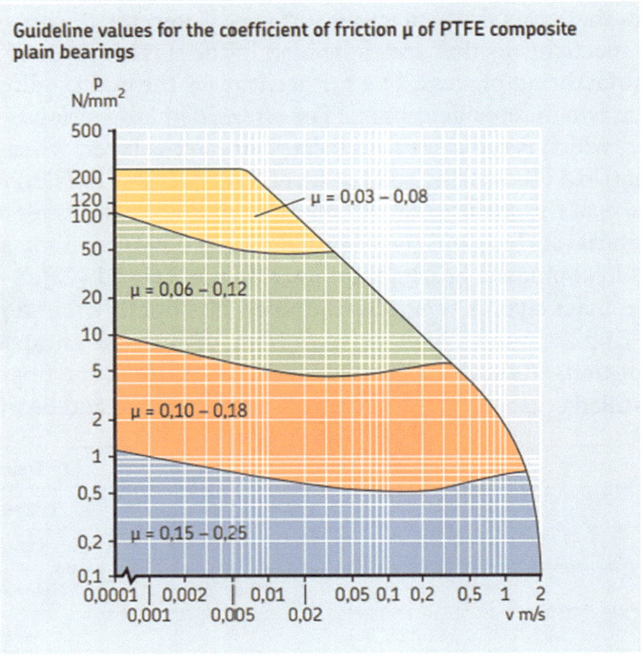

Figure 6.
Operating Conditions and Coefficient of Friction [16].

are designed to have an initial film of lubrication, and they do not require any further application for the life of the bearing. Both the PTFE and POM composite bearings have similar friction coefficients, which vary between 0.03 and 0.25, depending on external factors and operating conditions. It is important to note that POM composite bearings tend to have lower friction coefficients, due to their initial application of lubricant. **Figure 6** shows how operating conditions can influence the sliding contact friction coefficients. Under high loads and at low interface velocities, composite bearings show minimal rolling friction between the contact surfaces. The coefficient of friction rises with increasing speeds, although under low loads and high velocity conditions, or any other unfavorable conditions, such as inadequate surfaces or improper alignment, the coefficient of friction for a composite bearing is low [16].

4.3 Strength and loads

Composite bearings excel in high static load environments and also perform well under low-speed dynamic loads. Bearing manufacturers typically offer a full spectrum of specialized products that can meet the requirements of each application, whether that be large loads, high speeds, or extreme temperatures. As shown in **Table 2**, composite bearings tend to meet or exceed competitive traditional designs in most

Material	Maximum dynamic capacity [<0.025 m/s (5 sft-min)]		Maximum temperature		Thermal expansion rate – Hoop		Specific gravity
Cast Bronze*	41	6000	71	160	18.0	10	8.80
Porous Bronze**	28	4000	71	160	18.0	10	7.50
Alloyed Bronze*	69	10,000	93	200	28.8	16	8.10
Steel-Backed Bronze*	24	3500	93	200	14.4	8	8.00
Hardened Steel*	276	40,000	93	200	12.6	7	7.90
Zinc Aluminum*	38	5500	93	200	27.0	15	5.00
Fabric-Reinforced Phenolic*	41	6000	93	200	36.0	20	1.60
Reinforced PTFE	14	2000	260	500	99.0	55	2.00
GAR-MAX®	140	20,000	160	320	12.6	7	1.87
GAR-FIL®	140	20,000	205	400	12.6	7	1.96
HSG	140	20,000	160	320	12.6	7	1.87
MLG	140	20,000	160	325	12.6	7	1.87
HPM	40	20,000	160	320	12.6	7	1.87
HPMB®	140	20,000	160	325	12.6	7	1.87
HPF® Sliding Plate	140	20,000	140	285	10.8***	6***	1.90
GGB MEGALIFE™ XT	140	20,000	175	350	12.6***	7***	1.85
MULTIFIL	35	5000	280	540	—	—	2.37

**With lubrication*, ***Oil impregnated*; ****Lengthwise*

Table 2.
Properties comparison between composite and traditional material bearings [14].

bearing metrics, especially in maximum dynamic bearing capacity. Almost all bearing designs offered commercially by GGB by Timken have higher maximum dynamic bearing capacities than traditional materials like bronze or aluminum [14]. Only hardened steel is stronger than the composite material, but it is important to note that steel bearings require lubrication and are not recommended for corrosive surroundings. Composite materials do not require any external lubrication and will not corrode, even within the most reactive environments. The bearing performance given in **Table 2** limits the maximum dynamic bearing capacities to only be applicable below 0.025 m/s [14]. Composite bearings struggle at high rotational speeds, and many manufacturers only offer bearings rated for speeds below specific limits, as shown in the product comparison in **Table 3** [14].

This limitation on rated speeds may seem restrictive, but there remain countless potential applications where composite bearings could be helpful, particularly in the realm of off-road vehicles. Composite bearings are not suitable for applications involving high speed shafts, like crankshafts in an engine or axle shafts within an automotive differential, but they can be an excellent choice for low-speed cyclic loads like those found in an excavator arm or power steering mechanism.

The combination of load and speed is an important consideration to a designer, and this metric is the primary factor in approximating bearing life. An engineer will want to work closely with potential bearing suppliers and provide them with appropriate values for use in their design calculations. Generally, it is important to minimize the specific load and sliding contact speed to maximize bearing life. It is also important to size the bearing appropriately. Increasing the bearing length will reduce the specific load, but it will also introduce additional, and often significant, new loads if the shaft deflects inside the bearing [14]. There can be contradictory requirements in design, and compromises must often be made to best fit individual applications with appropriate choice in composite bearings.

4.4 Thermal properties

Composite bearings provide impressive temperature resistance, and in some cases, better thermal expansion rates than traditional bearing materials. Many traditional bearing materials, like bronze and steel, require lubrication to function properly. Even if the bearing material itself may be able to sustain higher tem-peratures, oils and greases will quickly deteriorate at high temperatures. Composite bearings, as shown in **Table 1**, generally allow for higher maximum sustained temperatures than traditional materials. They are typically needed in assemblies that see elevated operating temperatures above 90°C. High temperatures are a limiting factor on applicability, as the bearing surface will soften under prolonged exposure to extreme thermal stresses above 150°C [14]. In some cases, the composite bearing may still function, but it will have a lower load capacity and a shorter lifespan. In fact, as detailed in **Figure 7**, a decreased expected bearing life begins at 65°C [14]. Even though composite bearings have high maximum temperatures capabilities, sustained high temperatures are not a preferred operational environment, and elevated temperatures will have a detrimental effect on bearing life.

Composite bearings also show excellent thermal expansion rates. For a bearing, thermal expansion in the 'hoop' direction should be minimized, as much as possible. Thermal expansion can place additional stresses on the shaft, increasing friction and generating additional heat, leading to continued thermal expansion. This cycle feeds on itself continuously, creating accelerated wear and eventual bearing destruction. **Table 1** compares the thermal expansion of many composite bearings with traditional

Physical properties	Units	GAR-MAX®	GAR-FIL®	HSG	MLG	HPM	HPMB®	HPF® Sliding Plates	GGB MEGALIFE™XT	Mulitifil
Ultimate Compressive Strength	N/mm^2	414	379	621	414	345	414**	379	207	—
	psi	60,000	55,000	90,000	60,000	50,000	60,000	55,000	30,000	—
Static Load Capacity	N/mm^2	210	140	415	210	210	210	140	140	70
	psi	30,000	20,000	60,000	30,000	20,000	30,000	20,000	20,000	10,000
Maximum Dynamic Load Capacity	N/mm^2	140	140	140	140	140	140	140	140	35
	psi	20,000	20,000	20,000	20,000	20,000	20,000	20,000	20,000	5000
Maximum Relative Surface Speed	m/s	0.13	2.5	0.13	0.13	0.13	0.13	2.5	0.5	2.5
	fpm	25	500	25	25	25	25	500	100	500
Maximum pU Factor	$N/mm^2 \times m/s$	1.05	1.23	1.05	1.05	1.23	1.23	1.23	1.23	0.32
	psi × fpm	30,000	35,000	30,000	30,000	35,000	35,000	35,000	35,000	9000
Maximum Operating Temperature	°C	160	205	160	160	160	160	140	175	280
	°F	320	400	320	320	320	320	285	350	540
Minimum Operating Temperature	°C	−195	−195	−195	−195	−195	−195	−195	−195	−200
	°F	−320	−320	−320	−320	−320	−320	−320	−320	−330
Thermal Expansion Rate - Hoop	$10^{-6}/K$	12.6	12.6	12.6	12.6	12.6	12.6	10.8*	12.6*	—
	$10^{-6}/F$	7.0	7.0	7.0	7.0	7.0	7.0	6.0*	7.0*	—
Thermal Expansion Rate - Axial	$10^{-6}/K$	27.0	27.0	27.0	27.0	27.0	27.0	—	—	—
	$10^{-6}/F$	15.0	15.0	15.0	15.0	15.0	15.0	—	—	—
Specific Gravity	—	1.87	1.96	1.87	1.87	1.87	1.87	1.90	1.85	2.37

*Lengthwise, **For details contact GGB Applications Engineering department*

Table 3.
Properties of GGB bearings [14].

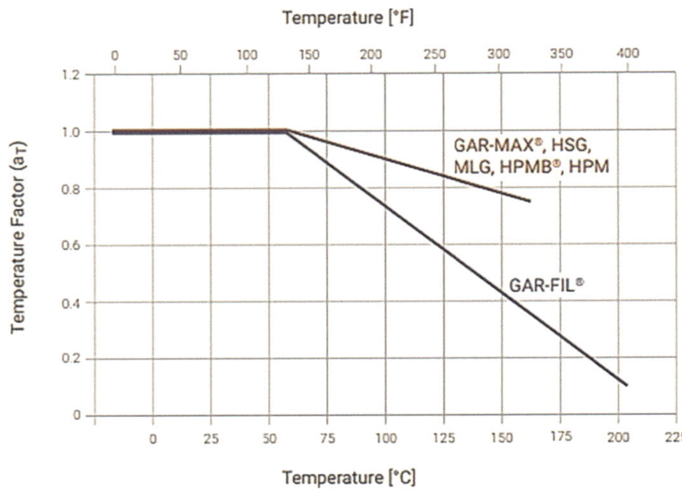

Figure 7.
Temperature factors for bearing life calculations [14].

bearing materials. The composite offerings meet or exceed the traditional materials in this metric. Additionally, **Table 2** clearly illustrates that composite bearings have a greater thermal expansion capability in the axial direction.

4.5 Sizing and tolerances

Composite bearings are offered in similar sizes as traditional bearings, to work with standard shafts. Composite bearing selection follows the same systematic design process as metallic bearing selection. Manufacturers provide catalogs that include critical dimensions and part drawings for their various bearing selections. Important dimensions to consider are the inside diameter (ID), outside diameter (OD), and the bearing length (w). The bearing manufacturer will also provide tolerances and clearances for the bearing housing. ID tolerances are critical to ensuring that the bearing fitment is suitable for the application, without being too tight or too loose. In the past, composite material-based bearings have been limited in applicability by the inability to provide tightly-toleranced parts, due to the composite material manufacturing processes and filament winding processes used to produce composite bearings. Most metallic bearings are machined from steel. The ability to apply modern high-precision machining processes allows metallic bearings to be manufactured with extremely tight tolerances. Composite bearings are made in an additive manner by winding filaments and resins, rather than removing material to size with a computer numerical controlled (CNC) machine tool. Previously this introduced uncertainty into the composite manufacturing process, but advancements in fiber construction for self-lubricating materials used in the assembly of composite bearings have allowed the production of bearings with similar dimensional tolerances to metallic bearings [20].

Composite bearing manufacturers are very transparent about their specified tolerances. The specifications outlined in product catalogs and design guides include ID and OD tolerances, recommended bearing housing bore diameters, press-fit tolerances, pin diameter recommendations for sleeve bearings, running clearances, and bearing length tolerances. **Table 4** presents an example of a typical design guide used for composite bearing selection [18].

Bearing part number	Nominal ID	ID	OD	Recommended housing bore	Press fit	Recommended pin diameter	Running clearance	Length tolerance
PWR08M16	1/2	0.5058	1.0025	1.000	0.0005	0.5000	0.0013	±0.010
		0.5088	1.0045	1.002	0.0045	0.4980	0.0103	
PWR10M18	5/8	0.6309	1.1276	1.125	0.0005	0.6250	0.0013	±0.010
		0.6339	1.1296	1.127	0.0046	0.6230	0.0104	
PWR12M20	3/4	0.7560	1.2526	1.250	0.0006	0.7500	0.0014	±0.010
		0.7590	1.2546	1.252	0.0046	0.7480	0.0104	
PWR14M22	7/8	0.8811	1.3777	1.375	0.0006	0.8750	0.0014	±0.010
		0.8841	1.3797	1.377	0.0047	0.8730	0.0105	
PWR16M24	1	1.0062	1.5027	1.500	0.0007	1.0000	0.0015	±0.010
		1.0092	1.5047	1.502	0.0047	0.9980	0.0105	
PWR18M26	1-1/8	1.1314	1.6278	1.625	0.0007	1.1250	0.0016	±0.010
		1.1344	1.6298	1.627	0.0048	1.1230	0.0107	
PWR20M28	1-1/4	1.2564	1.7528	1.750	0.0008	1.2500	0.0016	±0.010
		1.2594	1.7548	1.752	0.0048	1.2480	0.0106	
PWR22M30	1-3/8	1.3816	1.8779	1.875	0.0008	1.3750	0.0017	±0.010
		1.3856	1.8799	1.877	0.0049	1.3730	0.0117	

Table 4.
Bearing selection charts [18].

Proper sizing calculations and load considerations must be made to achieve optimal bearing performance. Bearing sizes are driven by the radial forces acting on the shaft. When determining the radial and axial forces being applied to the shaft, there are three main considerations: static loads, applied loads, and drive loads. Static loads are most commonly loads produced by the weight of the components. Applied loads are those derived from normal operating conditions on the component, such as thrust loads exerted onto a shaft. Drive loads are generally torques transmitted by the power source to the load. Although there are many steps to ensure that a bearing is properly sized, the calculation of radial force for belt, chain, or gear driven shafts is the same for both metallic and composite bearings [1] and is presented as Eq. (1):

$$F_r = \frac{(P)(19.1 \times 10^6)(K_t)}{(PD)(\Omega)} \tag{1}$$

where:
F_r = radial force on shaft (N).
P = power transmitted (kW).
PD = pitch diameter (mm).
Ω = shaft speed (rpm).
K_t = drive tension factor (1 for chains & gears, 1.5 for belt drives).

5. Composite bearings in heavy off-road equipment

Off-road equipment design provides numerous constraints and criteria that could be interpreted as advantageous toward the use of composite bearings by an off-road vehicle manufacturer. These bearings minimize the need for continuing maintenance as the machinery is operated. They provide a positive effect on the field capacity of the equipment, and they are beneficial to the overall life of the equipment in corrosive and abrasive environments. A few illustrative case studies have been provided, showing the potential use of composite bearings in the design of agricultural, construction, forestry, and mining vehicles.

5.1 Reduction of maintenance

Off-road vehicles require regular service on traditional metallic bearings at specific operating intervals to keep the equipment operating at its optimal performance. These service intervals include grease applications for the multitude of bearings and other moving mechanical components. Grease intervals can range from daily to seasonal, and it is now standard to label the grease fitting location with a specific decal featuring the recommended number of operational hours between servicing. The heavy equipment industry uses National Lubricating Grease Institute (NLGI) #2 grease for most greaseable parts, unless otherwise specified. Grease is rated with an NLGI number, which represents the consistency of the lubrication and is derived using the American Society for Testing Materials (ASTM) standard D217 [24]. This standard outlines the testing methods for the cone penetration of lubricating grease at 25°C [25]. The grease for use on the operating equipment must be purchased, and labor must be allotted for the lubrication process. Replacing traditional metallic bearings with composite bearings eliminates both of these expenses for the converted point of application.

5.2 Cost-effectiveness

Regular servicing is time-consuming for operators to complete, but a necessary step to continuing trouble-free daily operation. Minimizing the amount grease-requiring components on off-road vehicles would be highly beneficial for a machine's field capacity, but it might not be feasible for every specific function on a machine. However, replacing traditional metallic, grease-requiring bearings with greaseless composite bearings could eliminate the need for daily service in many locations, and it would allow more time per shift for equipment productivity. The cost of grease can be a sizable annual maintenance cost for a variety of different industries, especially for those that are managing large fleets of equipment. Designing equipment with more self-lubricating composite bearings could significantly reduce the continuing costs of ownership and help minimize equipment down-time for maintenance.

Vehicle maintenance typically falls under the vague accounting charges of indirect labor and miscellaneous supplies. To illustrate the somewhat hidden expense of grease and routine equipment service in maintenance budgets, calculations were performed for the cost to service all 25 h grease fittings on a standard Class VIII combine for one harvest season in the midwestern region of the United States. An axial combine used for corn, soybean, and cereal grain harvest in the Midwest has approximately twenty grease fittings around the machine, requiring service every 25 h of engine operation. Additionally, there are around seven grease fittings that require service every machine operating day, along with numerous others having different intervals. The number of grease fittings on combines vary between manufacturers, but they are all comparable. Each requires routine maintenance that creates downtime for the machine. Most grease fittings need three to five pumps of grease per fitting, unless specified otherwise by the manufacturer. A standard, manual-lever grease gun supplies about 1.3 g of grease per pump [21]. Harvest seasons in midwestern United States are typically around 8 weeks long. It was assumed that 75% of the total days for the completion of harvest season are combine-operating-days, since it is not reasonable to assume a combine would operate continuously through the length of harvest season. It was also assumed that a 25 h combine operating interval is equivalent to approximately 2 days. **Table 5** and Eqs. (2)-(4) estimate the cost of grease to service one standard combine for a single harvest season. Ten standard 14 oz tubes of #2 NIGL multi-purpose grease were used to calculate the average price used in Eq. (2). The cost of the 25 h service interval for the harvest season was roughly US$60.

$$\text{Total Grease for One, 25 h Service} = 20\,(1.3\,g * 4\,pumps) = 104\,g \tag{2}$$

Qty. of 25 h grease fittings	20	fittings
Flow rate/pump	1.3	g
Avg. qty. of pumps/fitting	4	pumps
Avg. length of harvest	56	days
Estimated combine operating days	42	days

Table 5.
Inputs for grease cost analysis calculations.

$$Total\ Cost\ of\ One, 25\,h\ Service = \left(\frac{104\,g}{14\,oz.}\right) * \left(\frac{1\,oz}{28.3\,g}\right) * \$10.98 = \$2.82 \quad (3)$$

$$Total\ Cost\ for\ One\ Harvest\ Season = \$2.82\left(\frac{42}{2}\right) = \$59.50 \quad (4)$$

Service cost estimations become proportionally larger for agricultural and construction operations with multiple pieces of equipment and various implements requiring regular servicing within their fleets. **Table 4** and Eqs. (2)-(4) are for a single machine and a single specific service interval. In addition to the 25 h service interval, combines will typically also have 10 h, 50 h, 75 h, 100 h, and 200 h grease fitting service intervals, with their own annual charges. Maintenance neglect at any of these service intervals will cause a premature failure of bearings and potentially other larger mechanical components. The total service cost for each unit would be the sum of the service intervals, and the total cost to an entity would be multiplied by the number of vehicles in their operational fleet.

5.3 Resistance to harsh environments

Providing resistance to resist debris entrance and contamination is one of the greatest challenges to overcome when designing heavy equipment for the agricultural, construction, forestry, and mining industries. Engines, mechanical motion parts, fluid power, and electronics are highly sensitive to dust, debris, and other contaminants found within the environment that heavy equipment spends most of its operational life. While metallic bearings and composite bearings both should be protected from contamination, neglecting bearing seal integrity for metallic bearings will cause a premature failure at a more rapid rate in these types of environments. The threat of bearing failure caused by contamination and lack of service is minimal for a self-lubricating, composite bearing. This feature also makes composite bearings an excellent choice in difficult-to-access locations. Some bearings on heavy equipment require the removal of several parts to gain access to grease fittings for service. This often causes operators to not apply the proper amount of grease to the bearing or ignore its required service intervals altogether. A common solution to this problem is the installation of remote grease fittings on a bulkhead in an accessible location, connected to the bearing housings with steel lines or hoses. However, remote grease fittings require more grease initially, because their lines must be filled with grease. Hard-to-reach bearing locations outfitted with self-lubricating composite bearings are an excellent solution to issues with impeded service accessibility and the potential for premature bearing failure, if neglected.

5.4 Case studies

Some real-world equipment issues that were resolved through the application of composite bearings in heavy equipment designs are summarized in this section. Many of these solutions included maintenance reductions, increased wear resistance in corrosive environments, and economic advantages in both design and manufacturing. These case studies were published by Polygon Composites to summarize collaborations between Polygon and various equipment manufacturers looking to resolve problematic designs deriving from metallic bearings [26].

5.4.1 Livestock feed mixer

Polygon and Kuhne Industrie BV, a Netherlands-based manufacturer of composite bearing materials, collaborated to produce a custom, composite bearing to work with an agricultural equipment manufacturer's tandem-axle, feed mixing cart. The equipment manufacturer had tested several bearing types, all of which caused excessive friction and premature wear on the shafts, due to the high, uneven load on the bearings during operation. In addition to not supporting the known load demands of the equipment, properly-sized bearings were difficult to procure. Polygon and Kuhne worked closely to design a composite material solution for the feed mixing cart manufacturer. This led to the development of a custom composite bearing that withstood the high, uneven loads, minimized the need for maintenance, and reduced noise caused by the rapid oscillation of the assembly held by the bearing [27].

5.4.2 Reduced maintenance in bulldozer designs

Polygon provided composite bearings and bushings to three major US construction equipment manufacturers, as a greaseless bearing alternative that minimized chassis maintenance requirements. Bearing service in heavy equipment applications can be very time consuming and difficult, depending on the location of the bearing to be serviced. Due to these common challenges, one of the leading bulldozer manufacturers in the US approached Polygon for a composite bushing to replace a traditional, metallic bushing located in a position that was difficult for operators and maintenance technicians to service with grease. This bushing was designed as part of an assembly that connected the bulldozer track assembly to the machine's chassis. To solve this critical design challenge, Polygon developed a self-lubricating, composite bushing with an industrial wear surface that permanently mated with the internal diameter of the bushing. Due to the development of this wear surface, this bearing does not need to be greased or serviced for the design-life of the equipment and will not corrode, due to the non-metallic construction of the bushing. Since this initial technology introduction, the utilization of this composite bushing has become essential in bulldozer designs by all major US construction equipment manufacturers [28].

5.4.3 Composite bushings in earthmoving equipment

Polygon has designed composite bushings for K-Tec earthmoving equipment, a Canadian-based supplier of mining equipment. K-Tec used Polygon's bushings in their hydraulic cylinders and for other critical mechanical pivot points on their machinery. Mining equipment is operated in extremely harsh environments with many contaminants and debris present in the ambient air, water, and soil of the mine. Equipment running in this industry must be rugged, able to withstand high loads, and perform reliably with minimal maintenance. Leveling and scraping equipment uses hydraulic actuation to adjust the depth of the working edge for various digging and scraping duties. These hydraulic leveling systems must withstand high amounts of strain and stress, and they must operate accurately on a daily basis. To reduce the need for daily lubrication service, Polygon developed custom, greaseless bushings for the hydraulic linkage pivot points that can withstand the severe-duty requirements of the equipment and maintain their functionality in the debris and contaminate-laden environment of the mines [29].

6. Testing & validation

Reputable manufacturers of composite bearing products complete rigorous testing to ensure their products' capabilities and evaluate them against other competitors' products. Common bearing tests include rotational wear, linear wear, pivoting wear, and friction coefficient measurements. Igus has published several of their product experiments, as well as the test reports completed for their iglide® composite bearing line [19]. Their test lab has researched low load pivot wear, composite bearings versus metal bearings on different shaft types, frictional wear on composite bearings over many cycles, and plain bearing and brass bearing wear on various shafts [30]. In one circumstance, several variations of iglide® composite bearings were tested against a customer's brass bearing, which was used on a rotating shaft. During operation, the brass sleeve bearing was subject to 'tumbling' load stresses. This term describes a shaft that wobbles, creating uneven wear patterns. The shaft does not wear uniformly, due to the tumbling-effect, and typical metallic bearings in these applications have low lives. Igus tested two different variations of composite bearings, each with two different lubrication configurations, along with two different configurations of a competitor's brass bearing. All experimental components were tested with and without grease. The purpose of the testing was to discover the optimal bearing material for the specific intended application that reduced both shaft and bearing wear over time. It was demonstrated that an average of 490.4 μm/km of bearing wear was measured for the dry-running iglide® bearings, compared to the measured wear of 1325.0 μm/km for dry-running brass bearings. An average of 425.4 μm/km of bearing wear was measured for the grease-applied iglide® bearings, compared to 497.6 μm/km of measured wear for the grease-applied brass bearing. However, the most optimal bearing, when subjected to the unique tumbling stress was the grease-applied iglide® Q bearing, which showed the least amount of measured wear at 241.2 μm/km, compared to the customer's brass bearing [30]. The results for each bearing type and lubricant configuration are shown in **Figure 8** [30].

Composite bearing manufacturers typically use state-of-the-art testing facilities to validate their product's performance. Obtaining high-confidence testing data for a specific application is critical for product design and allows for comparisons to be drawn between the product and other competitive products for marketing purposes. These tests are designed to simulate the working conditions of the equipment in a variety of applications, and they are primarily conducted to measure wear and

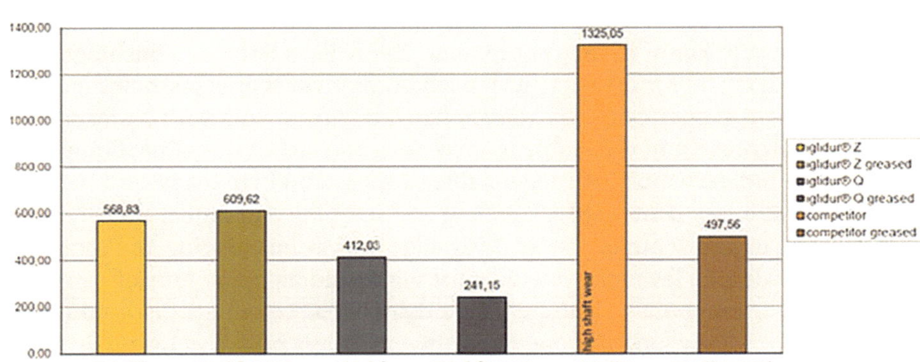

Figure 8.
Measured wear (μm/km) (y-axis) test results for iglide® composite bearing in tumbling motion [30].

DOI: http://dx.doi.org/10.5772/intechopen.1004436

friction. The design guides that many bearing manufacturers provide can help determine the optimal bearing for an intended design and application. Specific results with the data collected during well-designed product testing and validation program would be needed, prior to any full commitment to convert a design to composite bearings.

7. Design considerations

The overall design of a composite bearing joint proceeds exactly like a traditional metallic bearing joint [1]. All motions and loads must be accurately categorized, and performance criteria should be specified. Any specific constraints for the application should be noted. Design guides from specific manufacturers will now be useful. It is always wise for an engineer to explore their own solutions with a design guide, prior to contacting a company, to familiarize themselves with the company's key parameters and processes. It may be useful to utilize multiple potential vendors. After an introduction, a prototype product for the specific application should be developed as a joint project. Testing and validation should be planned and reviewed, before committing to a long-term partnership. These steps will be explored in the following sections in greater detail.

7.1 Design guides

Design guides provide engineers and technologists with the details and specifications necessary to evaluate which products are best suited for the specific design. These guides are published by bearing manufacturers and are available on-line. Often, there is an abundance of information in these product design guides, which should be appreciated by modern engineers and technologists. Obtaining critical details regarding a product can sometimes be difficult, due to proprietary constraints and lack of transparency about design. Bearing design guides contain specifications on mechanical properties, thermal properties, dimensions, and sizing recommendations for working stresses. Different bearing classifications relative to the direction of motion in the intended application may also be included. Many of these design guides also incorporate specific design calculations to aid with bearing sizing and selection within the manufacturer's product line. **Figure 9** provides calculations and recommendations for pressure and velocity in Polygon's bearing selection and design guide [15]. **Figure 10** shows the physical forces and geometry of the bearing defined in their preferred terminology [15]. The sample example calculation uses a 1.9 cm (0.75 in) diameter pin, turning at 200 rpm, supported by a sleeve bearing subject to 380 N (85 lb_f) of load. The pin diameter and rotational speed of the pin are utilized in the

CALCULATING SLEEVE BEARING PV LIMIT

Example: .750" Pin Dia @200 rpm , 85.0 lb. total load, bearing length .750"	A = .750 (Pin) x .750 (bearing length) = .562 in.^{2}
V = 0.262* x rpm x diameter = 0.262 x 200 x .750 = 39.3 ft/min	P = 85.0 lbs. / .562 in.^{2} = 151.2 psi
P = total load / projected area (A)**	PV = 151.2 psi x 39.3 ft/min = 5,942 psi* ft/min

Figure 9.
Pressure and volume limit calculations [15].

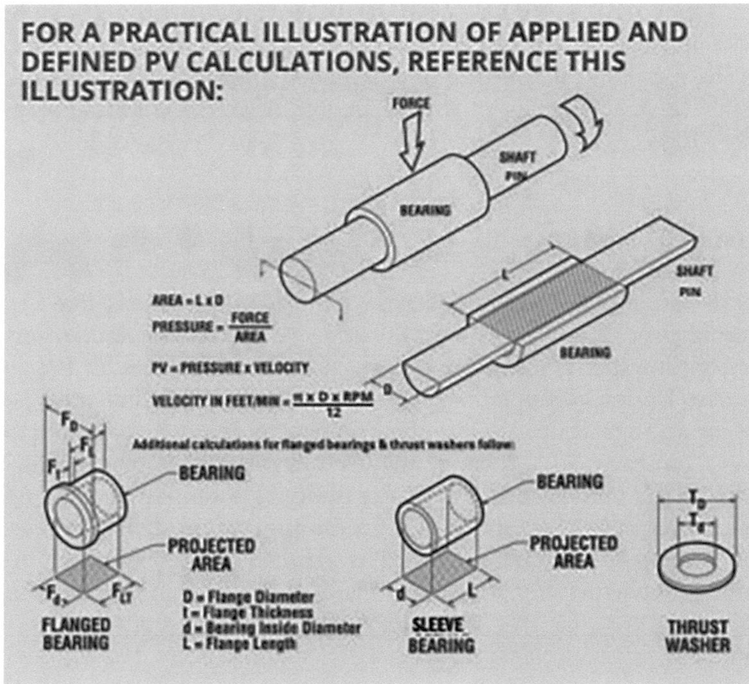

Figure 10.
Geometry and forces present in sleeve bearing calculations [15].

velocity calculation, while the load on the bearing over the projected contact area is used in the bearing pressure calculation. The solutions to 'P' and 'V' are multiplied to equate a singular, quantified metric for comparison against other sleeve-type bearings. Pressure and velocity limit calculations are standard methods for comparing bearing performance [15]. Design engineers use 'PV' calculations to determine safety limits with respect to loading and operating speed for the application.

7.2 Basic design and manufacturing

There are standard design practices associated with the design and manufacturing of all products. Bearing assemblies are no different. For composite bearings, some design practices that should be utilized are to ensure the correct material choice for the bearing, press-fit interference tolerance in the assembly, minimized post-curing effects in the backing layers, proper overall length, and correct surface finish. Bearing material selection is important, because most composite materials have a greater thermal expansion coefficient than metallic bearings and will require larger clearances to properly fit across all operational temperatures. Proper mating materials in the assembly are also very important, because a harder mating material can result in a longer bearing life [31]. Interference tolerances are important in bearing design, because proper secure fitment is critical to bearing life. During the post-curation process for the composite material, it is recommended the bearing be baked at 140–150°C for four hours or annealed in SAE 20 oil at 107–130°C [31]. This provides the best material properties for long-term bearing life. A best design practice regarding the length of the bearing is for the L:D ratio to not exceed 3.5 [31]. Unexpected forces in a bearing can develop on rotating joints with slender pins. The composite bearing surface finish should be molded, if possible, to provide the least amount of wear in motion and lowest possible coefficient of friction [31]. These common design best practices have been listed to provide a general understanding of the variety of different processes that manufacturers use to measure and control to provide critical performance properties in composite bearings.

7.3 Product availability and procurement

When designing an off-road vehicle, parts that are not produced in-house or are being prototyped, require vendor sourcing. Vendor relationships and shared business processes are critical for the successful manufacturing and maintenance of composite bearing systems in sustained, heavy-duty applications. The collaborative themes discussed previously will be expanded in this section.

The three ways to determine the necessary product fitment for components in the manufacture of heavy-duty off-road-vehicles are through ordering existing parts from a manufacturer of the desired component, purchasing mass-produced parts from an authorized retailer, or partnering with a company to create a new specific product for use through joint research and development efforts. When purchasing existing components, it is generally advised for engineers to look for vendors that already supply products to their company, due to the history of past transactions and an established business partner relationship. If these vendors do not have the needed product, and they cannot or will not help produce a new product to fit the specifications, then it is advisable to seek new vendors. This can be an involved process, since a new business-to-business relationship is being developed, as opposed to a technical one.

Existing products are manufactured items that have a history of usage behind them and are generally available in large quantities, due to their established demand. Many times, these products can be found through wholesalers or parts houses. These products typically have extensive testing and engineering support proving the product works as claimed, which means these types of components are good selections for use in a production design. When an engineer seeks a new product, research on the manufacturer or vendors is critical to determining if the desired product from this particular vendor is suitable for use. For example, one can go to the iglide® website, enter the “Plain Bearings” tab, and search for a component that would fit most products' specifications. Many, if not all, commercial websites have filter applications available for easy site navigation, and it is advisable to utilize these useful tools, when selecting components for a larger design.

If there are not products on the market that are suitable for a specific design, then one must consider manufacturing a new part to complete the assembly. Generally, it is more cost-effective and time-efficient to contact a manufacturer already in the business to have a prototype designed for the specific new assembly, rather than attempt to make one in-house. If the component is something that is related to a trade secret or is need of intellectual property protection, then the necessary legal protections must be in place before divulging any information to another corporation. If these protections are needed and adequately secured, then a manufacturer who deals in the desired product area can be contacted. These organizations are typically more than happy to work with customers and assist them with finding the right part for the job. New design prototypes will cost more than an existing part, due to the need to create new tooling, casting fixtures, and other equipment to manufacture test batches, before full production can be made. However, this is a one-time cost. If the new prototypes are to be produced for a long period of time, then the added costs will be negligible, amortized over the entire production run.

8. Conclusion

The advantages, unique properties, and design characteristics of composite material bearings have been illustrated by this design overview in effort to convey the state-of-the-art for this technology to practicing engineers and technologists working on all types of designs for off-road heavy-equipment. The off-road vehicle industry should be able to take advantage of many types of composite bearings to incorporate near-frictionless, maintenance-free, and contaminate-resistant motion into their equipment designs. Metallic bearings, which use grease or oil to create a near-frictionless barrier between mechanical motion parts, are preferred in many heavy-equipment applications, but these components have continuing financial and time costs associated with their use. Traditional metallic bearings feature high load factors, durability, and low operating noise. Metallic bearings function very well, when maintained properly in accordance with specific design specifications and recommended service intervals, which typically require the application of grease or oil. Maintenance is critical to bearing life and avoidance of premature bearing failure. Composite bearings maintain critical functions and reduce the need for continuing maintenance.

Servicing every day or once per week may seem like a minute challenge. However, for large commercial construction, forestry, mining or agricultural operations, maintenance is very costly in two primary ways. The cost of consumables, tools, and labor adds significant cost to properly maintaining equipment. A numerical experiment on a modern combine, where the costs of grease and labor were evaluated for only one specific service interval, demonstrated the potential of these expenses to grow and become significant. For many machines, there are often five to seven different service interval periods with many different grease fittings and associated intervals. The additional costs that stem from the equipment downtime during the maintenance period reduce the profitability of the machinery. Equipment downtime is detrimental to machine efficiency, field effectiveness, and return on investment. These factors are always analyzed in detail by equipment and project managers to accurately predict operating costs, making a composite bearing-equipped machine a potential superior consumer choice. Another design challenge that might encourage composite bearing use includes accessibility issues for a specific bearing location. Off-road equipment utilizes many complex chassis designs with tough-to-reach areas. Many of these inaccessible locations have bearings that require regular service or occasional replacement. Sometimes a remote grease fitting is not an option, depending on the bearing type. Additionally, some heavy equipment operates in corrosive environments that can attack metallic bearings. Composite bearings can be used to increase the useful life and maintenance intervals of such a vehicle.

Composite material bearings offer a variety of solutions to each of these unusual challenges. Case studies were reviewed on the application of composite bearings to unique situations, where metallic bearings were problematic or were not an option. A leading bulldozer manufacturer significantly reduced downtime for maintenance through the integration of a composite bushing that did not require regular lubrication to maintain optimal bearing performance and would not corrode in harsh environmental conditions. This improvement was provided by utilizing a bearing that could last the design life of the equipment. This is a prime example of the advantages associated with composite bearings in heavy-duty off-road vehicles. The self-lubricating features provided by PTFE and other special coatings eliminate the need for applied lubrication and regular access. Additionally, composite resins and materials are non-corrosive, making them extremely advantageous in adverse conditions, where many types of off-road equipment spend most of their life cycles.

Many types of bearings and configurations are applied in many types of machinery designs to provide a frictionless barrier, smooth motion, and minimal operating noise for working machine parts. The heavy-equipment industry, specifically the agricultural, construction, forestry, and mining segments, could potentially integrate many composite bearings into their designs for the basic vehicles and auxiliary implement functions on their machines. The specific heavy-duty mechanical components designed for specific tasks and operations differentiate off-road equipment from the automotive industry and other on-road equipment. Composite bearings will allow these machine and chassis designs to improve through the elimination of accessibility and continuing service requirements, providing significant cost reductions for consumers and equipment managers.

9. Recommendations

The scope of this paper was to introduce composite bearing technologies and educate design engineers and technicians on the advantages of using non-traditional composite bearing types in mobile and industrial equipment design applications. Emphasis was placed on the off-road vehicle industry in the agricultural, construction, mining, and forestry equipment manufacturing sectors, given the off-road industry's unique ability to take advantage of the anti-corrosive and maintenance-free characteristics of the composite bearings. There are numerous other reasons and advantages to incorporate composite bearings into designs. It is recommended that designers consider composite bearings in their research and development processes by reviewing and utilizing the detailed design guides provided by several composite technology manufacturers, some of which were described. It should be expected that composite bearing modules will appear in computer aided design software soon, to aid in embedding these new bearings in different designs. Similar to the ability to import any component data from a supply house for use in a new design, this inclusion will speed-up the introduction of this technology into common use.

Many composite technology companies have exceptional technical service engineers and are willing to create custom solutions for incorporation into a variety of design applications. The emphasis on custom bearing applications is greater in the composite technology industry than with traditional metallic bearing manufacturers. Many composite bearing manufacturers have a systematic engineering process for developing custom solutions for unique applications requiring their product. This begins with a preliminary study and direct conversations with a client to formulate the background and scope of the problem. This initial step is typically followed by a joint design process that eventually leads to product validation and testing. Once testing has been completed, and a solution has been optimized for a specific application, mass production can take place for the inclusion of the composite bearing into a new off-road vehicle design. There are many resources available to take advantage of in the composite bearing industry, and the product transparency offered by these companies will be greatly appreciated by engineers and technicians looking to take advantage of these innovative products.

Acknowledgements

The Fall 2023 Design of Off-Road Vehicles class at Purdue University's School of Agricultural and Biological Engineering is acknowledged for their contributions to the structure and content of this technical chapter. Dr. Carol S. Stwalley is graciously thanked for her editorial work with the manuscript and in the final formatting of the document.

Author details

Thomas Barnes[1], Brecken Beyer[1], Wyatt Griffey[2], Caleb Huffmeyer[1], Clayton Ness[1], Owen Nifong[1], Tyler J. McPheron[1] and Robert M. Stwalley III[1*]

1 Purdue University Agricultural and Biological Engineering, West Lafayette, IN, USA

2 Purdue University Mechanical Engineering, West Lafayette, IN, USA

*Address all correspondence to: rms3@purdue.edu

References

[1] Krutz GW, Schueller JK, Claar PW II. Machine Design for Mobile and Industrial Applications. Warrendale, PA: Society of Automotive Engineers, Inc; 1994

[2] saVRee. Plain Bearing [Online]. 2023. Available from: https://savree.com/en/encyclopedia/plain-bearin [Accessed 27 December 2023]

[3] Forsthoffer M. Journal (Radial) Bearings. In: Forsthoffer's Component Condition Monitoring. Oxford, UK: Butterworth-Heinemann; 2019

[4] GGB by Timken. How does a Self-Lubricating Bearing Work? [Online]. 2019. Available from: https://www.ggbearings.com/en/why-choose-ggb/faq/bearings-faq/what-self-lubricating-bearing [Accessed 27 December 2023]

[5] Gibson RF. Principles of Composite Material Mechanics. New York, NY: McGraw-Hill; 1994

[6] Karandikar PM, Kharde RR, Bhoyar SB, Kadu RL. Study the tribological properties of PEEK/PTFE reinforced with glass fibers and solid lubricants at room temperature. International Journal of Current Engineering and Technology. 2014;**4**(4): 2401-2404. Available from: https://inpressco.com/wp-content/uploads/2014/07/Paper192401-2404.pdf

[7] Yu HN, Kim SS, Lee DG. Optimum design of aramid-phenolic/glass-phenolic composite journal bearings. Composites Part A: Applied Science and Munufacturing. 2009;**40**(8):1186-1191. DOI: 10.1016/j.compositesa.2009.05.001

[8] Burris DL, Sawyer WG. A low friction and ultra low wear rate PEEK/PTFE composite. Wear. 2006;**261**(3-4): 410-418. DOI: 10.1016/j.wear.2005.12.016

[9] Henninger F, Friedrich K. Thermoplastic filament winding with on-line impregnation. Part A: Process technology and operating efficiency. Composites Part A: Applied Science and Manufacturing. 2002;**33**(11):1479-1486. DOI: 10.1016/S1359-835X(02)00135-5

[10] Henninger F, Hoffmann J, Friedrich K. Thermoplastic filament winding with on-line impregnation. Part B: Experimental study of processing parameters. Composites Part A: Science and Manufacturing. 2002;**33**(12): 1684-1695. DOI: 10.1016/S1359-835X(02)00136-7

[11] Jadhav GR, Zoman DB, Mahajan DR. Analysis of composite journal bearing. International Engineering Research Journal. 2025;**1**:1998-2002. Available from: https://www.ierjournal.org/pupload/mitpgcon/1998-2002.pdf

[12] Jadhav G, Badguiar T, Zorman DB, Jadhav M. Tribological performance analysis of composite materials for journal bearing. International Journal of Modern Trends in Engineering and Research. 2016;**3**:131-138. DOI: 10.21884/ijmter.2016.3017.apwts

[13] Sengsri P, Marsico MR, Kaewunruen S. Base isolation fibre-reinforced composite bearings using recycled rubber. IOP Conference Series: Materials Science and Engineering. 2019; **603**:1-10. DOI: 10.1088/1757-899X/603/2/022060

[14] GGB by Timken. Fiber Reinforced Composite Bearing Handbook [Online]. 2023. Available from: https://www.ggbearings.com/sites/default/files/

2023-08/GGB-Fiber-Reinforced-Composite-Bearings-High-Load-Self-Lubricating-Bearings-Bushings_0.pdf [Accessed 27 December 2023]

[15] Polygon Composites Technology. Bearings Design Guide [Online]. 2023. Available from: https://polygoncomposites.com/design-guide-bearings/ [Accessed 27 December 2023]

[16] SKF. SKF Composite Plain Bearings [Online]. 2012. Available from: https://cdn.skfmediahub.skf.com/api/public/0901d19680229dfc/pdf_preview_medium/0901d19680229dfc_pdf_preview_medium.pdf [Accessed 27 December 2023]

[17] Igus Motion Plastics. Igus Engineer's Toolbox Plain Bearing Design Guide [Online]. 2023. Available from: https://toolbox.igus.com/design-guides/iglide-plain-bearings-design-guide [Accessed 27 December 2023]

[18] Polygon Composites Technology. Industries [Online]. 2023. Available from: https://polygoncomposites.com/industries/ [Accessed 27 December 2023]

[19] Igus Motion Plastics. Bearings are Rigorously Tested within the Iglide® Test Laboratory [Online]. 2023. Available from: https://www.igus.com/info/plain-bearings-testlaboratory [Accessed 27 December 2023]

[20] GGB by Timken. Composite Bearing Design with Improved Tribology. GGB, Thorofare, NJ [Online]. 2020. Available from: https://www.ggbearings.com/sites/default/files/inline-files/GGB-Whitepaper-Composite-Bearings-for-Aggressive-Applications.pdf [Accessed 27 December 2023]

[21] Fitch B. The Grease Gun: Applications, Uses, and Benefits [Online]. 2023. Available from: https://www.machinerylubrication.com/Read/29356/grease-gun-anatomy [Accessed 27 December 2023]

[22] Carrera A. How do Self-Lubricating Bearings Lubricate? [Online]. 2023. Available from: https://www.tstar.com/blog/qa-how-do-self-lubricating-bearings-lubricate [Accessed 27 December 2023]

[23] Bushing MFG. PTFE Bearings [Online]. 2023. Available from: https://bushingmfg.com/ptfe-bearings/ [Accessed 27 December 2023]

[24] National Lubricating Grease Institute. Grease Glossary [Online]. 2017. Available from: https://www.nlgi.org/grease-glossary/nlgi-grade/ [Accessed 27 December 2023]

[25] American Society for Testing Materials. Standard Test Methods for Cone Penetration of Lubricating Grease [Online]. 2021. Available from: https://www.astm.org/d0217-21a.html [Accessed 27 December 2023]

[26] Polygon Composites Technology. Case Studies [Online]. 2023. Available from: https://polygoncomposites.com/case-studies/ [Accessed 27 December 2023]

[27] Polygon Composites Technology. Polygon and Kuhne Industrie delivers composite Bearings for Demanding Mixer Cart Application [Online]. 2023. Available from: https://polygoncomposites.com/resources/kuhne-industrie-delivers-composite-bearings-mixer-cart-application/ [Accessed 27 December 2023]

[28] Polygon Composites Technology. Polygon Composite Bushings Ensure reduced Maintenance in Bulldozer Designs [Online]. 2023. Available from: https://polygoncomposites.com/resources/composite-bushings-ensure-red

uced-maintenance-in-bulldozer-designs/ [Accessed 27 December 2023]

[29] Polygon Composites Technology. Polygon Composite Bushings Simplify Maintenance in K-Tec heavy-duty Earthmoving Equipment [Online]. 2023. Available from: https://polygoncomposites.com/resources/polygon-composite-bushings-simplify-maintenance-in-heavy-duty-equipment/ [Accessed 27 December 2023]

[30] Igus Motion Plastics. iglide® Q: Able to Withstand High Loads, even When Subjected to the Strain of Tumbling [Online]. 2023. Available from: https://www.igus.com/info/plain-bearings-test-iglide-q [Accessed 27 December 2023]

[31] Franklin Fibre. Essential Design Practices for Composite Bearings and Bushings [Online]. 2023. Available from: https://www.franklinfibre.com/blog/essential-design-practices-for-composite-bearings-and-bushings#:~:text=For%20best%20design%2C%20the%20length,cause%20edge%20concentration%20of%20loading.&text=%E2%80%8DThe%20smoother%20the%20surface%20of,16RMS%20or%20bet [Accessed 27 December 2023]

Section 2

Non-Metallic Composites

Chapter 6

Mechanical Constitutive Models of Fiber Reinforced Plastics for Finite Element Analysis

Changfang Zhao

Abstract

This chapter is devoted to introducing some basic constitutive models of fiber reinforced plastics (FRPs), presenting the rationale for modifying and improving the constitutive model in accordance with the actual physical mechanism, and proposing the application ideas of the constitutive model for material subroutines in conjunction with the finite element (FE) method. These constitutive models can serve as a foundation for the establishment and development of new constitutive models. As we know, the advent of fiber reinforced plastics (FRPs) has furnished robust reinforcement for aerospace and other sophisticated equipment, thereby facilitating and enhancing human exploration of the sea, land, air, and sky. This kind of material has the characteristics of multi-phase distribution, which allows artificial control of its properties, qualifying it as an advanced composite material. However, as the scope of FRP applications continues to expand, a number of new challenges have emerged, including those related to physical fields, boundary conditions, material limitations, and more. This underscores the necessity for further development of some existing theories. Finite element analysis (FEA), as the third paradigm of scientific research, has been instrumental in facilitating significant advances in numerous engineering design projects. At present, the application of FEA to FRPs remains a complex, time-consuming, and pivotal process, which suggests that this chapter will provide useful insights.

Keywords: fiber reinforced plastics, constitutive models, finite element analysis, mesoscopic mechanics, interfacial mechanics, failure criteria, damage evolution, material subroutines

1. Introduction

Over the past six decades, fiber reinforced plastics (FRPs), a category of advanced composite materials, have undergone significant advancement, particularly in the area of carbon fiber reinforced plastics (CFRPs). In comparison with conventional engineering materials such as steel, FRPs are distinguished by relatively low density, enhanced strength and stiffness, and flexible designability [1, 2], which has facilitated their extensive utilization in a wide range of sectors, including aerospace, automotive, drones, and marine engineering [3, 4]. The strong designability and multi-phase characteristics of FRPs present significant challenges to the comprehensive structural design method, which must consider failure, damage, and other factors [5, 6]. Consequently, consensus on a common, accurate, and simple failure criterion has yet to be reached. Nevertheless, this highlights the necessity for further investigation into the mechanics of FRPs.

FRPs are man-made, inhomogeneous, non-isotropic, and interfacial material systems consisting of more than two phases, each of which has complete and independent properties [2, 3]. Consequently, the studies of mechanical and other properties of FRPs can be divided into two fields and four levels from the geometric scale [7]. These are the micro-mechanics and meso-mechanics levels in the material field, as well as the macro-mechanics and structural mechanics levels in the structural field. Micro-mechanics encompasses the study of phases, crystals, and even molecules and atoms. In contrast, the domain of meso-mechanics focuses on the analysis of the particle or fiber diameter, which serves as the minimum geometric characteristic, within the context of a representative volume element (RVE). Additionally, macro-mechanics employs a single layer or a larger domain comprising of numerous RVEs as the subject of investigation, whereas structural mechanics focuses on the characteristics of laminates or composite structures. Typically, the aforementioned four aspects are integrated into the process of material selection/design for structure design and application. This approach is commonly referred to as cross-scale analysis, and the corresponding theory is known as hybrid composite mechanics theory [8, 9].

FRPs are typically anisotropic materials [10, 11]. However, under specific circumstances, the unidirectional single layer can be considered to exhibit transversely isotropic characteristics [12], and the laminate can be analogous to a quasi-isotropic material when subjected to specific layup rules [13]. The primary cause of the anisotropy observed in FRPs can be attributed to the angular distribution of the fibers within a mixed material system, such as short-cut CFRPs. Moreover, the stacking method and fiber orientation of the fiber layers in multi-layer laminates can also result in anisotropy. It is noteworthy that the matrix is typically a continuous isotropic material, which does not significantly contribute to the anisotropy of FRPs. In addition, the fiber-matrix interface not only exerts a slight influence on the anisotropy of FRPs, but also plays a significant role in failure [14, 15].

Furthermore, it is essential to consider additional factors when examining the mechanical properties of FRPs. To illustrate, in the context of high-speed impact loading, the external force acting on the material is instantaneously converted to strain energy, frictional energy, and kinetic energy [16, 17]. This phenomenon is also referred to as the adiabatic temperature, which is associated with thermodynamic behavior (namely mechanical-thermal coupling). Moreover, the strain rate effect is now recognized as a crucial factor influencing the mechanical behavior of materials [18, 19], which must be considered under impact loading [20]. It has recently been reported that CFRPs exhibit initial nonlinearity, which is attributed to matrix and interface defects [16, 18, 21] and the buckling of fibers [7]. Therefore, it is imperative that these identified issues, namely strain rate, initial nonlinearity, and adiabatic temperature, be taken into account when designing FRPs in the new era. This is crucial for the advancement of lightweight and high-strength structures for aerospace and other equipment, which necessitates the formulation of a fundamental constitutive model.

This chapter addresses the aforementioned issues by illustrating the mechanical constitutive models for FRP composites. This provides a foundation for developing more effective models that consider new factors. The chapter is divided into six sections. The second section presents the basic elastic mechanical constitutive models, which includes continuum medium mechanics, anisotropic models, and transverse isotropic models. The third section outlines the basic meso-mechanical models, which considers average stress, elastic modulus, and Poisson's ratio. The fourth section discusses the interlaminar interface constitutive models, which addresses damage evolution. The fifth section reviews several commonly used failure criteria, such as the Mises, Hill, and Hashin criteria. The sixth section presents a stiffness degradation model based on the effective load-bearing area. These models provide a mathematical framework for the design and prediction of structural properties within the domain of traditional mechanics, particularly in the context of finite element analysis (FEA) and user-defined material subroutines.

2. Basic anisotropic elastic constitutive models for FRPs

2.1 Continuum medium mechanics model for FRPs

The macro-mechanics of elastomer are typically founded upon the principles of homogeneity and continuity, which fall within the purview of continuum mechanics. However, the deformation theory associated with elastomers is only deemed valid under specific conditions. Fibers, matrices, and even FRPs are all examples of elastomers, and their constitutive models must satisfy the following assumptions [22]:

(1) Continuous medium assumption: It is assumed that the material occupies a continuous space, ensuring continuity of deformation and internal force at each material point (also called a substance point) when the material is deformed by an external force.

(2) Small deformation assumption: It is assumed that the deformation of a continuum under external force is negligible compared to its own geometric dimensions, such that dimensional changes due to deformation can be disregarded.

(3) Prestress-free assumption: It is postulated that the continuum is originally in a stress-free natural state, ensuring that the stress at any material point of the continuum is zero when it is not subjected to external forces.

(4) Isotropic assumption: It is assumed that each material point in the continuum has the same mechanical state and thermo-mechanical properties, allowing for the modeling of each material point in an identical manner.

In accordance with the aforementioned assumptions, consider the adjacent material points $\boldsymbol{p}$ and $\boldsymbol{q}$ within a three-dimensional Euler space, subjected to an external force and representing a FRP. If the distance between the two points is (dx, dy, dz), and the material point $\boldsymbol{p}(x_1, y_1, z_1)$ is taken as the reference point, then the material point $\boldsymbol{q}$ can be expressed as $\boldsymbol{q}(x_1{+}dx, y_1{+}dy, z_1{+}dz)$. Therefore, following deformation due to an external force, the material point $\boldsymbol{p}$ is transformed into $\boldsymbol{p}_1(x_2, y_2, z_2)$, as shown in **Figure 1(a)**. The spatial displacement of the material point $\boldsymbol{p}$ due to the external force can be expressed by the three-dimensional space vector $\boldsymbol{r}(u, v, w)$, that is,

$$p_1 = p + r \Rightarrow \begin{cases} x_2 = x_1 + u(x,\ y,\ z) \\ y_2 = y_1 + v(x,y,z) \\ z_2 = z_1 + w(x,y,z) \end{cases} \tag{1}$$

In the meantime, the displacement of the nearby material point $\boldsymbol{q}$ is determined with the material point $\boldsymbol{p}$ serving as the reference point. The three-dimensional spatial displacement vector of the material point $\boldsymbol{q}$ due to the force is obtained by expanding the material point $\boldsymbol{q}(x_1{+}dx, y_1{+}dy, z_1{+}dz)$ in accordance with the Taylor series expansion and omitting higher order infinitesimals. Moreover, the displacement vector is expressed in third-order symmetric tensor form using gradient notation, obtaining the geometric equation [23]. The geometric equation represents the specific form of what is often referred to as the Cauchy strain tensor. This represents the spatial deformation path of the material point, which corresponds to the deformation gradient in the finite element (FE) algorithm. The Cauchy strain tensor $\boldsymbol{\Gamma}$ and the corresponding strain matrix are expressed in matrix component form as follows:

$$\boldsymbol{\Gamma} = \begin{bmatrix} \dfrac{\partial u}{\partial x} & \dfrac{1}{2}\left(\dfrac{\partial v}{\partial x} + \dfrac{\partial u}{\partial y}\right) & \dfrac{1}{2}\left(\dfrac{\partial w}{\partial x} + \dfrac{\partial u}{\partial z}\right) \\ \dfrac{1}{2}\left(\dfrac{\partial v}{\partial x} + \dfrac{\partial u}{\partial y}\right) & \dfrac{\partial v}{\partial y} & \dfrac{1}{2}\left(\dfrac{\partial v}{\partial z} + \dfrac{\partial w}{\partial y}\right) \\ \dfrac{1}{2}\left(\dfrac{\partial w}{\partial x} + \dfrac{\partial u}{\partial z}\right) & \dfrac{1}{2}\left(\dfrac{\partial v}{\partial z} + \dfrac{\partial w}{\partial y}\right) & \dfrac{\partial w}{\partial z} \end{bmatrix} = \begin{bmatrix} \varepsilon_{11} & \gamma_{12} & \gamma_{13} \\ \gamma_{21} & \varepsilon_{22} & \gamma_{23} \\ \gamma_{31} & \gamma_{32} & \varepsilon_{33} \end{bmatrix} \tag{2}$$

where ε_{ii} denotes the principal strain and γ_{ij} denotes the shear strain, and $(i, j = 1, 2, 3)$.

In the field of elastic mechanics, the volume micro-element (VME) is a more prevalent concept. It represents the finite differential element of a given material, analogous to the mesh element in FEA. Consider a hexahedral VME, as shown in **Figure 1(b)**, where the material point is at the center of the VME, and the length of each side of the VME is known. In a given direction along the VME, an external force vector $\boldsymbol{f}(f_x, f_y, f_z)$ acts upon the system, and the force equilibrium equations can be derived. In the event that the hexahedron undergoes a deformation at the material point subsequent to the application of a force, the equilibrium equations between the stress and the external force $\boldsymbol{f}_i$ can be derived as follows [22, 23]:

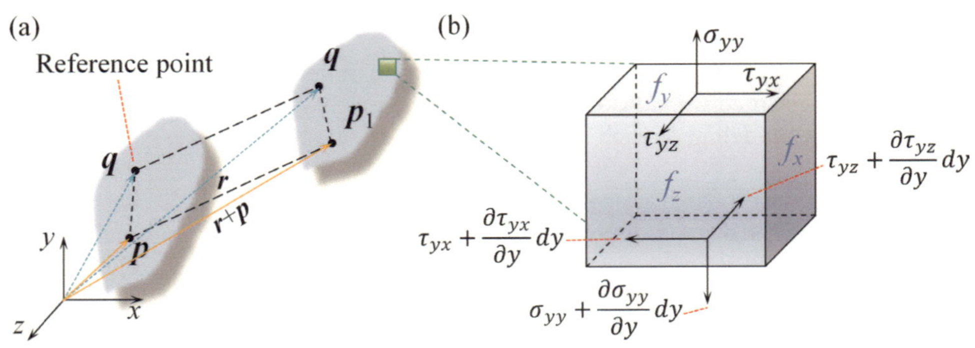

Figure 1.
Deformation diagram of material points within an Euler space (panel a) and within a volume micro-element (panel b), where panel b shows only one stress pair.

$$\begin{cases} \dfrac{\partial \sigma_{xx}}{\partial x} + \dfrac{\partial \tau_{xy}}{\partial y} + \dfrac{\partial \tau_{xz}}{\partial z} + f_x = 0 \\ \dfrac{\partial \tau_{yx}}{\partial x} + \dfrac{\partial \sigma_{yy}}{\partial y} + \dfrac{\partial \tau_{yz}}{\partial z} + f_y = 0 \\ \dfrac{\partial \tau_{zx}}{\partial x} + \dfrac{\partial \tau_{zy}}{\partial y} + \dfrac{\partial \sigma_{zz}}{\partial z} + f_z = 0 \end{cases} \tag{3}$$

where σ_{ii} denotes the principal stress, σ_{ij} denotes the shear stress, and $(i, j = 1, 2, 3)$.

The geometric equation (Eq. (2)) combines the deformation and strain of a material point under external force, while the equilibrium equation (Eq. (3)) combines the force and stress of a VME under external force. However, the stress-strain relationship has not been represented. The stress-strain relationship, frequently designated as the constitutive relationship, is the basic criterion for describing the mechanical behavior of VMEs. It is also a fundamental theory in the context of FE simulation.

The derivation of the constitutive model is based on the principles of thermodynamics. Based on the fact that the work done on a VME by an external force and the heat input from the outside are equal to the internal energy increment of the VME, it can be concluded that the VME is in a thermodynamic equilibrium state at all times. Assumption: there is no other source of entropy, the density of the VME remains constant, the internal energy is only associated with strain and entropy, and therefore, the free energy is only related to strain and temperature. According to the second law of thermodynamics, the Clausius inequality can be established through a Taylor series expansion and omitting the higher-order infinitesimals, which is the initial constitutive relationship. The derivation process is outlined as follows [22, 23]:

$$\begin{gathered} dU - TdS - \boldsymbol{\sigma} : d\boldsymbol{\Gamma} \geq 0 \\ \Downarrow \\ \boldsymbol{\sigma}_{ij} = \frac{\partial U}{\partial \boldsymbol{\gamma}_{ij}} \end{gathered} \tag{4}$$

where dU is the increment of the VME's internal energy U, dS is the increment of the VME's entropy S (which is equal to zero in the absence of biochemical process), T is the absolute temperature, $\boldsymbol{\sigma}$ and $\boldsymbol{\Gamma}$ are the stress and strain tensors, and $\boldsymbol{\gamma}_{ij}$ denotes the shear stress tensor. The first term on the left-hand side of the inequality denotes the internal energy, the second denotes the increase in entropy, and the third denotes the strain energy.

Further, for both adiabatic and isothermal processes, there are internal energies that only manifest as strain energies, namely $W = U$ and $W = U\text{-}TS$. Consequently, Eq. (4) can be expressed as follows:

$$\boldsymbol{\sigma}_{ij} = \frac{\partial W}{\partial \boldsymbol{\gamma}_{ij}} \tag{5}$$

where W is the strain energy.

By expanding Eq. (5) in terms of the Taylor series and approximating higher-order infinitesimals, we can obtain the generalized Hooke's law, which is the most common form of the constitutive equation. The representation in the common form of the tensor index is as follows:

$$\boldsymbol{\sigma}_{ij} = \boldsymbol{E}_{ijks}\boldsymbol{\gamma}_{ks} \tag{6}$$

where $\boldsymbol{E}_{ijks}$ is the elasticity constant.

As can be observed from the aforementioned derivation process, the three equations were obtained through the utilization of a Taylor series expansion, with the exclusion of the higher-order infinitesimal (small deformation assumption). Additionally, it should be noted that there is no initial stress and the material point is continuous. From the perspective of incremental analysis, the three equations encapsulate the fundamental concept of FEA: first, an object is discretized into a continuous number of VMEs (mesh elements). Subsequently, according to the specified boundary conditions, the stress and strain data of the individual VMEs are calculated incrementally. Finally, the boundary continuity of all VMEs is employed to integrate the mechanical response for the macroscopic structure. In conclusion, FEA is a cross-scale calculation process based on VMEs to the macro-structure.

2.2 Anisotropic constitutive model for single-layer FRPs

Section 2.1 presents the derivation of the three fundamental equations governing the behavior of a continuum medium. Additionally, it offers a mathematical expression of the deformation mechanism of elastomers subjected to force and introduces the concept of FE. In this section, the theory related to continuum elastomers is applied to FRPs. Laminated FRPs are anisotropic materials due to the variation in layup angles and exhibit a general anisotropic stress-strain relationship. The laminate can be considered to be the stacked product of single layers, and its mechanical behavior or response can be assembled and characterized by the single layer. Therefore, the constitutive model of single-layer FRPs is discussed first. Since the geometric and equilibrium equations have been incorporated into the FE software, it is now possible to focus on establishing the constitutive equations that will describe the mechanical behavior of the target material. For the sake of convenience in writing in the matrix relationship, the symbols for stress $\boldsymbol{\sigma}_{ij}$ and strain $\boldsymbol{\varepsilon}_{ij}$ will henceforth be abbreviated as follows [24]:

$$\begin{cases} \sigma_{11} = \sigma_1, \quad \sigma_{22} = \sigma_2, \quad \sigma_{33} = \sigma_3 \\ \sigma_{12} = \tau_{12} = \sigma_4, \quad \sigma_{23} = \tau_{23} = \sigma_5, \quad \sigma_{31} = \tau_{31} = \sigma_6 \end{cases} \tag{7}$$

$$\begin{cases} \varepsilon_{11} = \varepsilon_1 \\ \varepsilon_{22} = \varepsilon_2 \\ \varepsilon_{33} = \varepsilon_3 \\ \varepsilon_{12} = \dfrac{1}{2}\gamma_{12} = \varepsilon_4 \\ \varepsilon_{23} = \dfrac{1}{2}\gamma_{23} = \varepsilon_5 \\ \varepsilon_{31} = \dfrac{1}{2}\gamma_{31} = \varepsilon_6 \end{cases} \tag{8}$$

where $i = j$ stands for the principal state, whereas $i \neq j$ represents the shear state.

Let $\boldsymbol{C}_{ij}$ be the stiffness matrix. According to classical laminate mechanics [25], the anisotropic constitutive relationship in matrix form, which contains 36 elastic constants, can be expressed as follows based on the Eq. (6). That is,

$$\begin{Bmatrix} \sigma_1 \\ \sigma_2 \\ \sigma_3 \\ \sigma_4 \\ \sigma_5 \\ \sigma_6 \end{Bmatrix} = \begin{bmatrix} C_{11} & C_{12} & C_{13} & C_{14} & C_{15} & C_{16} \\ C_{21} & C_{22} & C_{23} & C_{24} & C_{25} & C_{26} \\ C_{31} & C_{32} & C_{33} & C_{34} & C_{35} & C_{36} \\ C_{41} & C_{42} & C_{43} & C_{44} & C_{45} & C_{46} \\ C_{51} & C_{52} & C_{53} & C_{54} & C_{55} & C_{56} \\ C_{61} & C_{62} & C_{63} & C_{64} & C_{65} & C_{66} \end{bmatrix} \begin{Bmatrix} \varepsilon_1 \\ \varepsilon_2 \\ \varepsilon_3 \\ \varepsilon_4 \\ \varepsilon_5 \\ \varepsilon_6 \end{Bmatrix} \tag{9}$$

Besides, Eq. (9) can be expressed in the tensor form as follows:

$$\boldsymbol{\sigma}_{ij} = \mathbf{C}_{ij}\boldsymbol{\varepsilon}_{ij}(i,j = 1,\ 2,\ 3,\ 4,\ 5,\ 6) \tag{10}$$

Further, as the material strain energy density is a function of the six strain components, it can be demonstrated that the stiffness matrix is symmetric. Therefore,

$$\mathbf{C}_{ij} = \mathbf{C}_{ji}(i,j = 1,\ 2,\ 3,\ 4,\ 5,\ 6) \tag{11}$$

On this basis, the symmetric stiffness matrix reduces the number of elastic constants from 36 to 21. The stiffness matrix is related to the elastic coefficients, such as elastic modulus E_{ii}, Poisson's ratio v_{ij}, and shear modulus G_{ij}. Nevertheless, it remains a challenge solves the constitutive equation with 21 independent elastic constants. In the case of unidirectional single-layer FRP, it is reasonable to assume that there are three principal axes parallel to the three directions of the mechanical coordinate system. In such a case, the number of independent elastic constants is reduced from 21 to 9, and the constitutive equation can be further simplified [26], that is,

$$\begin{Bmatrix} \sigma_1 \\ \sigma_2 \\ \sigma_3 \\ \sigma_4 \\ \sigma_5 \\ \sigma_6 \end{Bmatrix} = \begin{bmatrix} C_{11} & C_{12} & C_{13} & 0 & 0 & 0 \\ C_{12} & C_{22} & C_{23} & 0 & 0 & 0 \\ C_{13} & C_{23} & C_{33} & 0 & 0 & 0 \\ 0 & 0 & 0 & C_{44} & 0 & 0 \\ 0 & 0 & 0 & 0 & C_{55} & 0 \\ 0 & 0 & 0 & 0 & 0 & C_{66} \end{bmatrix} \cdot \begin{Bmatrix} \varepsilon_1 \\ \varepsilon_2 \\ \varepsilon_3 \\ \varepsilon_4 \\ \varepsilon_5 \\ \varepsilon_6 \end{Bmatrix} \tag{12}$$

Moreover, since the mechanical properties of fibers in the transverse and normal directions are basically the same, unidirectional FRPs can be regarded as belonging to the category of transverse isotropic materials. From another perspective, the matrix is an isotropic material, meaning it is uniformly distributed in an ideal state. Therefore, this will not affect the mechanical properties of unidirectional fibers. In addition, the transverse isotropy remains unaffected if the interface between the fibers and the matrix is ignored. Based on Eq. (12), the independent elastic constants of unidirectional FRPs are reduced from 9 to 7. Consequently, the simplified stress-strain constitutive equation is as follows [27]:

$$\begin{Bmatrix} \sigma_1 \\ \sigma_2 \\ \sigma_3 \\ \sigma_4 \\ \sigma_5 \\ \sigma_6 \end{Bmatrix} = \begin{bmatrix} C_{11} & C_{12} & C_{13} & 0 & 0 & 0 \\ C_{12} & C_{22} & C_{23} & 0 & 0 & 0 \\ C_{13} & C_{23} & C_{33} & 0 & 0 & 0 \\ 0 & 0 & 0 & C_{44} & 0 & 0 \\ 0 & 0 & 0 & 0 & 0.5(C_{11} - C_{12}) & 0 \\ 0 & 0 & 0 & 0 & 0 & C_{44} \end{bmatrix} \cdot \begin{Bmatrix} \varepsilon_1 \\ \varepsilon_2 \\ \varepsilon_3 \\ \varepsilon_4 \\ \varepsilon_5 \\ \varepsilon_6 \end{Bmatrix} \tag{13}$$

Additionally, the stiffness coefficients utilized in the transverse isotropic constitutive model are calculated based on four basic elastic constants, namely,

$$\begin{cases} C_{11} = \dfrac{1\text{-}v_{23}v_{32}}{E_{22}E_{33}\Delta}, \ C_{22} = \dfrac{1\text{-}v_{13}v_{31}}{E_{11}E_{33}\Delta}, \ C_{33} = \dfrac{1\text{-}v_{12}v_{21}}{E_{11}E_{22}\Delta} \\ C_{12} = \dfrac{v_{12}+v_{13}v_{32}}{E_{11}E_{33}\Delta}, \ C_{13} = \dfrac{v_{13}+v_{12}v_{23}}{E_{11}E_{22}\Delta}, \ C_{23} = \dfrac{v_{23}+v_{21}v_{13}}{E_{11}E_{22}\Delta} \\ \Delta = \dfrac{1}{E_{11}E_{22}E_{33}}(1 - v_{12}v_{21} - v_{23}v_{32} - v_{13}v_{31} - 2v_{21}v_{32}v_{13}) \end{cases} \tag{14}$$

where Δ is an operator, and $C_{44} = G_{12}, \ C_{55} = G_{23},$ and $C_{66} = G_{31}$.

This section has established the elastic constitutive model of single-layer FRPs, which includes the initial anisotropic constitutive model and the simplified transverse isotropic constitutive model. Nevertheless, it remains unsuitable for predicting laminates, which are the typical format in engineering practice. To this end, an effective method will be made in the next section.

2.3 Anisotropic constitutive for laminated FRPs

In the case of orthotropic anisotropic laminated FRPs with different single-layer material principal axes, the total constitutive equation can be derived through a coordinate transformation. Firstly, each single layer is transformed into a Cartesian orthogonal coordinate system. Subsequently, laminate theory is employed to analyze the behavior of the stacked multi-layers.

As illustrated in **Figure 2**, the principal axes of the FRP layer are defined as directions 1 (along the fiber direction), 2 (along the fiber transverse direction), and 3 (along the fiber normal direction). The Cartesian orthogonal coordinate directions, which serve as the principal axes of the FRP laminate, are designated as x, y, and z. In the case of a single-layer FRP, whose axes 1 and 2 are not aligned with the x and y axes, the z axis (axis 3) can be used as a rotation axis to convert the original axes 1 and 2 to the Cartesian coordinate system (x, y). This can be achieved by applying the appropriate transformation matrix.

Let us consider the angles between the material axis 1 and the orthogonal axis x, and between the material axis 2 and the orthogonal axis y, which we will denote by θ. Using the left transformation matrix $\boldsymbol{T}$, we can obtain the stress-strain constitutive relationship in the orthogonal coordinate system, that is,

$$\boldsymbol{T} = \left[\cos^2\theta \sin^2\theta - 2\sin\theta\cdot\ \cos\theta \sin^2\theta \cos^2\theta 2\sin\theta\cdot\ \cos\theta \sin\theta\cdot\ \cos\theta -\ \sin\theta\cdot\ \cos\theta \cos^2\theta -\ \sin^2\theta\right] \tag{15}$$

$$\begin{Bmatrix} \sigma_x \\ \sigma_y \\ \sigma_{xy} \end{Bmatrix} = [T]\cdot \begin{Bmatrix} \sigma_1 \\ \sigma_2 \\ \sigma_{12} \end{Bmatrix}, \begin{Bmatrix} \varepsilon_x \\ \varepsilon_y \\ \varepsilon_z \end{Bmatrix} = [T]\cdot \begin{Bmatrix} \varepsilon_1 \\ \varepsilon_2 \\ \varepsilon_{12} \end{Bmatrix} \tag{16}$$

In accordance with classical laminate theory, the stress in the stacking direction (z direction) is markedly inferior to that in the x and y directions. Furthermore, the out-of-plane shear strain prior to deformation is equal to zero, indicating that the in-plane mechanical response is the sole factor that necessitates consideration. Consequently, the mechanics of single-layer FRPs with varying fiber orientations are initially

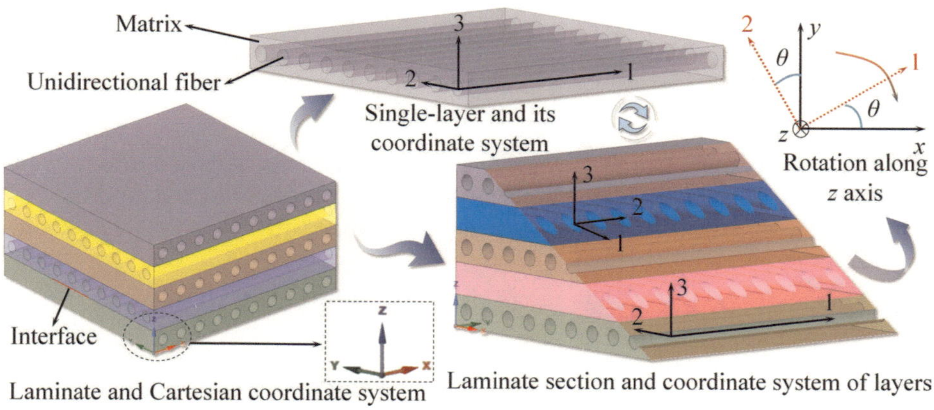

Figure 2.
Diagram of stacking characteristics and coordinate transformation relationships for a laminate and its single layer.

transformed into the laminate coordinate system. Subsequently, the converted mechanics of the single layer are synthesized with the objective of obtaining the elastic mechanics of the laminate. In accordance with Eq. (12), if we let the two-dimensional stiffness matrix be $\mathbf{Q}$, then the in-plane constitutive relationship of a single-layer FRP can be expressed as follows [28]:

$$\begin{Bmatrix} \sigma_1 \\ \sigma_2 \\ \sigma_4 \end{Bmatrix} = \begin{bmatrix} Q_{11} & Q_{12} & 0 \\ Q_{21} & Q_{22} & 0 \\ 0 & 0 & Q_{44} \end{bmatrix} \cdot \begin{Bmatrix} \varepsilon_1 \\ \varepsilon_2 \\ \varepsilon_4 \end{Bmatrix} \tag{17}$$

$$\begin{cases} Q_{11} = \dfrac{E_{11}}{1\text{-}v_{12}v_{21}}, \quad Q_{22} = \dfrac{E_{22}}{1\text{-}v_{12}v_{21}} \\ Q_{44} = G_{12} \\ Q_{12} = Q_{21} = \dfrac{v_{21}E_{11}}{1\text{-}v_{12}v_{21}} \end{cases} \tag{18}$$

Let $\tilde{\mathbf{Q}}$ be the reduced stiffness matrix after the coordinate transformation, then the constitutive relationship of the laminate can be obtained by transforming the principal direction of the single-layer FRP to the Cartesian coordinate system of the laminate. That is,

$$\begin{Bmatrix} \sigma_x \\ \sigma_y \\ \sigma_{xy} \end{Bmatrix} = \tilde{\mathbf{Q}} \cdot \begin{Bmatrix} \varepsilon_x \\ \varepsilon_y \\ \varepsilon_{xy} \end{Bmatrix} = \begin{bmatrix} \tilde{Q}_{11} & \tilde{Q}_{12} & \tilde{Q}_{14} \\ \tilde{Q}_{21} & \tilde{Q}_{22} & \tilde{Q}_{24} \\ \tilde{Q}_{41} & \tilde{Q}_{42} & \tilde{Q}_{44} \end{bmatrix} \cdot \begin{Bmatrix} \varepsilon_x \\ \varepsilon_y \\ \varepsilon_{xy} \end{Bmatrix} \tag{19}$$

The transformation of the two-dimensional stiffness matrix $\mathbf{Q}$ to the reduced stiffness matrix $\tilde{\mathbf{Q}}$ is given by

$$\tilde{\mathbf{Q}} = \mathbf{T}^{-1}\mathbf{QRTR}^{-1} \tag{20}$$

where $\boldsymbol{R}$ is the Reuter matrix, that is,

$$\boldsymbol{R} = \begin{bmatrix} 1 & 0 & 0 \\ 0 & 1 & 0 \\ 0 & 0 & 2 \end{bmatrix} \tag{21}$$

Further, the stiffness matrix $\boldsymbol{C}_{\mathrm{L}}$ of the laminate can be decomposed into the tensile stiffness matrix $\boldsymbol{A}$, the tensile and bending coupling stiffness matrix $\boldsymbol{B}$, and the bending stiffness matrix $\boldsymbol{D}$. Thus, the stress-strain constitutive relationship for the laminate can be given by [29, 30].

$$\boldsymbol{\sigma} = \mathbf{C}_L \cdot \boldsymbol{\varepsilon} = \begin{bmatrix} \boldsymbol{A} & \boldsymbol{B} \\ \boldsymbol{B} & \boldsymbol{D} \end{bmatrix} \cdot \boldsymbol{\varepsilon} \tag{22}$$

$$\boldsymbol{C}_L = \begin{bmatrix} \boldsymbol{A} & \boldsymbol{B} \\ \boldsymbol{B} & \boldsymbol{D} \end{bmatrix} = \begin{bmatrix} A_{11} & A_{12} & A_{14} & B_{11} & B_{12} & B_{14} \\ A_{21} & A_{22} & A_{24} & B_{21} & B_{22} & B_{24} \\ A_{41} & A_{42} & A_{44} & B_{41} & B_{42} & B_{44} \\ B_{11} & B_{12} & B_{14} & D_{11} & D_{12} & D_{14} \\ B_{21} & B_{22} & B_{24} & D_{21} & D_{22} & D_{24} \\ B_{41} & B_{42} & B_{44} & D_{41} & D_{42} & D_{44} \end{bmatrix} \tag{23}$$

As illustrated in **Figure 3(a)**, for a laminate comprising an even number of single layers (2 k layers), if each layer has the same thickness t, the global Cartesian coordinate system is positioned at the center of symmetry of the geometric plane (within the interface between the layers). This configuration allows a stiffness matrix calculation method for a laminate consisting of k consecutive single layers [31], namely,

$$A_{ij} = \sum_{k=1}^{k}\left[\left(\tilde{Q}_{ij}\right)_k t\right],\ B_{ij} = \frac{1}{2}\sum_{k=1}^{k}\left[\left(\tilde{Q}_{ij}\right)_k (2k+1)t^2\right],\ D_{ij} = \frac{1}{3}\sum_{k=1}^{k}\left[\left(\tilde{Q}_{ij}\right)_k (3k^2 - 3k + 1)t^3\right] \tag{24}$$

Similarly, for a laminate comprising an odd number of single layers ($2k$+1 layers), as illustrated in **Figure 3(b)**, if the global Cartesian coordinate system is situated on the neutral plane of the middle layer in the out-of-plane direction (within a layer), the stiffness matrix is

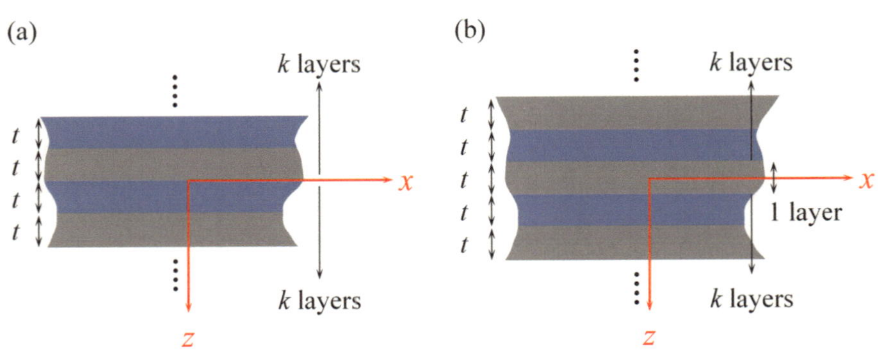

Figure 3.
Diagram of the laminate sectional areas with 2k *layers (panel a) and* 2k + 1 *layers (panel b).*

$$\begin{cases} A_{ij} = \sum_{k=2}^{k} \left[\left(\tilde{Q}_{ij} \right)_k t \right] + \frac{1}{2} \left(\tilde{Q}_{ij} \right)_1 \frac{t}{2} \\ B_{ij} = \frac{1}{2} \sum_{k=2}^{k} \left[\left(\tilde{Q}_{ij} \right)_k (2k+1) t^2 \right] + \frac{1}{4} \left(\tilde{Q}_{ij} \right)_1 \frac{t^2}{4} \\ D_{ij} = \frac{1}{3} \sum_{k=2}^{k} \left[\left(\tilde{Q}_{ij} \right)_k (3k^2 - 3k + 1) t^3 \right] + \frac{1}{6} \left(\tilde{Q}_{ij} \right)_1 \frac{t^3}{8} \end{cases} \tag{25}$$

where the thickness of the initial layer is $t/2$, while the subsequent layer is t; thus, the initial half layer must be computed individually, and the remaining layers should be arranged in accordance with the k layer sequence. In the above equation, the subscripts situated outside the parentheses of $\tilde{Q}$ indicate the position of the layer in question.

3. Mesoscopic mechanical constitutive models for FRPs

In Section 2.1, it is assumed that the single-layer FRPs containing fibers and matrix are blended, homogeneous, and isotropic in the fiber transverse direction. In other words, the elastic constitutive relationship is based on such homogeneous materials, without considering the differences in meso-scale mechanical properties between the fibers and the matrix, which results in the generation of stresses and damages in the entire FRPs. However, when the scale of observation is reduced to the mesoscopic level, it becomes evident that FRPs are non-uniform media, comprising fibers and matrix; additionally, they are non-continuous media, exhibiting the presence of voids; furthermore, they are often mixed with other reinforcing phases, such as flame retardant particles [32, 33].

Indeed, there is an interaction between fibers and matrix, and the mechanical properties of the two are markedly disparate. The former is a brittle material, whereas the latter is an elastic-viscous-plastic material [34]. The function of the fiber is to bear the load and provide strength, whereas the function of the matrix is to safeguard and secure the fiber, and to transfer the load to the fiber in the form of interfacial shear at the fiber ends [7]. Accordingly, the macro-mechanical constitutive equations are insufficient for fully capturing the multi-phase characteristics of composites. Typically, FRPs comprise three distinct phases: the reinforcing phase (fiber), the continuous phase (matrix), and an interfacial phase between the two. The failure of the fiber and matrix often begins at the interface, as microscopic cracks typically originate at this point and subsequently propagate along the axial direction of the fiber or the transverse direction of the fiber section. This highlights the significance of a mesoscopic model for understanding the mechanics of fiber-matrix interactions.

The meso-scale mechanical constitutive is frequently employed in fine modeling, particularly in the investigation of RVE models. Here, the subscripts f, m, h, and ψ are used to denote the parameters associated with fibers, the matrix, voids or holes, and the volume share of fibers, matrices, and voids, respectively. The classical mixed law can be employed to compute the common elastic mechanical parameters of FRPs, including density, elastic modulus, shear modulus, Poisson's ratio, and strength [35].

In theory, the density of voids is equivalent to the air density, but in practice it does not contribute much to the weight of FRPs. Therefore, the mesoscopic density ρ is calculated as follows:

$$\begin{gathered} \rho = \rho_f \psi_f + \rho_m \psi_m + \rho_h \psi_h \\ \Downarrow (\rho_h = 0) \\ \rho = \rho_f \psi_f + \rho_m \psi_m \end{gathered} \tag{26}$$

In the case of RVE i with a volume V, denoting the stress field by σ_i, the average volume stresses $\overline{\sigma}_i$ are given by

$$\begin{cases} \overline{\sigma}_i = \dfrac{1}{V}\displaystyle\int_V \sigma_i dV = \overline{\sigma}_{fi}\psi_f + \overline{\sigma}_{mi}\psi_m + \overline{\sigma}_{hi}\psi_h \\ \overline{\sigma}_{fi} = \dfrac{1}{V_f}\displaystyle\int_{v_f} \sigma_i dV, \text{(Avarage volume stress of fibers)} \\ \overline{\sigma}_{mi} = \dfrac{1}{V_m}\displaystyle\int_{v_m} \sigma_i dV, \text{(Avarage volume stress of matrix)} \\ \overline{\sigma}_{hi} = \dfrac{1}{V_h}\displaystyle\int_{v_h} \sigma_i dV, \text{(Avarage volume stress of holes)} \\ V = V_f + V_m + V_h \end{cases} \tag{27}$$

Similarly, the average volume strain of RVE i and its average strain $\overline{\varepsilon}_i$ and strain energy density $\overline{W}_i$ are

$$\overline{\varepsilon}_i = \overline{\varepsilon}_{fi}\psi_f + \overline{\varepsilon}_{mi}\psi_m + \overline{\varepsilon}_{hi}\psi_h \tag{28}$$

$$\overline{W}_i = \frac{1}{2V}\int_V \overline{\sigma}_i \overline{\varepsilon}_i dV = \frac{1}{2}\overline{\sigma}_{fi}\overline{\varepsilon}_{fi} + \frac{1}{2}\overline{\sigma}_{mi}\overline{\varepsilon}_{mi} + \frac{1}{2}\overline{\sigma}_{hi}\overline{\varepsilon}_{hi} \tag{29}$$

As previously stated, unidirectional single-layer FRPs are typically assumed to exhibit identical mechanical properties in both the transverse (2 direction) and normal (3 direction) fiber directions. Additionally, the resin is assumed to be an isotropic material, exhibiting no differential effect on the transverse isotropic mechanics. Based on these assumptions, a series of transverse isotropic elastic constants can be derived. There are [36],

Elastic modulus E:

$$\begin{cases} E_1 = E_{f1}\psi_f + E_m\psi_m \\ E_3 = E_2 = E_{f2}E_m / \left[E_m\psi_f + E_{f2}\psi_m\left(1 - \psi_m^2\right)\right] \end{cases} \tag{30}$$

Poisson's ratio μ:

$$\begin{cases} \mu_{13} = \mu_{12} = \mu_{f12}\psi_f + \mu_m\psi_m \\ \mu_{23} = \mu_{f23}\psi_f + \mu_m\psi_m \end{cases} \tag{31}$$

Shear modulus G:

$$\begin{cases} G_{13} = G_{12} = G_{f12}G_m / \left[G_m\psi_f + G_{f12}\left(1 - \psi_f\right)\right] \\ G_{23} = G_{f23}G_m / \left[G_m\psi_f + G_{f23}\left(1 - \psi_f\right)\right] \end{cases} \tag{32}$$

Among them, the relationship between volume fraction ψ and volume V is

$$\begin{cases} V = V_f + V_m + V_h \\ \psi_j = V_j/V, \ j = f, \ m, \ h \end{cases} \tag{33}$$

The mixed law of meso-mechanics is applicable not only to FRPs, but also to the majority of solid composites, including particle-reinforced metal matrix composites, doped-phase-reinforced composites, and others. Moreover, it can be applied to hybrid materials and structures, including hybrid cellular structures, hybrid porous materials, and laminated materials. Nevertheless, in practice, the meso-mechanical elastic constants must be compared with the experimental data in order to obtain the matching coefficients, thus enabling a more accurate description of the mechanical properties of the composites.

4. Constitutive models of interlaminar interface for FRPs

As stated in Section 3, FRPs undergo three distinct phases: a matrix phase, a fiber phase, and an interface phase. As is widely understood, laminates are created through the lamination of multiple single layers with varying fiber layup directions, with two adjacent layers connected by a matrix interface. In the context of actual finished material products, there are two interlaminar interfaces theoretically present between a matrix layer and two fiber layers. Accordingly, to obtain more accurate calculation results, it is necessary to consider the two interfaces attached to the matrix layer. In the case of large-scale FRP engineering structures, the focus is typically on the mechanical properties of the entire material. Consequently, the interlaminar interface is often simplified as a zero-thickness cohesive interface. Section 2.2 derives the elastic constitutive equation for single-layer FRPs, which is used to establish the constitutive model for laminates in Section 2.3. However, the interlaminar interface properties are not considered in this model. Delamination at the interlaminar interface of FRPs is a significant factor in failure, particularly under compression, bending, and torsion loading. Therefore, it is essential to consider the interfacial cohesive properties in the FEA.

The delamination damage in laminated composites is primarily evidenced by the formation of openings or closures and misalignment between layers. The former mode of damage, opening or closure, is an out-of-plane phenomenon that arises due to tensile or compressive stresses. In contrast, misalignment represents an in-plane phenomenon resulting from shear forces. The occurrence of delamination damage is typically attributed to crack propagation. In order to describe the damage mode of the interlayer, three distinct categories are typically employed: types I, II, and III. Type I pertains to out-of-plane normal opening/closure, type II to in-plane longitudinal shear, and type III to in-plane transverse shear. The deformation modes of the three types of delamination are illustrated in **Figure 4**. The form of failure of the interlaminar interface of laminates in the loading process may contain all three simultaneously, namely, the stress states of tension/compression and transverse/longitudinal shear. In the case of FRPs, the interlaminar interface is an isotropic material, and thus, the shear strengths of types II and III are approximately equal, which have been demonstrated by Panettieri et al. [37] and Lou et al. [38].

A plethora of efficacious constitutive models, based on experimental evidence, have been proposed to explain the delamination of the interlaminar interface.

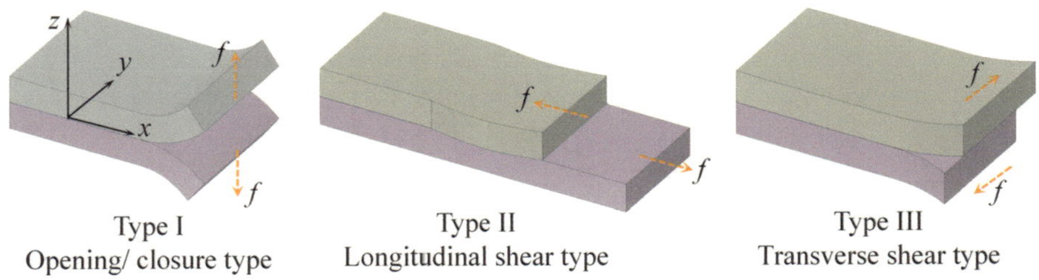

Figure 4.
Diagram of delamination modes on three distinct interlaminar interfaces, where f *represents the applied load.*

These models are typically described as cohesive zone models (CZMs). For instance, the trilinear CZM proposed by Tvergaard and Hutchinson [39] is a trilinear traction-separation law based on relative displacement and is also referred to as the trapezoidal traction-separation law, as illustrated in **Figure 5(a)**. The horizontal coordinate is defined as the ratio of the current propagation distance λ to the failure propagation distance λ_{fail}, which reaches a maximum value of 1 when the interface undergoes complete separation. λ is a dimensionless parameter that describes the separation displacements (δ_1, δ_2, δ_3) and their limits in the three directions. The vertical coordinate represents the traction function, $t(\lambda)$, which is a function of the propagation distance λ. The value of $\sigma_{\max}$ represents the peak traction or maximum traction. The model is divided into three segments: the first segment exhibits a linear increase in traction and reaches a maximum value; the second segment maintains the traction at the maximum value and then reaches a point of degradation; the third segment then decreases from the maximum value to the point of failure.

Moreover, there is a nonlinear CZM, which is based on a normal and tangential traction-separation displacement relationship with an irreversible loading/unloading process [40]. The constitutive curves for type I and type II mixed modes are illustrated in **Figure 5(b)**. The limiting separation displacement δ_F is calculated from the area enclosed by the corresponding normalized traction-separation displacement curves, mixing ratios, peak traction forces, fracture energies, and exponential terms, where the subscript m denotes the mixed state.

Of the above two models, the former is more applicable to the separation behavior of metal interlaminar interfaces, whereas the latter is more applicable to the

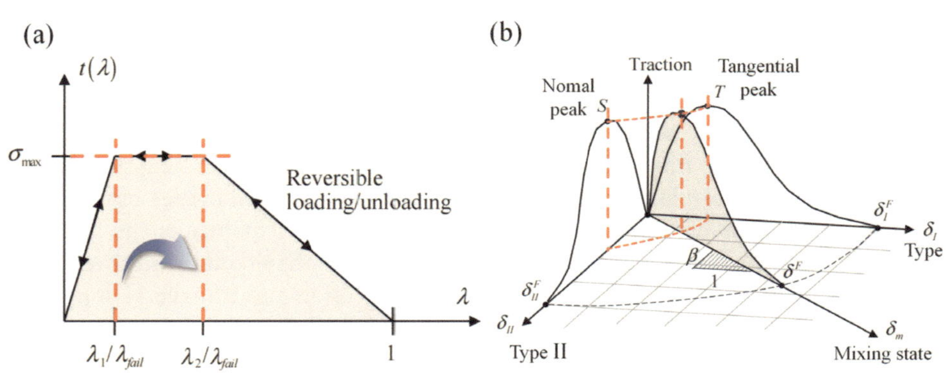

Figure 5.
Traction-separation displacement relationship for interlaminar interface. (a) Trilinear traction-separation law [39], and (b) nonlinear traction-separation law [40].

interlaminar separation behavior of FRPs [41]. However, the bilinear CZM improved by damage evolution [42], which synthesizes a linear increasing segment and a nonlinear damage degradation segment, as shown in **Figure 6(a)**, is the more commonly used interlaminar interface constitutive model for FRP laminates. **Figure 6(b)** illustrates the bilinear CZM curves for the type I and type II mixed states, where the subscripts I and II denote the respective type I and type II parameters respectively, m denotes the mixed parameter, 0 denotes the peak (the initial state in which delamination occurs), and F denotes the state of complete delamination failure. In regard to the mixed state of bilinear CZM with damage evolution, the failure point is typically determined using either Power's law [43] or the Benzeggagh-Kenane (B-K) law [44], the latter being more commonly used for FRPs.

Here, the bilinear CZM containing damage evolution is modeled below. The subscripts t, n, and s denote in-plane transverse shear, out-of-plane normal tension/compression, and in-plane longitudinal shear, respectively. Therefore, the bilinear interlaminar interface constitutive relationship is [45].

$$\begin{bmatrix} \sigma_n \\ \sigma_s \\ \sigma_t \end{bmatrix} = \begin{bmatrix} K_n & 0 & 0 \\ 0 & K_s & 0 \\ 0 & 0 & K_t \end{bmatrix} \begin{bmatrix} \zeta_n \\ \zeta_s \\ \zeta_t \end{bmatrix} \tag{34}$$

where with $\boldsymbol{\sigma}_i$ is the nominal stress, $\boldsymbol{K}_i$ is the stiffness matrix, $\boldsymbol{\zeta}_i$ is the separating displacement, and $i = n, s, t$.

Once the traction force attain its peak value, the continuous damage correction coefficient, designated as D_K, is introduced to perform a nonlinear correct of the $\boldsymbol{K}_i$ matrix. Let S denote the cohesive area and S_c the actual bearing area after failure. Then, the area where damage occurs is given by S-S_c. In this context, the mechanical property corresponding to the bearing area S_c after damage has occurred is referred to as the effective area, so the equilibrium relationship between the nominal stress $\boldsymbol{\sigma}$ and the effective stress $\hat{\boldsymbol{\sigma}}$ can be expressed as follows [46]:

$$\begin{gathered} \boldsymbol{\sigma} S = \hat{\boldsymbol{\sigma}} S_c \\ \Downarrow \\ \hat{\boldsymbol{\sigma}} = (S/S_c)\boldsymbol{\sigma} \end{gathered} \tag{35}$$

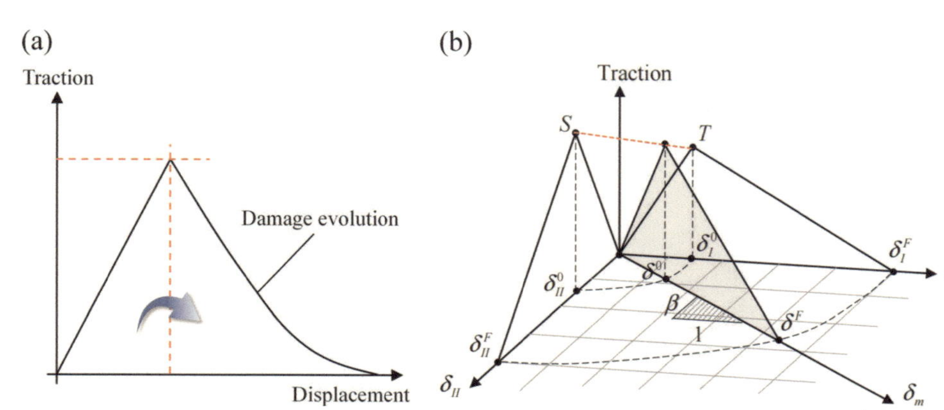

Figure 6.
Bilinear interlaminar cohesive relationship (a) with damage evolution and (b) without damage evolution [40].

By introducing the "Macaulay operator < >" and the "sign function sign ( )" the effective stress in the damage state can be constructed, namely,

$$\begin{cases} \sigma_n = [1 - D_K \cdot sign(\langle \widehat{\sigma}_n \rangle)]\widehat{\sigma}_n \\ \sigma_s = (1 - D_K)\widehat{\sigma}_s \\ \sigma_t = (1 - D_K)\widehat{\sigma}_t \end{cases} \tag{36}$$

$$\langle \widehat{\sigma}_n \rangle = \begin{cases} \widehat{\sigma}_n, & \widehat{\sigma}_n > 0 \\ \widehat{\sigma}_n, & \widehat{\sigma}_n = 0 \\ 0, & \widehat{\sigma}_n < 0 \end{cases} \tag{37}$$

$$sign\ (\langle \widehat{\sigma}_n \rangle) = \begin{cases} 1, & \langle \widehat{\sigma}_n \rangle > 0 \\ 0, & \langle \widehat{\sigma}_n \rangle = 0 \\ -1, & \langle \widehat{\sigma}_n \rangle < 0 \end{cases} \tag{38}$$

Thus, the stiffness matrix $\boldsymbol{K}_D$ of the interface in the damage state is

$$\boldsymbol{K}_D = \begin{bmatrix} [1 - D_K \cdot sign(\langle \widehat{\sigma}_n \rangle)]K_n & 0 & 0 \\ 0 & (1 - D_K)K_s & 0 \\ 0 & 0 & (1 - D_K)K_t \end{bmatrix} \tag{39}$$

where the damage correction factor D_K is derived from the effective separation displacement ζ_m in the mixed state, that is,

$$\begin{cases} D_K = \dfrac{\zeta_m^F (\zeta_m - \zeta_m^0)}{\zeta_m (\zeta_m^F - \zeta_m^0)} \\ \zeta_m = \sqrt{\langle \zeta_n \rangle^2 + {\zeta_s}^2 + {\zeta_t}^2} \end{cases} \tag{40}$$

where the superscript F for effective separation displacement indicates the failure value and 0 indicates the peak value. The Macaulay operator of ζ_n is expressed as follows:

$$\langle \zeta_n \rangle = \begin{cases} \zeta_n, & \zeta_n > 0 \\ \zeta_n, & \zeta_n = 0 \\ 0, & \zeta_n < 0 \end{cases} \tag{41}$$

As illustrated in **Figure 6(b)**, the interrelationships between the parameters in the aforementioned equations, classified as types I, II, and III, are as follows:

$$\begin{cases} \zeta_m \sim \delta_m \\ \zeta_m^0 \sim \delta^0 \\ \zeta_m^F \sim \delta^F \\ \zeta_n \sim \delta_I \\ \zeta_s \sim \delta_{II} \\ \zeta_t \sim \delta_{III} \end{cases} \tag{42}$$

In some condition, the final failure displacement ζ_m^F based on the traction stresses in the mixed mode can be calculated by the B-K law, that is,

$$\zeta_m^F = \begin{cases} \sigma_n \geq 0 & \dfrac{2}{\sqrt{\langle \sigma_n^0 \rangle^2 + \sigma_s^0 + \sigma_t^0}} \left[G_{IC} + (G_{IIC} - G_{IC}) \left(\dfrac{G_{IIC} + G_{IIIC}}{G_{IC} + G_{IIC} + G_{IIIC}} \right)^\eta \right] \\ \sigma_n < 0 & \sqrt{\left\langle \frac{2G_{IIC}}{T} \right\rangle^2 + \left(\frac{2G_{IIC}}{S} \right)^2} \end{cases} \tag{43}$$

where G_{IC}, G_{IIC}, and G_{IIIC} are the fracture energy, also called energy release rate and fracture toughness (unit: N/m) in the delamination state of types I, II, and III in that order, and T and S correspond to the nominal peak stresses in the normal and tangential directions, respectively. The assumption of fracture energy being the same in in-plane shear (i.e., $G_{IIIC} = G_{IIC}$ and $T = S$) leads to the conclusion that the delamination damage of types II and III has the same results in the initiation of separation displacement, the peak traction force, and the failure of separation displacement.

The delamination characteristic of the interlaminar interface exerts a profound influence on the overall mechanical behavior of the laminate, even with regard to the ply angle of the two adjacent layers. In this section, the actual bearing area was employed to construct the damage evolution matrix, which was then utilized to modify the bilinear CZM. This methodology provides a foundation for the development of CZM in FEA.

5. Common used failure criteria for FRPs

Once the constitutive model of single-layer FRPs has been established, the coordinate transformation relationship of the laminates and the interface constitutive model between the layers can be determined. The subsequent step is to ascertain the failure point of the material, thus obtaining the linear segment of the laminate (the first-half segment behavior). In the case of common elastic materials, the failure point can be determined by the maximum principal stress or maximum principal strain, once the constitutive relationship has been established. Alternatively, the failure state can be determined by the Mises and Tresca principles. Nevertheless, some studies have demonstrated that the maximum principal stress/strain failure criterion is inadequate for predicting the failure behavior of anisotropic FRPs, particularly for laminates with varying ply angles [47]. With regard to the laminate, it is important to note that in addition to interlaminar failure, the intralaminar interface also exerts a significant influence on the mechanical behavior. This primarily manifests as the generation of cracks at the fiber/matrix interface, which propagates along the fiber direction and other directions, leading to failure modes such as fiber/matrix cracking, fiber pull-out, and fiber debonding. Nevertheless, these failure modes can be incorporated into the failure criteria, underscoring the significance and complexity of establishing failure criteria.

The Mises equivalent stress, which integrates the principal stresses in the three directions, does not take into account the consideration of the material tensile/compressive strength limits and the shear strength. Consequently, Tsai [48] established a Tsai-Hill failure criterion based on the Mises failure criterion to consider the effects of tensile and shear strength. The development of FRPs has not formed a unified international failure judgment standard [49]. Instead, a multitude of proprietary failure criteria have emerged, the most commonly used being the Chang-Chang failure criterion [50–52], the Tsai-Wu failure criterion [53], the Hoffman failure criterion [54], the Puck failure criterion [55–58], the LaRC failure criterion [59], the Hashin failure criterion [60, 61], and others. In this section, a selection of these criteria is presented for illustration.

The effective yield limit of the material, designated as σ_e, is subject to a Von Mises-R Hill failure criterion [62], which can be expressed as follows:

$$(\sigma_1 - \sigma_2)^2 + (\sigma_2 - \sigma_3)^2 + (\sigma_3 - \sigma_1)^2 = 2\sigma_e^2 \tag{44}$$

Later, Stephen W. Tsai proposed the Tsai-Hill failure criterion based on Eq. (46). Additionally, he introduced the strength parameters F, G, H, L, M, and N, with the intention of further approximating for the failure of anisotropic unidirectional continuous FRP laminates. There are expressed as follows [53]:

$$H(\sigma_1 - \sigma_2)^2 + F(\sigma_2 - \sigma_3)^2 + G(\sigma_3 - \sigma_1)^2 + 2N\tau_{12}^2 + 2M\tau_{13}^2 + 2L\tau_{23}^2 = 1 \tag{45}$$

where

$$\begin{cases} G + H = \dfrac{1}{X_T^2}, F + H = \dfrac{1}{Y_T^2}, F + G = \dfrac{1}{Z_T^2} \\ L = \dfrac{1}{2S_{23}^2}, M = \dfrac{1}{2S_{13}^2}, N = \dfrac{1}{2S_{12}^2} \end{cases} \tag{46}$$

where X, Y, and Z represent the strength in Cartesian coordinates x, y, and z or principal axis 1, 2, and 3 directions, respectively. The subscript T denotes tensile strength, while S represents shear strength.

It is evident that the Tsai-Hill failure criterion solely considers tensile and shear strengths, yet this theory remains inadequate due to its significant inaccuracy in predicting material failure in compression. Accordingly, Hoffman introduces a correction term, which serves to further modify the Tsai-Hill failure criterion. The mathematical model of the Hoffman failure criterion, expressed as a function of the strength parameter C_i (i = 1 to 9), where the subscript C represents compression and T represents tension, is given by [54].

$$C_1(\sigma_1 - \sigma_2)^2 + C_2(\sigma_2 - \sigma_3)^2 + C_3(\sigma_3 - \sigma_1)^2 + C_4\sigma_1 + C_5\sigma_2 + C_6\sigma_3 + C_7\tau_{12}^2 + C_8\tau_{13}^2 + C_9\tau_{23}^2 = 1 \tag{47}$$

$$\begin{cases} C_1 = \dfrac{1}{2}\left(\dfrac{1}{X_T X_C} + \dfrac{1}{Y_T Y_C} - \dfrac{1}{Z_T Z_C}\right) \\ C_2 = \dfrac{1}{2}\left(\dfrac{1}{Y_T Y_C} + \dfrac{1}{Z_T Z_C} - \dfrac{1}{X_T X_C}\right) \\ C_3 = \dfrac{1}{2}\left(\dfrac{1}{Z_T Z_C} + \dfrac{1}{X_T X_C} - \dfrac{1}{Y_T Y_C}\right) \end{cases}, \quad \begin{cases} C_4 = \dfrac{1}{X_T} - \dfrac{1}{X_C} \\ C_5 = \dfrac{1}{Y_T} - \dfrac{1}{Y_C} \\ C_6 = \dfrac{1}{Z_T} - \dfrac{1}{Z_C} \end{cases}, \quad \begin{cases} C_7 = \dfrac{1}{S_{12}^2} \\ C_8 = \dfrac{1}{S_{13}^2} \\ C_9 = \dfrac{1}{S_{23}^2} \end{cases} \tag{48}$$

The Hoffman failure criterion addressed the issue of compression failure by referring to the Tsai-Hill tensile failure criterion. Despite the subsequent emergence of the Tsai-Wu tensor failure criterion, these theories did not elucidate the damage mechanism at the meso-mechanical level. Ultimately, the failure of FRPs is not an overall macroscopic phenomenon, but rather the result of the synthesis of tensile, compression, and shear failure of the fibers and matrix. In order to address this issue, Hashin put forth five distinct failure criteria that consider the mechanical properties of the matrix and the fibers, the three-dimensional form of which is as follows [60, 61, 63]:

$$\begin{cases} \text{fiber} \quad \text{failure} \begin{cases} \text{tension}: \ f_1^2 = \left(\frac{\sigma_{11}}{X_T}\right)^2 + \alpha\left[\left(\frac{\tau_{12}}{S_{12}}\right)^2 + \left(\frac{\tau_{31}}{S_{31}}\right)^2\right] \\ \text{compression}: \ f_2^2 = \left(\frac{\sigma_{11}}{X_C}\right)^2 \end{cases} \\ \text{matrix} \quad \text{failure} \begin{cases} \text{tension}: \ f_3^2 = \left(\frac{\sigma_{22}+\sigma_{33}}{Y_T}\right)^2 + \frac{{\tau_{12}}^2 + {\tau_{31}}^2}{{S_{12}}^2} + \frac{{\tau_{23}}^2 - \sigma_{22}\sigma_{33}}{{S_{23}}^2} \\ \text{compression}: \ f_4^2 = \left(\frac{\sigma_{22}+\sigma_{33}}{Y_C}\right)^2 \\ \text{shear}: f_5^2 = \left(\frac{\sigma_{31}+\sigma_{23}}{S_{23}}\right)^2 \end{cases} \end{cases} \tag{49}$$

where f_i is the damage variable, which takes the value between 0 and 1, with 0 representing no damage and 1 representing complete damage; σ_{ii} is the principal stress, τ_{ij} is the shear stress, and S_{12}, S_{23}, and S_{31} are the shear stresses of the single layer; X_T, X_c and Y_T, and Y_c are the tensile (subscripted as T) and compressive (subscripted as C) strengths in the x direction and y direction of the single layer.

Later, based on the Hashin failure criterion, Puck gave a failure criterion considering the uncertainty of the matrix cracking direction (i.e., interfacial phase cracking) [64], and established a theoretical model for calculating the stress and strain components at the fracture surface of single-layer FRPs. This further refines the evaluation criteria for matrix failure between fibers due to tensile and compressive stresses. However, a significant number of commercial software applications lack the provision of a pre-existing failure criterion, necessitating the development of a material subroutine in most cases. Recently, some modified failure criteria have been proposed [65], but the optimal, comprehensive, and straightforward failure criterion remains to be developed.

6. Damage evolution behavior for FRPs

In contrast to metal and polymeric materials, FRPs are typically brittle materials that do not exhibit a discernible plastic yielding phase or flow stress phase. Once the strength limit is reached, failure occurs, resulting in a loss of load-bearing capacity. Consequently, following the determination of the failure point of FRPs, it is essential to consider the mechanical behavior subsequent to failure (the second-half segment behavior), which is frequently designated the stiffness degradation or damage evolution stage. The stiffness degradation observed after failure varies depending on the matrix and interface in question. Some exhibit linear behavior, while others display nonlinear behavior, as illustrated in **Figure 7**.

As indicated in Section 4, the effective value is defined as the load-bearing area or volume that has been altered following a failure event. Consequently, the initial load-bearing area or volume calculations are insufficient for accurately representing the FRPs. In FEA, mechanical calculations for FRPs are based on the initial bearing area, that is, the area of a complete mesh. There is, however, no algorithm that reflects the change of load-bearing area through failure. As the load increases, damage

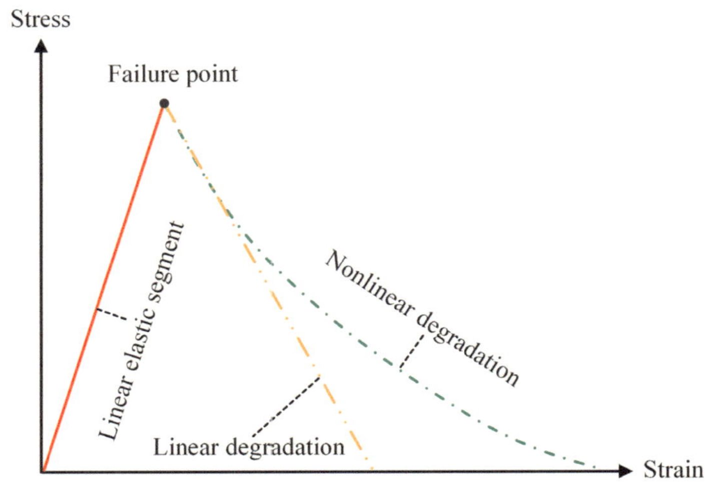

Figure 7.
Uniaxial theoretical stress-strain curve of FRPs containing damage evolution.

commences at zero and progresses to complete failure. The actual load-bearing area of the mesh declines in tandem, necessitating the establishment of a stiffness degradation matrix to account for the damage evolution behavior.

As stated in Section 5, damage to FRPs can be classified into three categories: fiber tension/compression, matrix tension/compression, and fiber/matrix shear. In this context, the term d_1 is used to denote fiber damage, while d_2 represents matrix damage. Both of these variables are expressed as values between 0 and 1, with 0 indicating no damage and 1 signifying complete damage. Similarly, the damage evolution matrix $\boldsymbol{M}_d$ can be established without considering the degradation of the out-of-plane normal stiffness and only considering the degradation of the in-plane stiffness, by establishing a relationship between the effective load-bearing area of the failed element and the load-bearing area of the undamaged element, according to the aforementioned principles [66]. That is,

$$\boldsymbol{M}_d = \begin{bmatrix} D_1 & 0 & 0 & 0 & 0 & 0 \\ 0 & D_2 & 0 & 0 & 0 & 0 \\ 0 & 0 & 1 & 0 & 0 & 0 \\ 0 & 0 & 0 & D_1D_2 & 0 & 0 \\ 0 & 0 & 0 & 0 & D_2 & 0 \\ 0 & 0 & 0 & 0 & 0 & D_1D_2 \end{bmatrix} \tag{50}$$

where $D_1 = 1 - d_1$, and $D_2 = 1 - d_2$.

In accordance with the postulates of strain equivalence and elastic strain energy equivalence in the effective and nominal states, the following conclusions can be drawn:

$$\begin{cases} \frac{1}{2}\boldsymbol{\sigma}^T\boldsymbol{C}^{-1}\boldsymbol{\sigma} = \frac{1}{2}\hat{\boldsymbol{\sigma}}^T\boldsymbol{C}^{M}{}_{-1}\hat{\boldsymbol{\sigma}} \\ \boldsymbol{\sigma} = M^{-1}\hat{\boldsymbol{\sigma}} \end{cases} \tag{51}$$

Subsequently, a transformation relation can be established. That is,

$$\begin{gathered}\frac{1}{2}\boldsymbol{\sigma}^T\boldsymbol{C}^{-1}\boldsymbol{\sigma}=\frac{1}{2}\widehat{\boldsymbol{\sigma}}^T\boldsymbol{C}^M{}_{-1}\widehat{\boldsymbol{\sigma}}\\ \Downarrow\widehat{\boldsymbol{\sigma}}=M\boldsymbol{\sigma}\\ \frac{1}{2}\boldsymbol{\sigma}^T\boldsymbol{C}^{-1}\boldsymbol{\sigma}=\frac{1}{2}\boldsymbol{\sigma}^T\left(M^T\boldsymbol{C}^M{}_{-1}M\right)\boldsymbol{\sigma}\\ \Downarrow\\ \boldsymbol{C}^{-1}=M^T\boldsymbol{C}^M{}_{-1}M=\left(M^{-T}\boldsymbol{C}_MM^{-1}\right)^{-1}\\ \Downarrow\\ \boldsymbol{C}=M^{-T}\boldsymbol{C}_MM^{-1}\\ \Downarrow\\ \boldsymbol{C}_M=M^T\boldsymbol{C}M\end{gathered} \tag{52}$$

Thereby, the stiffness matrix $\boldsymbol{C}_M$ with damage evolution is

$$\boldsymbol{C}_M=\begin{bmatrix} D_1{}^2C_{11} & D_1D_2C_{12} & D_1C_{13} & 0 & 0 & 0\\ D_1D_2C_{21} & D_2{}^2C_{22} & D_2C_{23} & 0 & 0 & 0\\ D_1C_{31} & D_2C_{32} & C_{33} & 0 & 0 & 0\\ 0 & 0 & 0 & D_1D_2{}^2G_{12} & 0 & 0\\ 0 & 0 & 0 & 0 & D_2{}^2G_{23} & 0\\ 0 & 0 & 0 & 0 & 0 & D_1D_2{}^2G_{31}\end{bmatrix} \tag{53}$$

The progressive failure behavior of FRPs containing damage evolution after failure can be obtained by replacing matrix **C** in Eq. (13) with the expression given in Eq. (55) above. The progressive damage observed in FRPs under compression is a distinctive feature that sets them apart from other materials. However, it is still necessary to control this phenomenon through secondary development for the purposes of FEA. In accordance with the aforementioned constitutive model, the mechanical behavior of FRPs can be described as follows: linear elasticity increases under anisotropic conditions. Following the attainment of the failure point, the effective load-bearing area diminishes and commences the damage evolution stage, ultimately resulting in the complete loss of load-bearing capacity, as illustrated in **Figure 7**.

The calculation of d_1 and d_2 in the damage evolution matrix is of paramount importance in determining the evolution path. In essence, the damage evolution matrix can be populated with linear factors, thereby facilitating linear degradation [67]. Nevertheless, the time-variant damage factors can result in nonlinear damage evolution behavior, which is a more accurate representation [68]. The nonlinear damage factors can be defined as follows:

$$\begin{cases} d_1=\dfrac{\varepsilon_1^f}{\varepsilon_1^f-\varepsilon_1^d}\left(1-\dfrac{\varepsilon_1^d}{\varepsilon_1}\right)\\ d_2=\dfrac{\varepsilon_2^f}{\varepsilon_2^f-\varepsilon_2^d}\left(1-\dfrac{\varepsilon_2^d}{\varepsilon_2}\right)\end{cases} \tag{54}$$

where the superscript f denotes the failure state, the superscript d denotes the state after the occurrence of damage, the subscript 1 denotes the 1 direction, and 2 denotes the 2 direction. In this context, where only damage accumulation in the 1 and 2 directions is considered, damage in the 3 direction is absent. Furthermore, the damage state in the 3 direction is regarded as identical to that in the 2 direction, contingent on the condition of transverse isotropy. The failure strain ε^f and the strain at which damage occurs ε^d are known. The current strain ε_i (i = 1, 2) is used as the denominator. Damage occurs when the current strain increases to the same level as the starting damage strain. The damage variable d_i (i = 1, 2) is set to zero at the onset of the damage evolution, whereupon it increases gradually.

It is observed that distinct FRPs exhibit disparate damage evolution behaviors, suggesting the necessity for calibration of the damage evolution matrix and damage variables through experimental means. Subsequently, the constitutive model incorporating damage evolution can be employed for this category of FRPs. Moreover, the choice of calculation method, such as the forward Euler method, the backward Euler method, the center difference method, and so on, can influence the expected outcome to a certain extent when the damage evolution is achieved through the use of a material subroutine [46, 68].

Furthermore, the FE incremental calculation method is introduced as a means of facilitating the developing of material subroutines. In each time increment, the mathematical models to be solved are linearly related, thus treating the nonlinear problem as a linear one [69, 70]. Based on the incremental FE method, the linearized stress increment and strain increment at the integration point within each mesh can be expressed as follows:

$$\Delta\boldsymbol{\sigma} = \mathbf{C}_\tau \Delta\boldsymbol{\varepsilon} \tag{55}$$

where C_τ is the stiffness matrix at time τ in a given time step/increment $[t^n, t^{n+1}]$. If the time increment is set to Δt^{n+1}, then the stress increment within this time increment can be obtained after integration, that is,

$$\begin{cases} t^{n+1} - t^n = \Delta t^{n+1} \\ \Delta\boldsymbol{\sigma} = \displaystyle\int_{t^n}^{t^{n+1}} \mathbf{C}_\tau d\boldsymbol{\varepsilon} = \int_{t^n}^{t^n+\Delta t^{n+1}} \mathbf{C}_\tau d\boldsymbol{\varepsilon} \end{cases} \tag{56}$$

Based on this, the total stress and strain in the time increment Δt^{n+1} are given by

$$\begin{cases} t^{n+1} = t^n + \Delta t^{n+1} \\ \boldsymbol{\varepsilon}^{n+1} = \boldsymbol{\varepsilon}^n + \Delta\boldsymbol{\varepsilon}^{n+1} \\ \boldsymbol{\sigma}^{n+1} = \boldsymbol{\sigma}^n + \Delta\boldsymbol{\sigma}^{n+1} \\ \boldsymbol{C}^{n+1} = \boldsymbol{C}^n + \boldsymbol{C}^{n+1} \\ \Delta\boldsymbol{\sigma}^{n+1} = \boldsymbol{C}_{n+1} : (\Delta\boldsymbol{\varepsilon}^{n+1} - \Delta\boldsymbol{\varepsilon}^n) \end{cases} \tag{57}$$

It is noteworthy that mesh integration points, which can also be referred to as material integration points, are points within meshes that are utilized for integration operations. The stress and strain state of the mesh is defined by the integration points, while the stress and strain distribution on the mesh is represented by interpolation of neighboring integrals. The aforementioned constitutive models can be developed further to predict the mechanical behavior of FRPs, thus enabling the prediction of the various engineering structures made of FRPs.

7. Discussions

FRPs are metamaterials in a broad sense, since they are man-made micro−/meso-structures with anomalous physical properties not found in nature [1, 71, 72]. From this point of view, the macroscopic phenomenological constitutive model in Section 2 is simple and needs further improvement, which has been demonstrated in many studies [9, 21, 65]. Considering fibers, matrices, and interfaces [14, 15, 36], a more accurate cross-scale model can be established to understand the mechanical behavior of FRPs at the micro-meso-macro level [7, 8, 32], and Sections 3 and 4 lay the foundation for this. Failure behavior directly determines the application scenarios of FRPs, and how to predict single loading [18], cyclic loading [11, 13] and impact loading [16, 19] with complex layup angles becomes a problem to be considered. Among them, one of the important fields is the damage tolerance analysis, which involves not only the failure criterion, but also the damage accumulation and evolution behavior before and after failure [35], which is related to whether the FRP structures can be further used or not. Sections 5 and 6 mention some of the failure criteria and damage evolution models, but there are still many issues to be taken into consideration for the real situation. It is well known that FRPs and their composite structures have a wide range of applications, such as load-bearing [31], energy absorption [4, 5], and impact resistance [17, 38], which requires the development of more intrinsic models, even in combination with deep learning [73]. This requires multi-dimensional studies. As a result, this chapter lays a solid foundation for this purpose (phenomenological constitutive models, cross-scale models, failure criteria, damage evolution, etc.) and even provides some approaches such as damage evolution considering changes in the effective load bearing area, and interfacial models considering damage evolution.

8. Conclusions

Fiber-reinforced plastics (FRPs) represent a category of advanced composite materials that facilitate the realization of a multitude of engineering designs, meeting the dual requirements of lightweightness and high strength, thereby enhancing the quality of human life. The significance and value of FRPs are widely acknowledged. In light of the above, this chapter has presented the constitutive models of FRPs from seven perspectives, with the aim of facilitating a more profound understanding of FRPs. Specifically, the following aspects are addressed: Firstly, the concept and origin of the constitutive equation were elucidated with reference to the basic geometrical equation, the equilibrium equation, and the thermodynamic equation. Secondly, the anisotropic constitutive model with 36 elastic constants was derived and simplified to yield the transverse isotropic constitutive equation, which contains only 7 elastic constants. Thirdly, a coordinate transformation matrix was employed to derive the constitutive equation for the laminate, based on the principles of classical laminate mechanics. Fourthly, with the fiber and matrix as the focus of investigation, the mesoscopic mechanical constitutive equations of FRPs were established. Finally, a cohesive zone model based on a bilinear traction-separation relationship for the interlaminar interface of laminates was presented, with consideration given to damage evolution behaviors. Sixthly, the most commonly used failure criteria, including the Mises, Tsai-Hill, and Hoffman criteria, were introduced, and particular attention was given to the three-dimensional Hashin failure criterion. Finally, the damage evolution behavior of FRPs was taken into account to achieve degradation after failure, and the stress and strain incremental relationships were provided for subroutine development. These models can be utilized for the development of a new constitutive model subroutine in the future, thereby offering substantial support for the prospective engineering application of FRPs.

Acknowledgements

The author gratefully acknowledged the financial support from the National Natural Science Foundation of China (12402458), the China Postdoctoral Science Foundation (2024M751640), and the Postdoctoral Fellowship Program of CPSF (GZC20231315), the China Scholarship Council (202106840033), the Postgraduate Research & Practice Innovation Program of Jiangsu Province (KYCX21_0342), and the Excellent Doctor Training Fund of Nanjing University of Science and Technology (2021–2023). Furthermore, the author would like to express his gratitude to Miss *Mengya Ai* for her support during the writing process, as well as Prof. *Guigao Le,* Prof. *Jie Ren,* Prof. *Jianlin Zhong,* Prof. *Heow Pueh Lee,* Prof. *Kheng Lim Goh,* Prof. *Chang Qing Chen*, and others for their valuable guidance in understanding the related knowledge.

Conflict of interest

The author declares no conflict of interest.

Nomenclature

B-K	Benzeggagh-Kenane
CFRPs	carbon fiber reinforced plastics
CZM	cohesive zone model
FE	finite element
FEA	finite element analysis
FRPs	fiber reinforced plastics
RVE	representative volume element
VME	volume micro-element
d_1	the damage to the fiber
d_2	the damage to the matrix
dS	the increment of the VME's entropy
$\boldsymbol{f}(f_x, f_y, f_z)$	an external force vector
$f_i(i = 1,2,3,4,5)$	the damage variable
k	the number of single layer
$\boldsymbol{p}(x_1, y_1, z_1)$	a material point
$\boldsymbol{p}_1(x_2, y_2, z_2)$	deformed material point
$\boldsymbol{r}(u, v, w)$	a three-dimensional space vector
sign( )	sign function
t	thickness of single layer
$t(\lambda)$	the traction function
γ_{ij}	the shear strain tensor
δ_F	the limiting separation displacement

DOI: http://dx.doi.org/10.5772/intechopen.1007980

ε_d the strain at which damage occurs
$\overline{\varepsilon}_i$ the average strain
θ angle between material axis and orthogonal axis
λ the current propagation distance
λ_{fail} the failure propagation distance
μ Poisson's ratio
v_{ij} Poisson's ratio
ζ_i the separating displacement
ζ_m the effective separation displacement
ζ_m^F the final failure displacement
ρ mesoscopic density
σ_e the effective yield limit of the material
$\widehat{\boldsymbol{\sigma}}$ the effective stress
$\overline{\sigma}_i$ the average volume stresses
ψ volume fraction
$\boldsymbol{A}$ the tensile stiffness matrix
$\boldsymbol{B}$ the tensile and bending coupling stiffness matrix
C_i the strength parameter
$\boldsymbol{C}_{ij}$ stiffness matrix.
$\boldsymbol{C}_{\mathrm{L}}$ the stiffness matrix of the laminate
$\boldsymbol{C}_M$ the stiffness matrix with damage evolution
C_τ the stiffness matrix at moment τ
$\boldsymbol{D}$ the bending stiffness matrix
D_K the continuous damage correction coefficient
E_{ii} elastic modulus
$\boldsymbol{E}_{ijks}$ the elasticity constant tensor
F, G, H, L, M, N the strength parameters
$G_{IC}, G_{IIC}, G_{IIIC}$ the fracture energy of type I, II, and III delamination
G_{ij} shear modulus
$\boldsymbol{K}_D$ the stiffness matrix of the interface in the damage state
$\boldsymbol{K}_i$ the stiffness matrix
$\boldsymbol{M}_d$ the damage evolution matrix
$\boldsymbol{Q}$ two-dimensional stiffness matrix
$\tilde{\boldsymbol{Q}}$ the reduced stiffness matrix
$\boldsymbol{T}$ left transformation matrix
V volume of RVE
W the strain energy
$\overline{W}_i$ the average strain energy density
$\boldsymbol{R}$ the Reuter matrix
S the cohesive area
S_c the actual bearing area after failure
S_{12}, S_{23}, S_{31} the shear stresses of the single layer
T, S the nominal peak stresses in the normal and tangential directions
X_c, Y_c the compressive strength in the x direction and y direction
X_t, Y_t the tensile strengths in the x direction and y direction
$< >$ macaulay operator
Δ an operator

Author details

Changfang Zhao
Department of Engineering Mechanics, CNMM and AML, Tsinghua University, Beijing, China

*Address all correspondence to: lackychang@tsinghua.edu.cn; lackychang@qq.com

References

[1] Talreja R, Waas AM. Concepts and definitions related to mechanical behavior of fiber reinforced composite materials. Composites Science and Technology. 2022;**217**:109081. DOI: 10.1016/j.compscitech.2021.109081

[2] Karim MA, Abdullah MZ, Deifalla AF, Azab M, Waqar A. An assessment of the processing parameters and application of fibre-reinforced polymers (FRPs) in the petroleum and natural gas industries: A review. Results in Engineering. 2023;**18**:101091. DOI: 10.1016/j.rineng.2023.101091

[3] Maiti S, Islam MR, Uddin MA, Afroj S, Eichhorn SJ, Karim N. Sustainable fiber-reinforced composites: A review. Advanced Sustainable Systems. 2022;**6**(11):2200258. DOI: 10.1002/adsu.202200258

[4] Zhao CF, Lee HP, Goh KL, Zhong JL, Zhang KB, Zhang ZD, et al. Preparation process and compression mechanics of carbon fiber reinforced plastics negative Poisson's ratio structure (CFRP+NPRS). Composite Structures. 2022;**292**:115667. DOI: 10.1016/j.compstruct.2022.115667

[5] Zhao CF, Goh KL, Lee HP, Yin C, Zhang KB, Zhong JL. Experimental study and finite element analysis on energy absorption of carbon fiber reinforced composite auxetic structures filled with aluminum foam. Composite Structures. 2023;**303**:116319. DOI: 10.1016/j.compstruct.2022.116319

[6] Wang B, Gao H. Fibre reinforced polymer composites. In: Shyha I, Huo D, editors. Advances in Machining of Composite Materials. Cham: Springer; 2021. pp. 15-43. DOI: 10.1007/978-3-030-71438-3_2

[7] Lin K, Wang ZL. Multiscale mechanics and molecular dynamics simulations of the durability of fiber-reinforced polymer composites. Communications Materials. 2023;**4**:66. DOI: 10.1038/s43246-023-00391-2

[8] Raju K, Tay TE, Tan VBC. A review of the FE2 method for composites. Multiscale and Multidisciplinary Modeling, Experiments and Design. 2021;**4**:1-24. DOI: 10.1007/s41939-020-00087-x

[9] Zhao CF, Zhou ZT, Zhao CX, Zhu HW, Zhang KB, Zhong JL, et al. Research on compression properties of unidirectional carbon fiber reinforced epoxy resin composite (UCFREP). Journal of Composite Materials. 2021; **55**(11):1447-1458. DOI: 10.1177/0021998320972176

[10] Fragoudakis R. Failure concepts in fiber reinforced plastics. In: Ali A, editor. Failure Analysis and Prevention. London, UK: IntechOpen; 2017. pp. 81-98. DOI: 10.5772/intechopen.71822

[11] Zhao CF, Zhou ZT, Zhang KB, Zhu HW, Zhong JL, Sun CB, et al. Experimental study on tensile mechanics of arrow combination structure with carbon fiber–epoxy resin composite. Arabian Journal for Science and Engineering. 2021;**46**:2891-2900. DOI: 10.1007/s13369-020-05202-1

[12] Gerbaud PW, Otero F, Bussetta P, Camanho PP. An invariant based transversely-isotropic constitutive model for unidirectional fibre reinforced composites considering the matrix viscous effects. Mechanics of Materials. 2019;**138**:103146. DOI: 10.1016/j.mechmat.2019.103146

[13] Hosoi A, Kawada H, Yoshino H. Fatigue characteristics of quasi-isotropic

CFRP laminates subjected to variable amplitude cyclic two-stage loading. International Journal of Fatigue. 2006; **28**(10):1284-1289. DOI: 10.1016/j.ijfatigue.2006.02.039

[14] Huang SL, Fu QN, Yan LB, Kasal B. Characterization of interfacial properties between fibre and polymer matrix in composite materials-a critical review. Journal of Materials Research and Technology. 2021;**13**:1441-1484. DOI: 10.1016/J.JMRT.2021.05.076

[15] Singh DK, Vaidya A, Thomas V, Theodore M, Kore S, Vaidya U. Finite element modeling of the fiber-matrix interface in polymer composites. Journal of Composites Science. 2020;**4**(2):58. DOI: 10.3390/jcs4020058

[16] Zhao CF, Zhong JL, Wang HX, Chen CQ. Complete constitutive model of CFRP including continuous damage in low strain rate compression and temperature generation in high strain rate impact. Polymer Composites. 2023; **45**(5):3965-3989. DOI: 10.1002/pc.28037

[17] Zhao CF, Zhong JL, Wang HX, Liu C, Li M, Liu H. Impact behaviour and protection performance of a CFRP NPR skeleton filled with aluminum foam. Materials & Design. 2024;**246**: 113295. DOI: 10.1016/j.matdes.2024.113295

[18] Zhao CF, Ren R, Sun CB, Ren J, Zhong JL, Zhang ZD. Compression mechanics for carbon-fiber reinforced epoxy resin composites under In-plane and out-of-plane quasi-static and dynamic loadings. Mechanics of Composite Materials. 2023;**59**:507-520. DOI: 10.1007/s11029-023-10112-y

[19] Zhao CF, Zhou ZT, Ren J, Xing CL, Zhang KB, He B, et al. Investigation of compression mechanics of strain rate-dependent forged/laminated carbon fiber-epoxy resin composites. Composites-Mechanics Computations Applications. 2020;**11**(4):341-367. DOI: 10.1615/compmechcomputapplintj.2020033979

[20] Tang ZR, Ma DW, Zou KD, Zhong JL, Ren J, Zhao CF, et al. The strain rate effect on the compression properties of basalt/carbon fiber reinforced composites. Applied Composite Materials. 2022;**29**(3):1007-1020. DOI: 10.1007/s10443-021-09997-6

[21] Zhong JL, Zhao CF, Ren J, Liu XX, Zhang ZD. A constitutive model for carbon fiber reinforced epoxy resin laminate under compression load: Considering the initial non-linearity. Applied Composite Materials. 2022;**29**: 629-649. DOI: 10.1007/s10443-021-09979-8

[22] Jones RM. Mechanics of Composite Materials. 2nd ed. Boca Raton, USA: CRC Press; 1999. 538 p. DOI: 10.1201/9781498711067

[23] Barber JR. Elasticity. 4th ed. Cham: Springer; 2022. 637 p. DOI: 10.1007/978-3-031-15214-6

[24] Altenbach H. Composite mechanics. In: Altenbach H, Johlitz M, Merkel M, Öchsner A, editors. Lectures Notes on Advanced Structured Materials. 1st ed. Cham: Springer; 2022. pp. 3-27. DOI: 10.1007/978-3-031-11589-9_1

[25] Gibson RF. Principles of Composite Material Mechanics. 4th ed. Boca Raton, USA: CRC Press; 2016. 700 p. DOI: 10.1201/b19626

[26] Tewary VK. Mechanics of Fibre Composites. 1st ed. New York, USA: John Wiley & Sons; 1978. 288 p

[27] Jiang YQ. Mechanics of Composites. 1st ed. Xi'an, China: Xi'an Jiaotong University Press; 1990. 222 p

[28] Shen GL, Hu GK, Liu B. Mechanics of Composite Materials. 2nd ed. Beijing, China: Tsinghua University Press; 2013. 365 p

[29] Yang QS. Mechanics of Composite Materials. 1st ed. Beijing, China: Beijing Science Press; 2020. 261 p

[30] Staab GH. Laminar Composites. 2nd ed. Waltham, USA: Butterworth-Heinemann; 2015. 466 p

[31] Giurgiutiu V. Stress, Vibration, and Wave Analysis in Aerospace Composites for SHM and NDE Applications. Chennai, India: Elsevier; 2022. 927 p. DOI:10.1016/C2016-0-04241-9

[32] Zhao CF, Ren R, Zhong JL, Goh KL, Zhang KB, Zhang ZD, et al. Intralaminar crack propagation of glass fiber reinforced composite laminate. Structure. 2022;**41**:787-803. DOI: 10.1016/j.istruc.2022.05.064

[33] Zhao CF, Ren R, Wei Y, Yang G, He B, Zhang KB, et al. Crack propagation for glass fiber reinforced laminates containing flame retardant: Based on single-edge tensile loading. Materiale Plastice. 2022;**59**(2):88-98. DOI: 10.37358/MP.22.2.5588

[34] Zhong JL, Ren R, Tang ZR, Ren J, Zhao CF. Analysis of nonlinear mechanical behavior of resin materials at low and medium strain rates. Polymer Composites. 2022;**43**(6):3699-3707. DOI: 10.1002/pc.26647

[35] Wang C, Roy A, Silberschmidt VV, Chen Z. Modelling of damage evolution in braided composites: Recent developments. Mechanics of Advanced Materials and Modern Processes. 2017;**3**: 15. DOI: 10.1186/s40759-017-0030-4

[36] Li X, Yu Y, Mu ZC, Hu YL. Meso-scale modeling for effective properties in continuous fiber-reinforced composites by state-based peridynamics. Acta Mechanica Solida Sinica. 2021;**34**: 729-742. DOI: 10.1007/s10338-021-00239-7

[37] Panettieri E, Fanteria D, Danzi F. A sensitivity study on cohesive elements parameters: Towards their effective use to predict delaminations in low-velocity impacts on composites. Composite Structures. 2016;**137**:130-139. DOI: 10.1016/j.compstruct.2015.11.011

[38] Lou X, Cai H, Yu P, Jiao F, Han X. Failure analysis of composite laminate under low-velocity impact based on micromechanics of failure. Composite Structures. 2017;**163**:238-247. DOI: 10.1016/j.compstruct.2016.12.030

[39] Tvergaard V, Hutchinson JW. The relation between crack growth resistance and fracture process parameters in elastic-plastic solids. Journal of the Mechanics and Physics of Solids. 1992; **40**(6):1377-1397. DOI: 10.1016/0022-5096(92)90020-3

[40] Hallquist J. LS-DYNA® Keyword user's Manual-VOLUME II Material Models. 13191st ed. California, USA: Livermore Software Technology Corporation; 2009. 1884 p

[41] Du C, Wang H, Zhao Z, Han L, Zhang C. A comparison study on the impact failure behavior of laminate and woven composites with consideration of strain rate effect and impact attitude. Thin-Walled Structures. 2021;**164**(3): 107843. DOI: 10.1016/j.tws.2021.107843

[42] Bouvet C, Castanié B, Bizeul M, Barrau JJ. Low velocity impact modelling in laminate composite panels with discrete interface elements. International Journal of Solids and Structures. 2009;**46**

(14–15):2809-2821. DOI: 10.1016/j.ijsolstr.2009.03.010

[43] Camanho PP, Davila CG, Moura MF. Numerical simulation of mixed-mode progressive delamination in composite materials. Journal of Composite Materials. 2003;**37**(16):1415-1438. DOI: 10.1177/0021998303034505

[44] Benzeggagh ML, Kenane M. Measurement of mixed-mode delamination fracture toughness of unidirectional glass/epoxy composites with mixed-mode bending apparatus. Composites Science and Technology. 1996;**56**(4):439-449. DOI: 10.1016/0266-3538(96)00005-X

[45] Shahriari M, Googarchin HS. Prediction of vehicle impact speed based on the post-cracking behavior of automotive PVB laminated glass: Analytical modeling and numerical cohesive zone modeling. Engineering Fracture Mechanics. 2020;**240**:107352. DOI: 10.1016/j.engfracmech.2020.107352

[46] Ren R. Study on gas pressure impact response characteristics of composite covers [thesis]. Nanjing, China: Nanjing University of Science & Technology; 2019

[47] Huang ZM. Mechanics theories for composite materials. Acta Materiae Compositae Sinica. 2023;**40**(6):3090-3114. DOI: 10.13801/j.cnki.fhclxb.20230117.007

[48] Tsai SW. Strength characteristics of composite materials. In: NASA Contractor Report. Newport Beach, UAS: Philco Corporation; 1965. 105 p

[49] Naser MZ, Hawileh RA, Abdalla JA. Fiber-reinforced polymer composites in strengthening reinforced concrete structures: A critical review. Engineering Structures. 2019;**198**:109542. DOI: 10.1016/j.engstruct.2019.109542

[50] Chang FK, Chang KY. A progressive damage model for laminated composites containing stress concentrations. Journal of Composite Materials. 1987;**21**(9):834-855. DOI: 10.1177/002199838702100904

[51] Chang KY, Llu S, Chang F. Damage tolerance of laminated composites containing an open hole and subjected to tensile loadings. Journal of Composite Materials. 1991;**25**(2):274-301. DOI: 10.1177/002199839102500303

[52] Chang FK, Lessard LB. Damage tolerance of laminated composites containing an open hole and subjected to compressive loadings: Part I-analysis. Journal of Composite Materials. 1991;**25**(1):2-43. DOI: 10.1177/002199839102500101

[53] Tsai SW, Wu EM. A general theory of strength for anisotropic materials. Journal of Composite Materials. 1971;**5**(1):58-80. DOI: 10.1177/002199837100500106

[54] Hoffman O. The brittle strength of orthotropic materials. Journal of Composite Materials. 1967;**1**(2):200-206. DOI: 10.1177/002199836700100210

[55] Puck A. Zur Beanspruchung und Verformung von GFK-Mehrschichtenverbund-Bauelementen: Teil 3. Versuche an Mehrschichtenverbunden. Kunststoffe. 1967;**57**(12):965-973

[56] Puck A. Calculating the strength of glass fiber/plastic laminates under combined load. Kunststoffe. 1969;**55**:780-787

[57] Puck A. A failure criterion shows the direction. Kunststoffe-German Plastics. 1992;**82**(7):29-32

[58] Puck A, Schürmann H. Failure analysis of Frp laminates by means of physically based phenomenological models. Composites Science and Technology. 1998;**62**(12–13):1633-1662. DOI: 10.1016/S0266-3538(01)00208-1

[59] Pinho ST, Davila CG, Camanho PP, Iannucci L, Robinson P. Failure Models and Criteria for FRP Under in-Plane or Three-Dimensional Stress States Including Shear Non-Linearity. NASA; 2005. p. NASA/TM-2005-213530

[60] Hashin Z. Failure criteria for unidirectional fiber composites. Journal of Applied Mechanics. 1980;**47**(2): 329-334. DOI: 10.1115/1.3153664

[61] Hashin Z. Analysis of cracked laminates: A variational approach. Mechanics of Materials. 1985;**4**(2): 121-136. DOI: 10.1016/0167-6636(85) 90011-0

[62] Shu DL. Mechanical Properties of Engineering Materials. 3rd ed. Beijing, China: China Machine Press; 2004. 256 p

[63] Hashin Z. Analysis of composite materials-a survey. Journal of Applied Mechanics. 1983;**50**(3):481-505. DOI: 10.1115/1.3167081

[64] Martin K. Analysis of Failure in Fiber Polymer Laminates: The Theory of Alfred Puck. 1st ed. Berlin, Heidelberg: Springer; 2014. 205 p

[65] Gu JF, Chen PH, Su L, Li K. A theoretical and experimental assessment of 3D macroscopic failure criteria for predicting pure inter-fiber fracture of transversely isotropic UD composites. Composite Structures. 2021;**259**:113466. DOI: 10.1016/j.compstruct.2020.113466

[66] Al-Rub R, Voyiadjis GZ. On the coupling of anisotropic damage and plasticity models for ductile materials. International Journal of Solids and Structures. 2003;**40**(11):2611-2643. DOI: 10.1016/S0020-7683(03)00109-4

[67] Lu X, Luo MJ, Zhao M, Shan ZZ. Thermal coupling analysis of lightning strike laminates. Journal of Aeronautical Materials. 2020;**40**(1):35-45. DOI: 10.11868/j. issn.1005-5053.2019.000095

[68] Ren R, Zhong JL, Le GG, Ma DW. Research on intralaminar load reversal damage modeling for predicting composite laminates' low velocity impact responses. Composite Structures. 2019;**220**:481-493. DOI: 10.1016/j. compstruct.2019.04.012

[69] Wang XC. Finite element method. 24th ed. Beijing, China: Tsinghua University Press; 2003. 776 p

[70] Simulia D. Abaqus Analysis user's Manual. 6.10 ed. Providence, USA: Dassault Systèmes; 2011

[71] Zhong JL, Zhao CF, Chen CQ, Lai WL, Wang Q. Mechanical behaviors of composite auxetic structures under quasi-static compression and dynamic impact. European Journal of Mechanics - A/Solids. 2025;**109**:105454. DOI: 10.1016/j. euromechsol.2024.105454

[72] Wegener M. Metamaterials beyond optics. Science. 2013;**342**:939-940. DOI: 10.1126/science.1246545

[73] Yang ZZ, Yu CH, Buehler MJ. Deep learning model to predict complex stress and strain fields in hierarchical composites. Science. Advances. 2021; 7(15):eabd7416. DOI: 10.1126/sciadv. abd7416

Chapter 7

Oxygen Vacancies in Spinels

Jiacheng Li

Abstract

Spinel oxides, with the general formula AB_2O_4, are a class of ternary transition metal compounds renowned for their diverse elemental composition and tunable electronic properties. Their cubic crystal structure, belonging to the Fd-3 m space group, features metal ions in both tetrahedral and octahedral coordination centers, which can be manipulated to achieve different catalytic functionalities. This review focuses on the impact of oxygen vacancies on the structural and catalytic properties of spinel oxides and the various strategies to introduce these vacancies. The formation of oxygen vacancies, facilitated through heat treatment, chemical reduction, etching, doping, and composite formation, has been shown to significantly enhance the conductivity, redox potential, and catalytic activity of spinel oxides. These vacancies act as active sites, promote electron/ion transfer, and lower activation energies for various catalytic reactions. Characterization techniques such as X-ray photoelectron spectroscopy (XPS), electron paramagnetic resonance (EPR), and density functional theory (DFT) calculations provide comprehensive insights into the distribution, concentration, and electronic structure modifications due to oxygen vacancies. Theoretical calculations, in particular, underscore the pivotal role of vacancies in stabilizing reaction intermediates and reducing energy barriers. This review synthesizes the current understanding of oxygen vacancies in spinel oxides and highlights the potential for defect engineering as a strategy to optimize catalysts for applications in energy storage, environmental catalysis, and electrochemical reactions.

Keywords: spinels, vacancies, catalysis, copper, spectroscopy

1. Introduction

Spinel, a ternary transition metal ion compound, boasts a rich elemental composition and a unique electronic structure, with a structure that is both variable and adjustable. It is represented by the chemical formula AB_2O_4, where A and B sites are occupied by transition metal elements, and it belongs to the Fd-3 m space group of the cubic crystal system [1]. Research on spinel transition metal oxides is extensive, focusing on the distribution of metal ions within the crystal structure, which are found in tetrahedral and octahedral coordination centers. The valence states of cations differ in these environments, while anions are situated at the vertices of the polyhedron. In the spinel geometry, only half of the octahedral sites and one-eighth of the tetrahedral sites are filled with metal cations, leaving ample space for the accommodation of external metal cations.

Spinel can exhibit three distinct crystal structures based on the coordination environments of A and B cations. A normal spinel structure occurs when all divalent

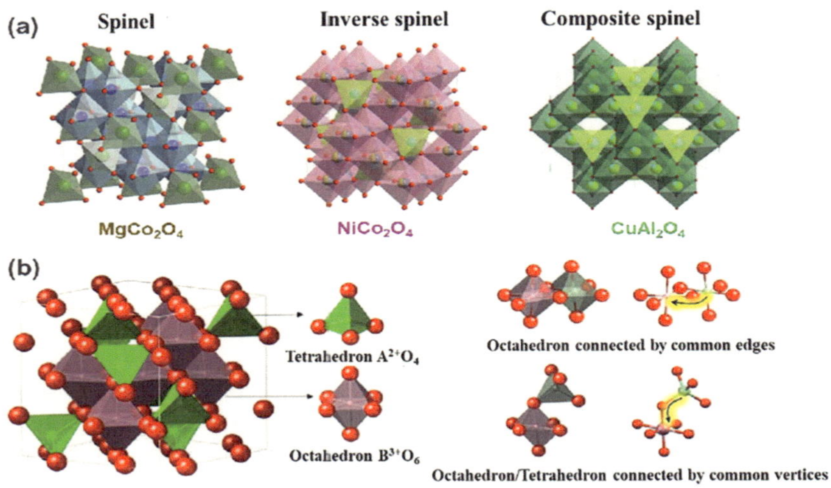

Figure 1.
Structural characteristics of spinel oxide. (a) Explain the categories of spinel oxides using several typical spinel oxides as representatives. Among them, $MgCo_2O_4$ *is spinel, with* Mg^{2+} *and* Co^{3+} *occupying tetrahedral and octahedral centers, respectively;* $NiCo_2O_4$ *is an inverse spinel, with half of* Co^{3+} *occupying tetrahedral centers and the other half of* Co^{3+} *and all* Ni^{2+} *occupying octahedral centers;* $CuAl_2O_4$ *is a composite spinel, with* Cu^{2+} *and* Al^{3+} *mixed to occupy tetrahedral and octahedral sites. (b) Coordination structure in spinel. Among them, octahedra are connected to each other through a common edge method, and octahedra and tetrahedral positions are connected to each other through a common vertex method.*

metal cations occupy the tetrahedral sites (A site) and all trivalent metal cations occupy the octahedral sites (B site). An inverse spinel structure is identified when half of the trivalent metal cations occupy the tetrahedral sites (A site), while the other half, along with divalent metal cations, occupy the octahedral sites (B site). A mixed spinel structure arises when both tetrahedral and octahedral sites are occupied by a mixture of divalent and trivalent metal cations. The connectivity of the Oh sites by common edges and the connection of Td and Oh sites by common vertices provide a theoretical foundation for modulating the electronic structure and catalytic performance of catalysts through the rational design of spinel oxide structures [2].

Typically, site A can be occupied by divalent cations such as Fe^{2+}, Mn^{2+}, Zn^{2+}, and Co^{2+}, while site B can be occupied by trivalent metal ions like Fe^{3+}, Al^{3+}, Cr^{3+}, Cu^{3+}, Ni^{3+}, etc. The unit cell, as depicted in **Figure 1(a)**, contains 24 metal ions (8 A^{2+} and 16 B^{3+}) and 32 O^{2-}. The O^{2-} ions are densely packed in a cubic form, creating a multitude of polyhedral gaps. With six oxygen ions forming octahedral gaps and four forming tetrahedral gaps, cations are positioned within these gaps in the crystal structure. However, the lattice contains as many as 96 gaps, which is significantly more than the number of cations. Consequently, not all gaps can be filled by cations, resulting in a multitude of vacant gaps. These excess gaps facilitate the regulation of metal ion diffusion and the formation of vacancies.

2. Definition and effects of oxygen vacancies on spinel

Perfect spinel crystals with stoichiometric ratios often display significant isotropic physical and chemical properties. The introduction of trace defects, such as oxygen vacancies, can profoundly alter these properties, including conductivity, band structure, adsorption/desorption, and thermodynamics. These defects can regulate the electronic density of states and surface properties of nanomaterials, which is particularly crucial for catalytic applications. Defect engineering creates unsaturated active sites for molecular adsorption by introducing oxygen vacancies, effectively enhancing reaction activity as demonstrated in various studies [3].

Oxygen vacancies in spinel oxides are formed by the absence of lattice oxygen on the surface of metal oxides. These vacancies possess abundant localized electrons that can influence the physical and chemical properties of the metal oxide, including redox capability, resistive switching, and magnetism. They can provide a large number of dangling bonds for intermediates in electrocatalytic reactions, thereby improving electrocatalytic performance through various mechanisms, such as influencing the evolution of active sites, altering metal valence states, regulating electronic structures, changing metal-oxygen bond lengths, and reducing reaction activation energies [4–6]. The presence of oxygen vacancies in spinel significantly enhances the catalyst's ability to adsorb and activate reactive species, accelerating the production of target products and thereby increasing the activity and selectivity of catalytic reactions. Moreover, oxygen vacancies can optimize the pore structure of the spinel, facilitating the in-situ formation of highly active sites on the catalyst surface.

In general, the introduction of oxygen vacancies in spinel materials can influence catalytic performance through various mechanisms, such as activating reactant molecules, enhancing active sites, improving electronic structure, and increasing catalytic selectivity. Therefore, designing and thoroughly analyzing the oxygen vacancy structure in spinel catalysts is crucial for optimizing catalytic efficiency.

3. Preparation methods and catalytic performance of oxygen vacancies

3.1 Heat treatment method

Heat treatment is a straightforward and common method for regulating oxygen vacancies. Xie et al. prepared ultra-thin nanosheet structured $NiCo_2O_4$ samples and obtained two types of nanosheets: one rich in oxygen vacancies (NiCo-r) and the other with few oxygen vacancies (NiCo-p) through calcination in air and oxygen atmospheres [7]. Electrochemical OER tests conducted in a 1 M KOH solution revealed that the NiCo-r sample had the lowest overpotential of 0.32 V and an exceptionally high current density of 285 $mA{\cdot}cm^{-2}$ at a potential of 0.8 V. In ultra-thin nanosheets, oxygen vacancies are structurally confined, which enhances the reactivity of active sites and effectively reduces the adsorption energy of water, promoting surface reactions and significantly improving OER activity.

Xu et al. controlled the synthesis temperature of spinel $MnCo_2O_4$ to obtain a series of samples with different Mn valence states [8]. The overpotential exhibited a volcanic trend with the valence state of the element, with the cation valence state in octahedral coordination playing a dominant role in catalytic activity. This was related to the electron filling of the oxide's e_g orbital. Research on the electron occupancy rate of e_g orbitals in spinel oxides indicates that the electron occupancy rate is a crucial indicator for catalyst selection and can effectively regulate the valence state of metal atoms by altering oxygen vacancies. Chen et al. successfully synthesized hollow $NiCo_2O_4$ nanoparticles with abundant oxygen vacancies and a porous structure by combining solvent heat and annealing [9]. The $NiCo_2O_4$-450-V_o, with its hollow porous structure, demonstrated dual functional electrocatalytic performance for

MOR and ORR under alkaline conditions. The enhancement in the performance of $NiCo_2O_4$-450-V_o is attributed to its morphological structure and active sites. Oxygen defect engineering can effectively modulate the electronic structure of the $NiCo_2O_4$ material, improving its intrinsic activity, while the hollow and layered porous structure can facilitate reaction kinetics.

3.2 Reduction method

Spinel samples rich in oxygen vacancies can be prepared using chemical reducing agents. Lu et al. reduced $CoFe_2O_4$ on carbon nanotubes with $NaBH_4$ and observed a significant increase in oxygen vacancy content [10]. OER test results indicated that the treated samples exhibited improved catalytic activity. Zheng et al. treated the spinel-type Co_3O_4 sample in a $NaBH_4$ solution for 1 hour and found that the sample's activity increased post-treatment [11]. The oxygen vacancies introduced altered the electronic structure of the oxide, enhancing the catalytic reaction activity. $NaBH_4$ could be utilized as a reducing agent to create a large number of oxygen vacancy defects in FeCo oxide nanosheets, which improved the material's conductivity, increased the number of surface active sites, and significantly enhanced the catalytic reaction activity [12]. Furthermore, $NaBH_4$ can effectively prepare reduced $CoFe_2O_4$ nanosheets (NS) rich in oxygen vacancies, leading to more active sites and enhanced conductivity compared to the pristine $CoFe_2O_4$ hollow nanosphere [13]. These nanosheets achieved a current density of 10 mA cm^{-2} at an overpotential of 320 mV in 1 M KOH. As expected, the $CoFe_2O_4$ hollow nanosphere, with a relatively higher surface area after $NaBH_4$ reduction, showed the most notable improvement in OER activity, and the corresponding reduced $CoFe_2O_4$ NS demonstrated the best OER performance compared to the reduced $CoFe_2O_4$ bulk and nanoparticles.

Hydrogen reduction treatment is also an effective approach. Du et al. treated the obtained mesoporous $NiFe_2O_4$ material in a mixed gas (40% H_2/N_2) for 1 hour and controlled different reduction temperatures to generate more oxygen vacancies on the catalyst surface [14]. The presence of a large number of oxygen vacancies and active sites promotes the catalytic reaction, and the increase in defect states enhances the oxide's conductivity.

Additionally, more defects can be introduced into the spinel structure through electrochemical reduction, improving the material's conductivity and electrochemical active sites. The potential on the cathode can be accurately adjusted, and the reduction process of the cathode can be precisely controlled without introducing strong reducing agents during the electrolysis process. Wang et al. induced oxygen vacancies by reduction at a potential of −0.9 V (NCO-0.9@CC) [15]. During the electrochemical reaction process, oxygen vacancies act as active sites, promoting electron/ion transfer rates and achieving the goal of enhancing energy storage.

3.3 Etching method

The etching method can directly control oxygen defects on the surface of spinel. Dai et al. treated the prepared sample with Ar plasma and found that the surface area of the treated sample increased significantly [16]. Moreover, the current density corresponding to 1.6 V VS RHE was 44.44 mA·cm^{-2}, with a lower starting potential. After plasma etching, the increased surface area and internal oxygen vacancy content of the sample contribute to the improved activity of the catalyst.

Oxygen vacancy-rich $FeCo_2O_4$ nanoparticles can also be prepared by laser treatment. These oxygen vacancies effectively reduce the thermodynamic energy barrier and expedite electron transfer, thereby simultaneously enhancing both the oxygen evolution and reduction reactions [17].

3.4 Doping method

Heteroatom doping is a low-cost and highly effective method for incorporating oxygen vacancies. N and P co-doping can cause the formation of oxygen vacancies and metal-nitrogen bonds, leveraging the use of ionic liquids as sources for N and P. This strategy results in superior catalytic performance in the activation of persulfate [18].

Stabilizing oxygen vacancies by filling with non-metallic elements with different atomic radii and electronegativities to compensate for the coordination number of metal elements is a viable strategy. For instance, using sulfur atoms to effectively stabilize oxygen vacancies and reduce the Bader charge of atoms near these vacancies is beneficial for the adsorption and desorption of intermediate products in the reaction [19]. It can also form cobalt-sulfur coordination with the cobalt active center adjacent to the oxygen vacancy, thereby regulating the electronic structure of the active center and increasing the H_2O adsorption energy on the cobalt active site.

Zeng et al. [20] introduced both oxygen vacancies and phosphorus ions into $NiCo_2O_4$ nanosheets, which initially had low conductivity and fewer active sites, resulting in significantly improved conductivity, rate performance, and capacity for the optimized P-$NiCo_2O_4$-x NS zinc-ion battery. Zhang et al. [21] constructed anionic defects on $NiCo_2O_4$ through a mechanochemical strategy involving Na-assisted milling for propane oxidation (**Figure 5a**). This process substantially improved the oxygen vacancies, active oxygen species, and oxygen mobility of the catalyst. Lu et al. [22] prepared a novel $CeO_2/MnCo_2O_4$ catalyst, with CeO_2 acting as a key substance for creating oxygen vacancies. The stability of methane catalytic oxidation in unmodified $MnCo_2O_4$ pure samples is poor, while the stability of oxidation–reduction in samples modified with CeO_2 is enhanced. This enhancement is attributed to the partial substitution of high-valence Ce^{4+} with low-valence cations (Mn^{3+}, Co^{2+}, and Co^{3+}), allowing them to be partially doped into the $MnCo_2O_4$ lattice and generate a large number of oxygen vacancies to maintain charge balance.

The incorporation of Ce into $CuFe_2O_4$ spinel lattice has been instrumental in facilitating electron exchange, bolstering the oxygen storage capacity, and fostering the formation of oxygen vacancies [23]. However, an excessive amount of Ce can diminish the specific surface area and occupy surface oxygen vacancies, leading to detrimental effects.

3.5 Composite method

The introduction of supplementary materials can also result in an elevation of oxygen vacancies within the spinel structure. Chuang et al. [24] synthesized a novel hybrid catalyst featuring $CoFe_2O_4$ nanorods encapsulated by single-walled carbon nanotubes (SWNTs). The synergistic effect between $CoFe_2O_4$ nanorods and SWNTs not only increases the oxygen vacancies and optimizes the electronic structure of the composite but also facilitates the adsorption and stabilization of hydroxyl groups, significantly improving both OER and HER performances.

The synergistic interaction between SWNTs and $CoFe_2O_4$ nanorods not only increases the oxygen vacancies and refines the electronic configuration of the composite, but also facilitates the adsorption and stabilization of hydroxyl groups, leading to a marked enhancement in both oxygen evolution reaction and hydrogen evolution reaction performances.

Additionally, the hollow $CoFe_2O_4$ nanocubes, which are carbon-coated and derived from Prussian blue analogs and feature an abundance of oxygen vacancies, have achieved remarkable results with an ammonia yield rate of 30.97 $\mu g \cdot h^{-1} \cdot mg_{cat.}^{-1}$ at −0.4 V versus the reversible hydrogen electrode in 0.1 M Na_2SO_4 [25].

4. Characterizations of oxygen vacancies

Various methods are available for characterizing oxygen vacancies. For instance, the distribution of oxygen vacancies can be directly observed through visualization techniques such as scanning tunneling microscopy (STM) and aberration-corrected scanning transmission electron microscopy (AC-STEM). Spectroscopic techniques like X-ray fine absorption spectroscopy (XAFS), electron paramagnetic resonance (EPR), X-ray diffraction (XRD), X-ray photoelectron spectroscopy (XPS), Raman spectroscopy, positron annihilation technique (PAT), and others can qualitatively or quantitatively provide information on oxygen vacancies. Moreover, density functional theory calculations (DFT) can theoretically predict the form and content of oxygen vacancies.

Determining the form, location, and concentration of defects in materials is essential for studying the influence of defect structure on the catalytic activity and stability of spinel. A range of physical and chemical characterization methods is required to study the location and content of defects, with spectroscopy being the primary method for defect characterization.

4.1 Raman spectroscopy

The application of Raman spectroscopy in oxygen defect analysis is extensive. Utilizing the Raman scattering effect, the technique analyzes the scattering spectra at various frequencies distinct from the incident light, thereby extracting insights into molecular vibrations and rotations and studying molecular structure. Raman spectroscopy can reveal structural changes in materials caused by oxygen vacancies or defects.

To verify the physical properties and further explore the structural defects of catalysts, Raman spectroscopy was performed. **Figure 2(a)** shows that $NiCo_2O_4$ exhibited typical Raman spectra at 655 cm^{-1} (A1g), 516 cm^{-1} (F2g), 468 cm^{-1} (Eg), and 186 cm^{-1} (F2g), consistent with the spinel vibrational modes. A noticeable blue shift of the main peak (644 cm^{-1}) for D-$NiCo_2O_4$, compared to $NiCo_2O_4$, sufficiently proves the presence of structural defects induced by lattice distortion in D-$NiCo_2O_4$ [26].

Raman spectroscopy was also used to study the structural characteristics of the $CuCo_2O_4$ spinel catalyst and surface defects, as shown in **Figure 2(b)**. The presence of two bands at 510 and 462 cm^{-1} is evident. The band at 510 cm^{-1} is attributed to the Cu-O stretching in the octahedral sites of Cu^{2+}-O^{2-}, while the Co-O stretching vibration in the tetrahedral sites of $Co^{3+} - O^{2-}$ is observed at 462 cm^{-1}. Meanwhile, the pronounced band at 651 cm^{-1} and the small band at 547 cm^{-1} are indicative of the $CuCo_2O_4$ spinel crystal structure [27].

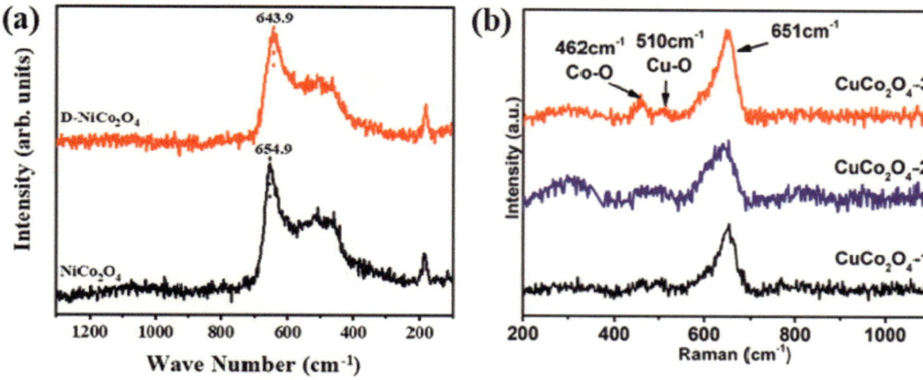

Figure 2.
(a) Raman spectra of $NiCo_2O_4$ and D-$NiCo_2O_4$; (b) Raman spectra of $CuCo_2O_4$-1, $CuCo_2O_4$-2, and $CuCo_2O_4$-3.

4.2 X-ray photoelectron spectroscopy (XPS)

XPS, a highly sensitive ultra-microscopic surface analysis technique, was commonly used in early research of defect chemistry to characterize defects in catalysts. It identifies defect types or concentrations by analyzing surface peak positions, valence bond changes, and bonding information [28, 29].

As shown in **Figure 3(a)**, the O 1 s spectrum of $NiCo_2O_4$ was fitted into O1, O2, and O3 with binding energies of 529.6, 531.2, and 532.6 eV, respectively. The O1 peak is attributed to the oxygen-metal bonds, specifically the interactions between oxygen and metals such as Co or Ni (O-Co/Ni). The O2 peak is likely associated with defective oxygen structures that exhibit low coordination numbers. Meanwhile, the O3 peak corresponds to oxygen species that are part of adsorbed water molecules [30]. It is evident that there is a significant increase in O2 for D-$NiCo_2O_4$, certifying the successful introduction of oxygen vacancies. The oxygen vacancy structure significantly influences the catalytic oxidation reaction by transforming O_2 molecules into active

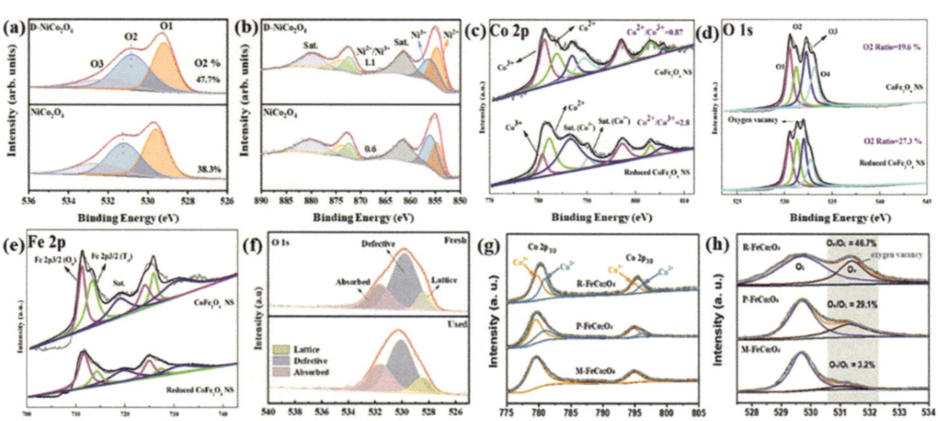

Figure 3.
XPS spectra of (a) O 1 s, (b) Ni 2p for $NiCo_2O_4$, and D-$NiCo_2O_4$; XPS spectra of (c) Co 2p, (d) O 1 s, and (e) Fe 2p for the pristine $CoFe_2O_4$ hollow nanospheres and the reduced $CoFe_2O_4$ nanosheet; (f) O 1 s XPS spectrum of CFB-2 before and after the reaction; the fitted (g) Co 2p and (h) O 1 s spectra of different $FeCo_2O_4$ samples.

surface oxygen and increasing oxygen mobility. In **Figure 3(b)**, the Ni 2p3/2 spectrum could be decomposed into the doublets of spin-orbits, with peaks corresponding to Ni^{2+} at 854.6 eV and Ni^{3+} at 855.9 eV, respectively [26]. The ratio of Ni^{2+}/Ni^{3+}, as well as Co^{2+}/Co^{3+}, in the defect-engineered D-$NiCo_2O_4$ is notably higher than that observed in the pristine $NiCo_2O_4$. The significant increase suggests that the defects engineering process could effectively reduce state of the metal ions. This correlates with the higher proportion of oxygen vacancies in D-$NiCo_2O_4$, created to balance the valence state. Interestingly, the Ni^{2+}/Ni^{3+} ratio exhibits a more pronounced increase compared to that of Co^{2+}/Co^{3+}, indicating that Ni^{3+} ions are more readily reduced than Co^{3+} ions.

The deficiencies in the reduced $CoFe_2O_4$ NS and its corresponding pristine $CoFe_2O_4$ hollow nanospheres were then characterized by XPS [13]. As shown in **Figure 3(c)**, the Co 2p spectra of hollow $CoFe_2O_4$ nanospheres and reduced $CoFe_2O_4$ NS are well fitted to 7 components, indicating the coexistence of Co^{2+} and Co^{3+} in $CoFe_2O_4$. The Co^{2+}/Co^{3+} ratio on the surface of the hollow $CoFe_2O_4$ nanospheres is 0.87, while that for the reduced $CoFe_2O_4$ NS is as high as 2.8. This significant increase suggests that the reduction process has led to a substantial enrichment of Co^{2+} species on the surface of the $CoFe_2O_4$ NS [12]. As evidenced by the detailed deconvolution of the O 1 s spectra, the reduction process involving $NaBH_4$ has been shown to generate a higher concentration of oxygen vacancies on the surface of $CoFe_2O_4$. **Figure 3(d)** illustrates the O 1 s region, which can be deconvoluted into four distinct oxygen peaks. The O1 peak, centered at approximately 529.7 eV, corresponds to metal-oxygen bonds. The O2 peak, observed at a higher binding energy of 531.2 eV, is characteristic of oxygen defect sites. The O3 peak, positioned at around 532.5 eV, is indicative of surface-adsorbed oxygen or hydroxyl groups. Lastly, the O4 peak at 533.4 eV is associated with adsorbed molecular water, as referenced in [12]. As shown in **Figure 3(e)**, the Fe 2p spectra are characterized by two shakeup satellites at 718.1 and 732.8 eV, along with two spin-orbit doublet peaks corresponding to Fe 2p3/2 and Fe 2p1/2. The XRD and XPS analyses suggest that the $NaBH_4$ reduction method is an efficient and reliable approach for preparing homogeneous metal oxides that are rich in oxygen vacancies.

As further observed in **Figure 3(f)**, there is a noticeable decrease in the relative peak area ratio between defect oxygen and lattice oxygen. This reduction primarily stems from the reaction of oxygen vacancies on the catalyst surface with dissolved oxygen in water, leading to the formation of metastable intermediates (O_V-O*) [31]. These intermediates serve as activation centers that facilitate the production of reactive oxygen species (ROS), particularly $\bullet O_2^-$ and $^1O_2$. The oxygen vacancies present on the surface of the $CoFe_2O_4$-loaded biochar composite (CFB-2) partially contribute to $^1O_2$ generation through their transformation into metastable intermediates with dissolved oxygen. To elucidate the enhanced performance of R-$FeCo_2O_4$, XPS measurements were conducted on three $FeCo_2O_4$ samples. **Figure 3(g)** presents the O 1 s spectra, which can be deconvoluted into two peaks at 529.7 and 531.3 eV. The peak at 529.7 eV is associated with lattice oxygen (O_L), while the peak at 531.3 eV corresponds to adsorbed oxygen on an oxygen vacancy (O_V) [17]. It is evident that the OV/OL ratio for R-$FeCo_2O_4$ (46.7%) is significantly higher than that of P-$FeCo_2O_4$ (29.1%) and M-$FeCo_2O_4$ (3.2%), indicating that laser fragmentation has generated a greater number of oxygen vacancies in R-$FeCo_2O_4$. The high-resolution Co 2p XPS spectra depicted in **Figure 3(h)** reveal that the majority of cobalt cations in the $FeCo_2O_4$ samples are in the Co^{3+} state. However, R-$FeCo_2O_4$ exhibits a more pronounced peak related to Co^{2+} ions compared to the other samples, and Co^{2+} ions have been identified as the catalytically active sites for the ORR.

4.3 Electron paramagnetic resonance (EPR) spectra

EPR is a magnetic resonance technique based on the magnetic moment of unpaired electrons, which is capable of both qualitatively and quantitatively detecting unpaired electrons within material atoms. It offers insights into the structural characteristics and immediate surroundings. EPR characterization is instrumental in confirming the presence of oxygen vacancies, providing valuable information about their formation and distribution within the material. As depicted in **Figure 4(a)**, D- $NiCo_2O_4$ exhibits a pronounced EPR signal, with a peak intensity approximately 4.1 times greater than that of $NiCo_2O_4$, at a g-value of 1.99975. This signal is indicative of paramagnetic oxygen vacancies within the spinel oxide structure, thereby robustly confirming that the mechanochemical process has successfully introduced a substantial number of oxygen vacancies into $NiCo_2O_4$ [26]. In **Figure 4(b)**, all $CuCo_2O_4$-1, $CuCo_2O_4$-2, and $CuCo_2O_4$-3 exhibited an EPR signal centered at approximately 3360, which can be ascribed to the surface electrons that are trapped at the oxygen vacancy sites. The highest intensity of EPR signal for $CuCo_2O_4$-3 suggests the most oxygen vacancies on the surface [27]. The oxygen vacancies present on the surface of the $CoFe_2O_4$-loaded biochar composite (CFB-2) play a partial role in generation of $^1O_2$ by transforming into a metastable intermediate in conjunction with dissolved oxygen.

The asymmetric stretching vibration of the Fe/Co-O bond, characteristic of the T2g mode, was detected at the tetrahedral position in the CFB-2 catalyst, signifying a substantial presence of oxygen vacancies (OVs). Concurrently, a pronounced asymmetric signal at g = 2.003 is distinctly visible in **Figure 4(c)**, which is indicative of the trapping of unpaired electrons within these oxygen vacancies. The underlying mechanism involves electron-rich groups on the biochar (BC) acting as Lewis bases to mediate electron transfer, thereby initiating a redox reaction that leads to the formation of singlet oxygen ($^1O_2$). The metastable intermediates created by OVs effectively reduce the O-O bond strength, resulting in a significant production of reactive oxygen species (ROS). Additionally, a high-valent metal ion formed through

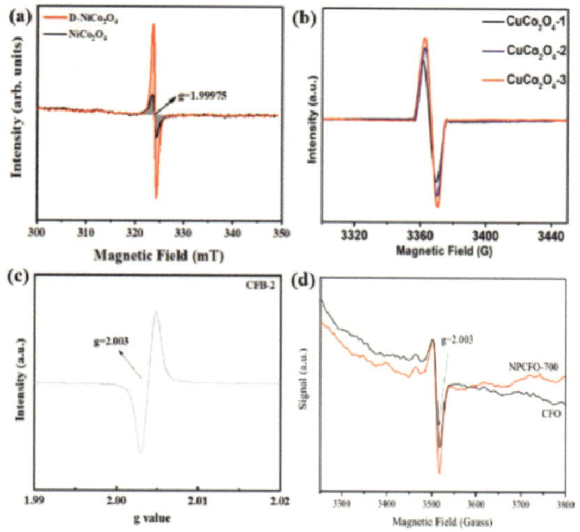

Figure 4.
(a) EPR spectra for $NiCo_2O_4$ and D-$NiCo_2O_4$; (b) EPR spectra of $CuCo_2O_4$-1, $CuCo_2O_4$-2, and $CuCo_2O_4$-3; (c) the EPR spectra of CFB-2 before and after the reaction; (d) EPR spectra of CFO and NPCFO-700.

a two-electron transfer process aids in the electron transfer to superoxide (•O_2^-), facilitating the generation of $^1O_2$.

In-situ EPR analysis was utilized to further elucidate the disparity in oxygen vacancy concentration between CFO and NPCFO-700, as shown in **Figure 4(d)**. It has been documented that the EPR resonance signal with a g value of 2.003 corresponds to the free spins induced by OVs and is directly proportional to the OV concentration [32]. In this context, the EPR signal intensity of NPCFO-700 was more intense than that of CFO, suggesting a higher concentration of oxygen vacancies in NPCFO-700 compared to other NPCFP-x samples.

5. Analysis of the mechanism of spinel oxygen defects

Oxygen vacancies can significantly adjust the electron density around the Fermi level, effectively forming and transforming reaction intermediates. They can effectively regulate the structure and electronic properties of spinel, adjust electronic configurations, and change the center position of the d-band of metal oxides, enhancing the surface adsorption/desorption ability of catalysts and optimizing electrocatalytic kinetics and catalytic activity. Theoretical studies have shown that vacancy modification increases the active sites and reduces the catalytic energy barrier, lowers the Gibbs free energy at the basic reaction level, and ultimately improves its electrocatalytic performance [33–38].

As shown in **Figure 5(a)**, to certify the enhancement of DCM dichlorination reaction activity on $CuCo_2O_4$ (311) and $CuCo_2O_4$-O_V, DFT calculations were carried out to investigate the active sites and catalytic mechanism. It is worth noting that the

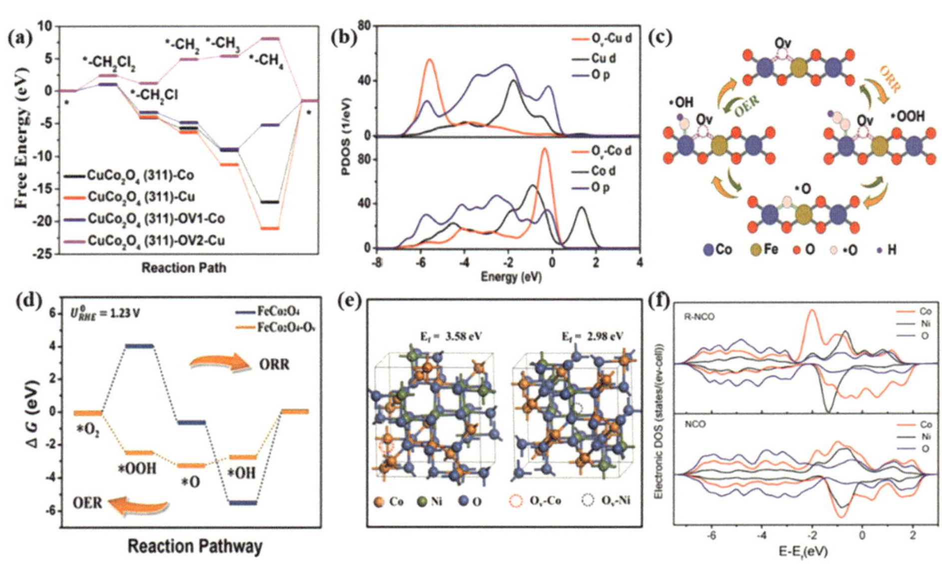

Figure 5.
(a) Reaction energy barrier for the dechlorination of DCM to CH_4 on $CuCo_2O_4$ catalysts; (b) Curves of projected density of states of Co d, Cu d, and O p orbitals for $CuCo_2O_4$ (311) and $CuCo_2O_4$ (311)-O_v; (c) The reaction mechanism of R-$FeCo_2O_4$ with O_V; (d) The DFT diagram of pristine $FeCo_2O_4$ and R-$FeCo_2O_4$; (e) O_v formation energy (E_f) induced by the reduction of Co, Ni, and the corresponding crystal structure; (f) Calculated DOS curves for pristine NCO and R-NCO.

second dechlorination step (from * – CH_2Cl to * – CH_2) is the potential limiting step for $CuCo_2O_4$ (311)-O_V-Cu, which shows the lowest energy barrier.

The results indicate that the $CuCo_2O_4$ (311)-O_V-Cu is the most active site for the dechlorination of CH_2Cl_2 into CH_4. The projected density of states (PDOS) plots, as depicted in **Figure 5(b)**, elucidate the intricate electronic interactions that occur between the adsorbates and the catalysts. Notably, they reveal a significant reduction in the d-band center for O_V-Cu, which leads to weaker binding between the *CH_4 intermediate and $CuCo_2O_4$ (311)-O_V-Cu, and further improvement in dechlorination performance [27].

Subsequently, they conducted DFT calculations to dissect the impact of oxygen vacancies on the catalytic mechanism. Both the OER and the ORR encompass four fundamental reaction steps, as illustrated in **Figure 5(c)**. During the ORR, O_2 molecules are adsorbed onto the catalyst surface and sequentially reduced to *OOH, *O, *OH, and ultimately H_2O. Conversely, the OER occurs in the opposite direction. The free energy diagrams derived from DFT calculations are portrayed in **Figure 5(d)**, where a substantial thermodynamic barrier (ΔG) is observed for the formation of *OOH on the pristine $FeCo_2O_4$ surface devoid of oxygen vacancies. Upon introducing oxygen vacancies into R- $FeCo_2O_4$, this ΔG value is markedly diminished, thereby enhancing the efficiency of both the ORR and OER processes. The substantially reduced ΔG values for the reduction of *O to *OOH and *OOH to O_2 indicate a higher catalytic efficiency for the four-electron dominant process, which is corroborated by the K-L plot results. Consequently, the DFT findings underscore the pivotal role of oxygen vacancies in augmenting the catalytic activities of $FeCo_2O_4$ for both ORR and OER [17].

The DFT calculations also assessed the formation energy (Ef) of oxygen vacancy structures (O_V-Ni and O_V-Co) associated with the reduction of Ni^{3+} and Co^{3+}, respectively. As depicted in **Figure 5(e)**, the formation energy for O_V-Ni is notably lower than that for O_V-Co, effectively substantiating the enhancement of reducibility and oxygen vacancy formation due to the incorporation of Ni into the spinel structure [26].

The extensive study, encompassing both experimental analyses and theoretical computations, has revealed the presence of single-site oxygen vacancies (SSOVs) and their role in enhancing the dissociation of N-O bonds in catalytic reactions. In the context of the rate-determining step for the NO+CO reaction on CO-$CuMn_2O_4$, we conducted density functional theory (DFT) calculations to compare the reaction pathways on CO-CuO and CO-Mn_2O_3.

As illustrated in **Figure 6(a)**, the computational models for SSOVs on mono-component catalysts are depicted as CO-CuO (111) and CO-Mn_2O_3 (111). On the CO-CuO (111) surface, CO and NO molecules are capable of adsorbing onto the copper site in proximity to the oxygen vacancy. However, the energy barrier for the transition state TS1' is exceedingly high at 2.31 eV, which confines the nitrogen atom (N*) to the surface, rendering it incapable of occupying the oxygen vacancy in the subsequent reaction step. A comparable situation arises in the reduction of NO by CO on the CO-Mn_2O_3(111), where the energy barrier for the transition state TS1" is also notably high at 2.20 eV. This barrier prevents the N* atom from directly filling the oxygen vacancy on the CO-Mn_2O_3(111) surface. Furthermore, as shown in **Figure 6(b-f)**, crystal orbital Hamilton population (COHP) analyses were performed for the CO-$CuMn_2O_4$(100), CO-CuO(111), and CO-Mn_2O_3(111) surfaces. In COHP analysis, negative values typically indicate bonding interactions, while positive values denote antibonding interactions. The analysis of the Integrated Crystal Orbital Hamilton Population (ICOHP) values for the N-O bonds in **Figure 6(b-d)** suggests that the difficulty in breaking the N-O bond should be ranked as CO-Mn_2O_3(111) > C

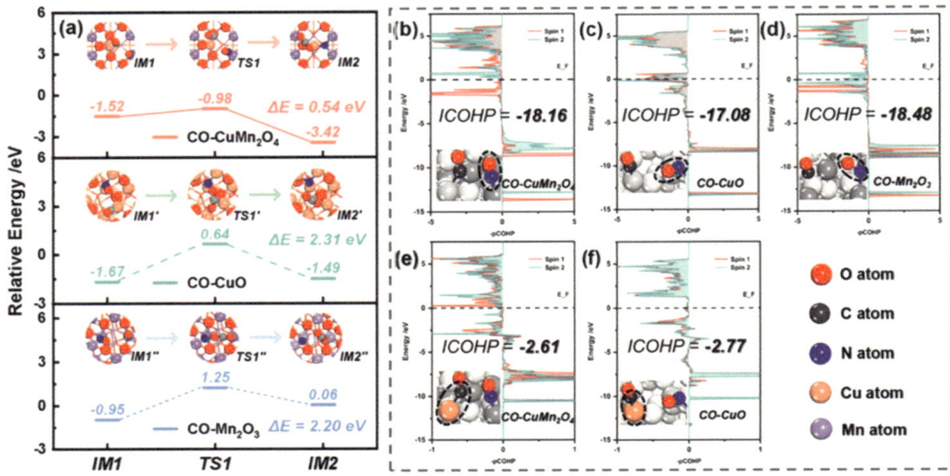

Figure 6.
(a) Energy comparison of key steps of the NO + CO reaction on CO-$CuMn_2O_4$ (100), CO-CuO(111), and CO-Mn_2O_3(111) surfaces, (b-d) the ICOHP results of the N-O bond over (b) CO-$CuMn_2O_4$ (100), (c) CO-CuO(111), (d) CO-Mn_2O_3(111) surfaces, (e, f) the ICOHP results of Cu and C atoms over, (e) CO-$CuMn_2O_4$(100), and (f) CO-CuO(111) surfaces.

O-$CuMn_2O_4$(100) > CO-CuO(111). Theoretically, a higher difficulty in breaking the N-O bond would impede the reaction progress. However, upon scrutinizing the energy barriers associated with the pivotal step (*N-O + Cu-*CO → *N + CO_2) as depicted in **Figure 6(a)**, it can be deduced that the reaction activity on the four catalysts follows the order of CO-$CuMn_2O_4$(100) < CO-Mn_2O_3(111) < CO-CuO(111).

In practice, for the NO + CO reactions on CO-CuO(111) and CO-$CuMn_2O_4$(100) surfaces, CO must undergo migration from Cu sites to the adsorbed NO, and the dissociation of the Cu-C bond during this transition could significantly influence the energy barrier of TS1. The ICOHP values presented in **Figure 6(e, f)** suggest that the strength of the Cu-C bond on the CO-$CuMn_2O_4$(100) surface is weaker than that on CO-CuO(111), which facilitates CO migration on the CO-$CuMn_2O_4$(100) catalyst surface and enhances its reactivity with NO. Consequently, the bond strength of the Cu-C is a pivotal factor that warrants consideration, in addition to the N-O bond strength of adsorbed NO. Moreover, the adsorption capacity for CO molecules is limited on the CO-Mn_2O_3(111) surface, which is detrimental to the subsequent NO + CO reaction.

The exhaustive investigation, encompassing both experimental characterizations and theoretical computations, has unveiled the presence of SSOVs and their pivotal role in facilitating the dissociation of N-O bonds during catalytic reactions. To gain insights into the rate-limiting step of the NO+CO reaction on CO-$CuMn_2O_4$, we conducted DFT simulations of analogous reaction pathways on CO-CuO and CO-Mn_2O_3. As depicted in **Figure 6(a)**, the computational models for single-component catalysts featuring SOVs are exemplified by CO-CuO(111) and CO-Mn_2O_3(111). Both CO and NO molecule adsorb preferentially at Cu sites adjacent to OVs on the CO-CuO(111) surface. However, the transition state of TS1' encounters a formidable energy barrier of 2.31 eV, which confines the N* atom to the surface, preventing it from occupying the oxygen vacancy in the subsequent steps. In similar, the reduction of NO by CO on CO-Mn_2O_3(111) mirrors this trend, with an equally steep energy barrier of 2.20 eV for TS1", hindering direct occupation of the oxygen vacancy by the N* atom.

Furthermore, **Figure 6(b-f)** presents crystal orbital Hamilton population (COHP) analyses for CO-$CuMn_2O_4$(100), CO-CuO(111), and CO-Mn_2O_3(111) surfaces. In COHP, negative values signify bonding interactions, whereas positive values indicate antibonding. Analyzing the ICOHP values for N-O bonds in **Figure 6(b-d)** suggests that the difficulty of N-O bond cleavage follows the order: CO-Mn_2O_3(111) > CO-$CuMn_2O_4$(100) > CO-CuO(111). Theoretically, this order implies a corresponding increase in reaction difficulty. Nevertheless, contrasting the energy barriers for the crucial step (N-O + Cu-CO → *N + CO_2) in **Figure 6(a)** reveals a contrasting trend: CO-$CuMn_2O_4$(100) < CO-Mn_2O_3(111) < CO-CuO(111). Notably, for reactions on CO-CuO(111) and CO-$CuMn_2O_4$(100), the migration of CO from Cu sites to the adsorbed NO is essential, and the disruption of the Cu-C bond during this migration can significantly modulate the energy barrier at TS1. The ICOHP values depicted in **Figure 6(e, f)** suggest that the Cu-C bond on the surface of CO-$CuMn_2O_4$(100) is comparatively weaker, facilitating CO migration and enhancing its reactivity with NO. Thus, the Cu-C bond strength, in addition to the N-O bond strength of adsorbed NO, is a crucial factor influencing the TS1 energy barrier. Lastly, the limited CO adsorption capacity on CO-Mn_2O_3(111) hinders the subsequent NO+CO reaction.

6. Conclusions

In summary, oxygen vacancies in spinel oxides are a crucial factor influencing their structural and functional properties. These vacancies, which can be introduced through various methods such as heat treatment, chemical reduction, etching, doping, and composite formation, significantly alter the material's physical properties like conductivity, redox capability, and catalytic activity. They provide active sites, enhance electron/ion transfer rates, and lower reaction activation energies, thereby improving the overall performance of spinel oxides in catalytic reactions. Characterization techniques provide insights into the distribution, concentration, and nature of oxygen vacancies. For example, XPS can reveal changes in the oxidation states of metal ions and the presence of defective oxygen structures, while EPR can confirm the presence of unpaired electrons associated with oxygen vacancies. Theoretical calculations, such as DFT, further support the experimental findings by illustrating the role of oxygen vacancies in modifying the electronic structure of spinel oxides and facilitating catalytic reactions.

Acknowledgements

We acknowledge the support of the Fundamental Research Funds for the Central Universities (Nos. 310400209508 and 311324240543) and the National Natural Science Foundation of China (No. 22472009) for their financial support of this work.

Author details

Jiacheng Li
School of Environment, Beijing Normal University, China

*Address all correspondence to: ljc@bnu.edu.cn

References

[1] Ming ZH, Jin YZ, Wen JZ. Synthesis of oxygen vacancy-enriched N/P co-doped $CoFe_2O_4$ for high-efficient degradation of organic pollutant: Mechanistic insight into radical and nonradical evolution. Environmental Pollution. 2021;**270**:116092-116103. DOI: 10.1016/j.envpol.2020.116092

[2] Song J, Wei C, Huang ZF, Chuntai L, Lin Z, Xin W, et al. A review on fundamentals for designing oxygen evolution electrocatalysts. Chemical Society Reviews. 2020;**49**:2196-2214. DOI: 10.1039/C9CS00607A

[3] Zhang N, Li X, Ye H, et al. Oxide defect engineering enables to couple solar energy into oxygen activation. Journal of the American Chemical Society. 2016;**138**:8928-8935. DOI: 10.1021/jacs.6b04629

[4] Liu Y, Zhang N, Yu C, et al. $MnFe_2O_4$@C nanofibers as high-performance anode for sodium-ion batteries. Nano Letters. 2016;**16**:3321-3328

[5] Li C, Han XP, Cheng FY, et al. Phase and composition controllable synthesis of cobalt manganese spinel nanoparticles towards efficient oxygen electrocatalysis. Nature Communications. 2015;**6**:7345

[6] Wang XT, Ouyang T, Wang L, et al. Surface reorganization on electrochemically-induced Zn-N-Co spinel oxides for enhanced oxygen electrocatalysis. Angewandte Chemie International Edition. 2020;**59**:6492-6499

[7] Bao J, Zhang XD, Fan B, et al. Ultrathin spinel-structured nanosheets rich in oxygen deficiencies for enhanced electrocatalytic water oxidation. Angewandte Chemie. 2015;**127**:7507-7512

[8] Wei C, Feng ZX, Scherer GG, et al. Cations in octahedral sites: A descriptor for oxygen electrocatalysis on transition-metal spinels. Advanced Materials. 2017;**29**:1606800

[9] Chen S, Huang D, Liu D, et al. Hollow and porous $NiCo_2O_4$ nanospheres for enhanced methanol oxidation reaction and oxygen reduction reaction by oxygen vacancies engineering. Applied Catalysis B: Environmental. 2021;**291**:120065

[10] Sahoo P, Tan JB, Zhang ZM, et al. Engineering the surface structure of binary/ternary ferrite nanoparticles as high-performance electrocatalysts for the oxygen evolution reaction. ChemCatChem. 2018;**10**:1075-1083

[11] Wang YC, Zhou T, Zheng GF. Reduced mesoporous Co_3O_4 nanowires as efficient water oxidation electrocatalysts and supercapacitor electrodes. Advanced Energy Materials. 2014;**4**:1400696

[12] Zhuang LZ, Ge L, Yang YS, et al. Ultrathin iron-cobalt oxide nanosheets with abundant oxygen vacancies for the oxygen evolution reaction. Advanced Materials. 2017;**29**:1606793-1606799

[13] Kai LY, Xiao S, Zhang L, Bin D, Shan SL, Jing QC, et al. A facile method for reduced $CoFe_2O_4$ nanosheets with rich oxygen vacancies for efficient oxygen evolution reaction. International Journal of Hydrogen Energy. 2017;**42**:24150-24158. DOI: 10.1016/j.ijhydene.2017.07.165

[14] Yue QD, Liu C, Wan YY, et al. Defect engineering of mesoporous nickel ferrite and its application for highly enhanced water oxidation catalysis. Journal of Catalysis. 2018;**358**:1-7

[15] Wang N, Zhao P, Zhang Q, et al. Monodisperse nickel/cobalt oxide composite hollow spheres with mesoporous shell for hybrid supercapacitor: A facile fabrication and excellent electrochemical performance. Composites Part B: Engineering. 2017;**13**:144-151

[16] Xu L, Jiang QQ, Xiao ZH, et al. Plasma-engraved Co_3O_4 nanosheets with oxygen vacancies and high surface area for the oxygen evolution reaction. Angewandte Chemie. 2016;**55**:5277-5281

[17] Kang WQ, Cong X, Yan Z, Rui ZL, Gu RS, Hui L, et al. Laser-induced oxygen vacancies in $FeCo_2O_4$ nanoparticles for boosting oxygen evolution and reduction. Chemical Communications. 2019;**55**:8579-8582

[18] Sun J, Guo N, Shao Z, Huang K, Li Y, He F, et al. A facile strategy to construct amorphous spinel-based electrocatalysts with massive oxygen vacancies using ionic liquid dopant. Advanced Energy Materials. 2018;**8**:1800980

[19] Wei X, Na Y, Xiao L, et al. 3D atomic-scale imaging of mixed Co-Fe spinel oxide nanoparticles during oxygen evolution reaction. Nature Communications. 2022;**13**:179

[20] Zeng YX, Lai ZZ, Lu XH, et al. Oxygen-vacancy and surface modulation of ultrathin nickel cobaltite nanosheets as a high-energy cathode for advanced Zn-ion batteries. Advanced Materials. 2018;**30**:1802396-1802403

[21] Zhang M, Sui X, Zhang X, Niu M, Li C, Wan H, et al. Multi-defects engineering of $NiCo_2O_4$ for catalytic propane oxidation. Applied Surface Science. 2022;**600**:154040

[22] Lu C, Deng R, Xu R, et al. Design of hybrid oxygen carriers with CeO_2 particles on $MnCo_2O_4$ microspheres for chemical looping combustion. Chemical Engineering Journal. 2021;**404**:126554

[23] Wang Y, Xue R, Zhao C, Liu F, Liu C, Han F. Effects of Ce in the catalytic combustion of toluene on $Cu_xCe_{1-x}Fe_2O_4$. Colloid. Surface. A. 2018;**540**:90-97

[24] Chuang W, Liang LG, Sheng You Q, Jian G, Yong Chao Z, Ke XW, et al. Modulating $CoFe_2O_4$ nanocube with oxygen vacancy and carbon wrapper towards enhanced electrocatalytic nitrogen reduction to ammonia. Applied Catalysis B: Environmental. 2021;**297**:120452-120462. DOI: 10.1016/j.apcatb.2021.120452

[25] Yu D, Jun Z, Wenqing Z, Jian Z, Xijie C, Fengchun Y, et al. Single-walled carbon nanotubes wrapped $CoFe_2O_4$ Nanorods with enriched oxygen vacancies for efficient overall water splitting. ACS Applied Energy and Materials. 2019;**2**:1026-1032

[26] Ming CZ, Xin S, Xin Z, Mang N, Cui QL, Hai QW, et al. Multi-defects engineering of $NiCo_2O_4$ for catalytic propane oxidation. Applied Surface Science. 2022;**600**:154040-154048. DOI: 10.1016/j.apsusc.2022.154040

[27] Jing W, Shi YF, Xin YL, Zhao DN, Zhi YL, Chun PB, et al. Rod-like nanostructured Cu-Co spinel with rich oxygen vacancies for efficient electrocatalytic dechlorination. ACS Applied Materials and Interfaces. 2023;**15**:12915-12923. DOI: 10.1021/acsami.2c19134

[28] Kumar M, Yun JH, Bhatt V, et al. Role of Ce^{3+} valence state and surface oxygen vacancies on enhanced electrochemical performance of single step solvothermally synthesized CeO_2 nanoparticles. Electrochimica Acta. 2018;**284**:709

[29] Ma L, Chen S, Pei Z, et al. Flexible waterproof rechargeable hybrid zinc batteries initiated by multifunctional oxygen vacancies-rich cobalt oxide. ACS Nano. 2018;**12**:8597-8605

[30] Bao J, Zhang X, Fan B, Zhang J, Zhou M, Yang W, et al. Ultrathin spinel-structured nanosheets rich in oxygen deficiencies for enhanced electrocatalytic water oxidation. Angewandte Chemie International Edition. 2015;**54**:7399-7404

[31] Shuo L, Ying NL, He SZ, Jun FN, Yoong KL, Duu JL, et al. Biochar loaded with $CoFe_2O_4$ enhances the formation of high-valent Fe(IV) and Co(IV) and oxygen vacancy in the peracetic acid activation system for enhanced antibiotic degradation. Bioresource Technology. 2023;**387**:129536-129546. DOI: 10.1016/j.biortech.2023.129536

[32] Gao P, Tian X, Nie Y, Yang C, Zhou Z, Wang Y. Promoted peroxymonosulfate activation into singlet oxygen over perovskite for ofloxacin degradation by controlling the oxygen defect concentration. Chemical Engineering Journal. 2019;**359**:828-839

[33] Hu Y, Lu W, Yan L, et al. Defect engineering of electrode materials towards superior reaction kinetics for high-performance supercapacitors. Journal of Materials Chemistry A. 2022;**10**:15267-15296

[34] Vazhayil A, Thomas N. Probing the electrocatalytic activity of hierarchically mesoporous M-Co_3O_4 (M= Ni, Zn, and Mn) with branched pattern for oxygen evolution reaction. Journal of Electroanalytical Chemistry. 2023;**934**:117298-117310

[35] Sheng JP, Feng G, Lin LL, De SY, Dong XJ, Tian RZ, et al. Necklace-like multishelled hollow spinel oxides with oxygen vacancies for efficient water electrolysis. Journal of American Chemical Society. 2018;**140**:13644-13653. DOI: 10.1021/jacs.8b05134

[36] Yu Z, Dan S, Yi XW, Xiao QL, Hong MS, Tong HC, et al. Facile electrochemically induced vacancy modulation of $NiCo_2O_4$ cathode toward high-performance aqueous Zn-based battery. Chemical Engineering Journal. 2023;**453**:139736-139745. DOI: 10.1016/j.cej.2022.139736

[37] Xiao LX, Xue QL, Long FM, Na NL, Shan Y, Hao L, et al. Construction of surface synergetic oxygen vacancies on $CuMn_2O_4$ spinel for enhancing NO reduction with CO. ACS Catalysis. 2024;**14**:3028-3040. DOI: 10.1021/acscatal.3c05337

[38] Fei X, Yi AW, Gui LX, Fei Y, Shang QZ, Cheng JS, et al. Fe–N–C boosts the stability of supported platinum nanoparticles for fuel cells. Journal of American Chemistry Society. 2022;**144**:20372-20384. DOI: 10.1021/jacs.2c08305

Chapter 8

Perspective Chapter: Modeling and Optimization Approaches of Physicomechanical Attributes for Biocomposite-Cultivating Pots

Manar E. Elashry, Elsayed G. Khater and Samir A. Ali

Abstract

This chapter explores the physicomechanical attributes of biocomposite materials used in cultivating pots, focusing on the modeling and optimization of filler compositions. It integrates analytical models to predict the behavior of biocomposites under varying conditions. The chapter also employs a hybrid TOPSIS-Taguchi method to optimize filler proportions, achieving an optimal balance between mechanical performance and sustainability. By investigating sustainable filler materials, the study advances the development of eco-friendly biocomposites, optimizing both functional properties and environmental impact. This approach bridges the gap between theoretical predictions and real-world applications, contributing to the innovative use of fillers and enhancing the durability and sustainability of biocomposite-cultivating pots.

Keywords: physical properties, mechanical properties, composite, cultivating pots, TOPSIS-Taguchi, fillers, fibers, theoretical analytical models

1. Introduction

In 2011, global bio-based polymer production was approximately 3.5 million tons, a stark contrast to the 235 million tons of traditional petrochemical-based polymers. By 2020, this figure was projected to rise to around 12 million tons annually, reflecting the increasing demand for sustainable materials [1]. Despite this growth, bio-based polymers still represent a fraction of the overall polymer market, underscoring the need for further innovations to mitigate the environmental consequences of petroleum-derived materials. Polylactic acid (PLA), produced from maize starch, exemplifies the environmental benefits of bio-based polymers, reducing oil consumption by 50% and greenhouse gas emissions by 60% compared to petrochemical-based alternatives. Fully bio-based and biodegradable polymers contribute to carbon neutrality by releasing sequestered carbon back into the atmosphere upon decomposition [2]. Drop-in bio-based polymers, such as bio-based polyethylene terephthalate, combine the recyclability of petrochemical-based polymers with the potential for renewable energy production *via* combustion. The advantages of bio-based resins over petrochemical resins include enhanced energy efficiency, improved safety during degradation, superior recyclability, and lower greenhouse gas emissions. However, bio-based resins typically incur a 10% higher production cost [3]. Green composites have gained prominence as environmentally sustainable alternatives in various applications, including biodegradable cultivating pots, which reduce plastic waste while promoting plant growth. The physical and mechanical properties of these pots, influenced by fillers such as sugarcane bagasse, compost, peat moss, vermiculite, and activated carbon, are crucial for optimizing their performance [4]. The mechanical attributes, such as tensile strength and elongation, enable the pots to endure mechanical stresses while maintaining structural integrity. Each filler type offers unique reinforcement mechanisms. Utilizing agricultural residues and by-products as fillers in composite materials not only addresses the shortage of wood resources but also holds promise for establishing a natural fiber industry. Blending these materials with polymers such as polypropylene, polyethylene, polyester, or PLA opens avenues for applications in the automotive, packaging, and construction sectors [5, 6]. In light of the growing demand for plastics and their adverse environmental impacts, developing eco-friendly alternatives such as biodegradable composites is essential. Numerous industrial by-products can be repurposed as fillers in green materials, offering both environmental and economic benefits. Composites, defined as combinations of two or more components bound in a matrix, often involve natural fibers due to their strong mechanical properties. These fibers are increasingly used as reinforcements in composite materials to meet the rising demand for sustainable alternatives to plastics. Sugarcane bagasse stands out as a particularly promising filler due to its favorable mechanical characteristics, low cost, and environmental sustainability [7]. Bagasse, the fibrous residue left after sugarcane juice extraction, is abundant and has been employed for various applications, including sustainable energy and environmental practices. Bagasse fibers are lightweight and possess good mechanical properties, such as high tensile strength and stiffness, making them suitable for reinforcing composite materials. They are biodegradable and eco-friendly, offering an effective substitute for synthetic fibers in composite applications [7]. Fillers such as sugarcane bagasse, compost, peat moss, and vermiculite enhance the physical and mechanical properties of biodegradable pots, contributing to water retention, nutrient delivery, and structural stability. Compost, often regarded as the cornerstone of sustainable agricultural systems, plays a critical role in conditioning the soil. With its higher organic matter content compared to synthetic fertilizers, compost enhances soil fertility, supports plant nutrition, and improves soil structure and water retention. The use of biodegradable pots, which can be planted directly into the soil with minimal waste, represents a further step toward sustainable horticultural practices. The selection of fillers, based on their physical attributes such as water absorption, density, and void space, determines the efficiency of water management and structural integrity. These attributes significantly affect root aeration and moisture retention, which are vital for plant growth. The mechanical properties, including tensile strength and elasticity, are equally important for ensuring the pots' durability and ability to withstand mechanical stresses. Novel biodegradable pots have been developed using fillers such as waste paper, peat, grape residue, tomato fibers, and hemp, each offering distinct reinforcement benefits [8]. The optimization of material formulations is key to enhancing the physicomechanical attributes of these biocomposite pots. The Taguchi method, integrated with the Technique for Order of Preference by Similarity to Ideal Solution (TOPSIS), has been applied to determine the optimal filler mixtures. This approach not only minimizes experimental costs but also reduces variability and improves material performance under specific conditions. The use of bio-based biodegradable polymers, which are derived from renewable resources and have a lower environmental footprint, is becoming increasingly widespread across various fields, including medicine and engineering. PLA, in particular, has emerged as a promising biopolymer due to its renewable origin from corn and sugar beets, making it an attractive alternative to petroleum-based polymers [9]. In composite materials, the matrix plays a crucial role by binding the reinforcing fibers, transferring loads uniformly, and resisting crack propagation. Extensive research has focused on identifying alternatives to petroleum-based matrices, driven by both environmental concerns and the finite nature of fossil fuel resources [4, 10, 11]. This chapter encompasses physical and mechanical analytical models, optimization techniques, multi-criteria decision-making (MCDM) methods, and validation processes, providing a comprehensive analysis of biocomposite filler selection and performance enhancement as pictorially presented in the mind map (**Figure 1**).

2. Physical properties modeling

2.1 Analytical models of moisture diffusion kinetics

Moisture diffusion is a key aspect in the study of biocomposites, particularly for applications such as biodegradable cultivating pots, biofertilization, and water preservation systems. Natural fibers, being hydrophilic, pose challenges in these areas, as moisture absorption and retention must be carefully controlled. Therefore, various predictive models have been developed to simulate experimental moisture diffusion curves.

2.1.1 Diffusion models

The moisture diffusion process in polymers is typically described by classical diffusion models, where Fick's law is the most widely used for isotropic diffusion. For more complex cases, like anisotropic or anomalous diffusion, dual-stage Fick's law and the Langmuir model are also applied [12, 13].

2.1.1.1 Governing equation of moisture diffusion

The governing equation for transient moisture diffusion in one dimension is expressed as:

$$\frac{\partial M}{\partial t} = D\frac{\partial^2 M}{\partial x^2} \tag{1}$$

where M represents the moisture content, t is the diffusion time, D is the moisture diffusion coefficient, and x is the coordinate along the diffusion direction.

This equation assumes moisture diffusion occurs in one direction, simplifying the analysis to a single dimension [14]. Fick's first law describes steady-state diffusion, where the material flux J across a unit area is proportional to the concentration gradient:

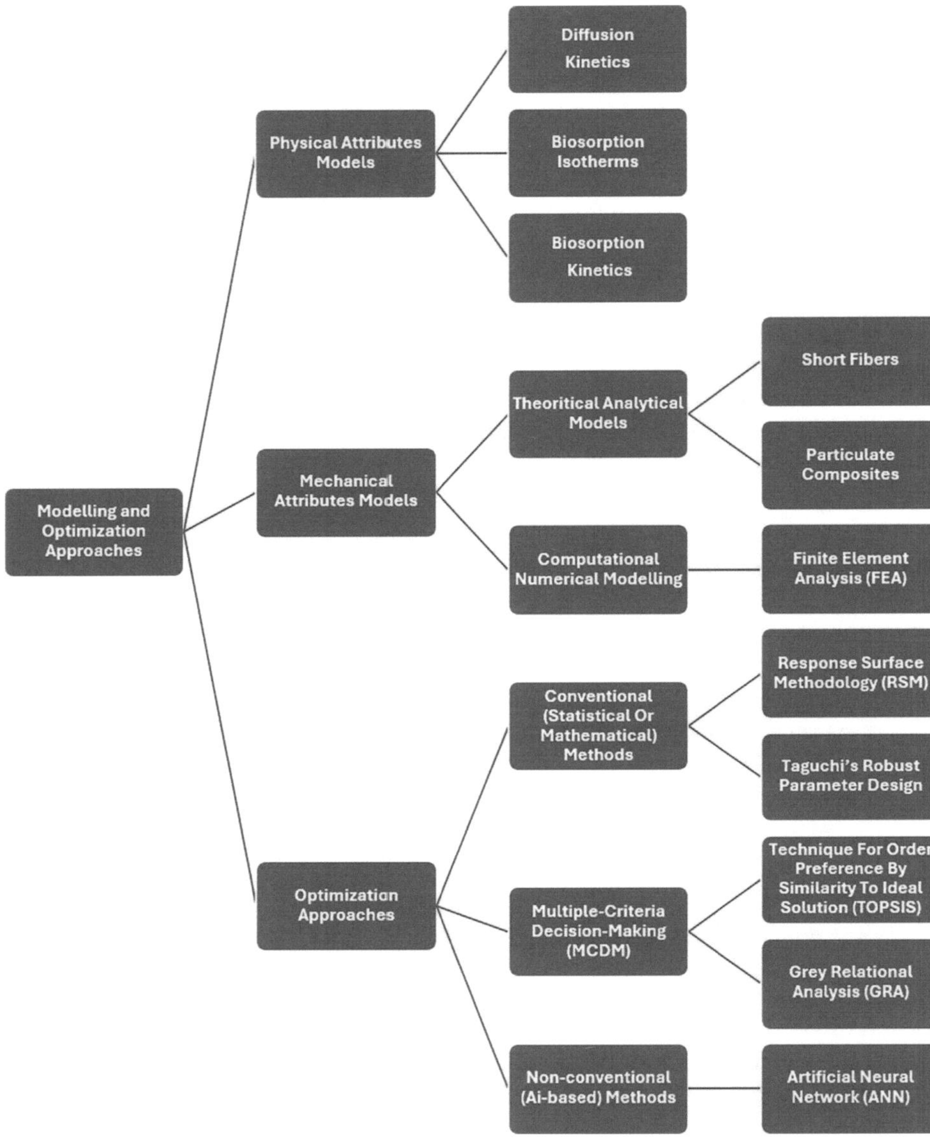

Figure 1.
Mind map deployed in modeling and optimization approaches of physicomechanical attributes for biocomposite cultivating pots.

$$J = -D\frac{\mathrm{d}C}{\mathrm{d}x} \tag{2}$$

where J is the diffusion flux, D is the diffusion coefficient, C is the concentration, and x is the diffusion path length. For unsteady-state diffusion, Fick's second law is applied, which incorporates time as a variable:

$$\frac{\partial c}{\partial t} = D\frac{\partial^2 c}{\partial x^2} \tag{3}$$

This equation describes how concentration changes over time at any point in the diffusion path.

In cylindrical coordinates, moisture diffusion for a radius r is governed by:

$$\frac{\partial c}{\partial t} = D\left(\frac{1}{r}\frac{\partial C}{\partial r} + \frac{\partial^2 c}{\partial r^2}\right) \tag{4}$$

This expression is particularly useful for studying moisture diffusion in cylindrical or spherical biocomposite structures [15]. Under certain boundary and initial conditions, analytical solutions to Fick's law can be derived. For example, in the case of a plane sheet with thickness h, where the initial moisture concentration is C_1 and surface moisture concentration is C_0, the solution is given by Crank [15]:

$$\frac{M_t}{M_\infty} = y(t) = 1 - \frac{8}{\pi^2}\sum\nolimits_{n=1}^{\infty}\frac{1}{(2n+1)^2}\,exp\left[\frac{-(2n+1)^2\pi^2}{h^2}Dt\right] \tag{5}$$

where M_t is the total amount of moisture absorbed at time t and M_∞ is the equilibrium moisture content. For simplification:

$$k = D\left(\frac{\pi}{h}\right)^2 \tag{6}$$

$$\frac{M_t}{M_\infty} = y(t) = 1 - \frac{8}{\pi^2}\sum\nolimits_{n=1}^{\infty}\frac{1}{(2n+1)^2}\,exp\left[-(2n+1)^2kt\right] \tag{7}$$

This solution helps estimate moisture uptake over time, useful for predicting swelling or deswelling behavior in biocomposites.

For a cylindrical body of radius r = a, the moisture uptake follows:

$$\frac{M_t}{M_\infty} = y(t) = 1 - \sum\nolimits_{n=1}^{\infty}\frac{1}{a^2\alpha^2}\,exp\left(-D\,\alpha_n^2 t\right) \tag{8}$$

where α_n are the roots of the Bessel function of the first kind. The water vapor transport mechanism can be simplified using the following relation for moisture absorption or release, described by three kinetic equations: power, exponential, and sigmoidal [16]:

$$\frac{M_t}{M_\infty} = kt^n \tag{9}$$

$$\frac{M_t}{M_\infty} = 1 - \;exp\,(-k\cdot t) \tag{10}$$

$$\frac{M_t}{M_\infty} = \frac{a-b}{1+\,exp\left(\frac{t-d}{c}\right)} + b \tag{11}$$

where k is a constant that depends on the system characteristics, and n indicates the transport mechanism. a, b, c, d are the sigmoidal equation constants. For Fickian diffusion, n = 0.5, while for non-Fickian or anomalous diffusion, 0.5 < n < 1 [17].

The diffusion of moisture, which may carry minerals, can be applied to slow-release fertilizers that consist of chemically or biologically decomposed materials characterized by high molecular weight, complex structures, and low solubility.

In contrast, controlled-release fertilizers regulate the release of nutrients through a polymer layer or membrane. Since nutrient uptake by plants follows a sigmoidal pattern, the design of biocomposite materials tailored for plant growth must account for both the sigmoidal diffusion of moisture and the decomposition of these materials [16].

$since, k = D_\gamma \left(\frac{\pi}{h}\right)^2$ the equation can be approximated by:

$$\frac{M_t}{M_\infty} = \frac{4}{h}\sqrt{\frac{Dt}{\pi}} \tag{12}$$

And for spherical device: [18].

$$\frac{M_t}{M_\infty} = 4\left[\frac{Dt}{\pi r^2}\right] \tag{13}$$

The following equation for moisture uptake beyond $M_t/M_\infty \geq 0.6$ [19].

$$\frac{M_t}{M_\infty} = 1 - exp\left[-7.3\left(\frac{Dt}{h^2}\right)^{0.75}\right] \tag{14}$$

The diffusion coefficient D is derived from the slope of the linear portion of the moisture absorption curve [20]. Moisture absorption in composite materials depends on several factors, including temperature, fiber orientation, void content (higher void content increases water absorption), fiber volume fraction, and interfacial properties [21]. For instance, Harper et al. [22] demonstrated that composites with a higher void content exhibit higher equilibrium moisture levels and may show non-Fickian behavior. Crank [15] summarized various diffusion behaviors in polymers and composites (**Figure 2**):

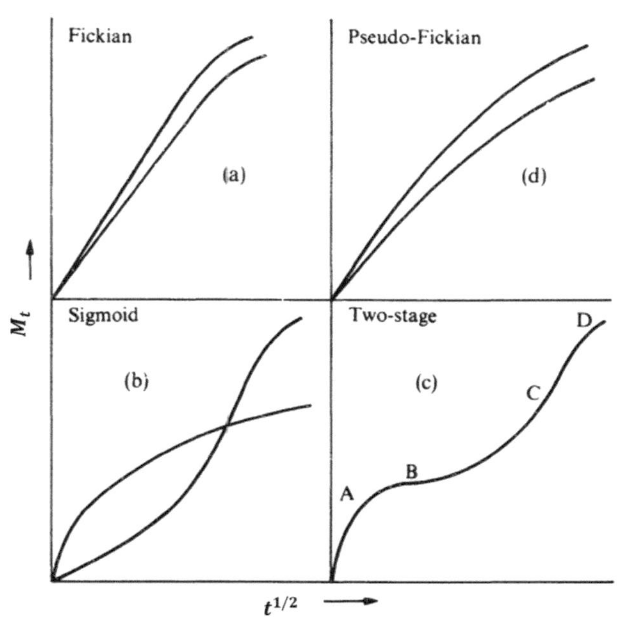

Figure 2.
Summarized various diffusion behaviors in polymers and composites [15].

1. Fickian (Case I): Controlled by the diffusion process.
2. Pseudo-Fickian: Shows short initial linear absorption.
3. Sigmoid: Curves have an inflection point with faster initial desorption.
4. Two-stage sorption: Initial rapid uptake followed by slow diffusion to equilibrium [23].

Understanding these behaviors helps optimize biocomposites for different applications, such as cultivating pots, hydrogels for irrigation, or biofertilization.

2.1.1.2 Dual-Stage Fick's Law

For systems exhibiting anomalous moisture uptake, Loh et al. [24] developed a dual-stage model that accounts for two Fickian diffusion processes occurring simultaneously. Each process has its own diffusion coefficient (D_1 and D_2) and saturation levels ($M_{\infty 1}$ and $M_{\infty 2}$). The total moisture absorption capacity of the specimen at steady state is the sum of the saturation levels from both diffusion stages [12]. The equation for total moisture uptake M_t is given by:

$$\begin{aligned} M_t = M_{\infty 1}\left(1-\frac{8}{\pi^2}\sum\nolimits_{n=1}^{\infty}\frac{1}{(2n+1)^2}\,exp\left[\frac{(2n+1)^2\pi^2}{h^2}D_1t\right]\right) \\ +M_{\infty 2}\left(1-\frac{8}{\pi^2}\sum\nolimits_{n=1}^{\infty}\frac{1}{(2n+1)^2}\,exp\left[\frac{(2n+1)^2\pi^2}{h^2}D_2t\right]\right) \end{aligned} \quad (15)$$

For cylindrical or radial samples, this equation is modified to reflect the radial geometry, but the overall approach remains the same, with the total saturation being the sum of the two independent saturation levels:

$$M_\infty = M_{\infty 1} + M_{\infty 2} \quad (16)$$

2.1.1.3 Langmuir Law (two-phase model of Carter and Kibler)

The Langmuir law or two-phase model, proposed by Carter and Kibler [25], builds on the Langmuir theory of adsorption to describe moisture absorption in polymers. In this model, absorbed moisture is divided into two phases: mobile molecules and bound molecules. Mobile phase molecules diffuse with a diffusion coefficient $D\gamma$, while bound molecules adhere to specific sites (such as polymer voids, hydrogen bonds, or heterogeneous morphologies). These bound molecules can return to the mobile phase with a probability per unit time γ, while molecules in the mobile phase can become bound with a probability β. The coupled differential equations governing this process are as follows:

For short fibers:

$$D\gamma\frac{\partial^2 n}{\partial x^2} = \frac{\partial n}{\partial t} + \frac{\partial N}{\partial t} \quad (17)$$

For spherical fillers:

$$\frac{\partial n}{\partial t} = D\left(\frac{\partial^2 n}{\partial r^2} + \frac{1}{r}\frac{\partial n}{\partial r}\right) + \frac{\partial N}{\partial t} \quad (18)$$

$$\frac{\partial N}{\partial t} = \gamma n - \beta N \tag{19}$$

where n is the number of mobile molecules per unit volume, N is the number of bound molecules per unit volume, Dγ is the diffusion coefficient for mobile molecules, and γ and β are the probabilities of binding and release, respectively. These equations can be numerically solved by finite different methods. The optimal values for Dγ, γ, and β are determined by minimizing the square difference between experimental and theoretical results using a least squares approach. The general solution for the moisture uptake, depending on the relation between γ, β, and t, when $2\gamma, 2\beta \leq k$ can be expressed as follows:

$$\frac{M_t}{M_\infty} = \left(\frac{\beta}{\gamma+\beta}\right)\mathrm{e}^{-\gamma t}y(t) + \left(\frac{\beta}{\gamma+\beta}\right)\left(\mathrm{e}^{-\beta t} - \mathrm{e}^{-\gamma t}\right) + \left(1 - \mathrm{e}^{-\beta t}\right) \tag{20}$$

For $k, t \leq 0.7$

$$\frac{M_t}{M_\infty} = \frac{4}{\pi^{3/2}}\left(\frac{\beta}{\gamma+\beta}\right)\sqrt{kt} \tag{21}$$

For $k, t \geq 1$

$$\frac{M_t}{M_\infty} = 1 - \left(\frac{\gamma}{\gamma+\beta}\right)\left(e^{-\beta t}\right) \tag{22}$$

This model is especially useful for describing deviations from Fickian behavior, as it considers how polymer structures change during sorption and desorption. The Langmuir-type diffusion model extends the basic Fickian model by classifying moisture molecules into mobile and bound phases. Importantly, the Fickian model can be viewed as a special case of Langmuir diffusion when only the mobile phase is considered [13]. In conclusion, dual-stage Fickian models and Langmuir-based two-phase models provide advanced frameworks for describing complex moisture absorption processes in biocomposites, particularly when non-Fickian behaviors are present. Both models highlight the intricate balance between diffusion and binding mechanisms that govern the overall moisture uptake.

The porosity characteristics of diverse fillers sourced from discrete origins, encompassing sugarcane bagasse, compost, peat moss, activated carbon, and vermiculite. The porosity values were quantified at 1.9%, 1%, 1.45%, 2.1%, and 2.8% for fillers derived from sugarcane bagasse, compost, peat moss, activated carbon, and vermiculite, respectively. It is noteworthy that vermiculite-based fillers exhibit the highest porosity at 2.8% due to their lightweight and highly porous structure. In contrast, fillers derived from compost demonstrate the lowest porosity at 1%, attributed to their particulate nature. Thus, the voids within each filler composite were minimal, ranging from 1 to 2.8%. Consequently, the disparities between the theoretical and actual density were negligible. According to Werber [26], the higher void content, the greater susceptibility to water dispersion, consequently affecting the strength properties. The void content may range between less than 1% for the good composite and 5% for the poor one [26]. The order of water absorption across different fillers was peat moss > activated carbon > vermiculite > sugarcane bagasse > compost, and water absorption values were 18.23, 15.56, 12.98, 10.73, and 8.74%, respectively. The theoretical and actual density values varied slightly, with sugarcane bagasse showing the highest densities (theoretical: 5.15 g/cm^3, actual: 5.05 g/cm^3) and compost showing the lowest (theoretical: 1.74 g/cm^3, actual: 1.72 g/cm^3). The regression analysis yielded the following equations to predict water absorption, porosity (voids), and theoretical and actual density based on filler types [27]:

$$\text{WA} = 13.2620 + 4.96800\ (\text{AC}) - 4.53200\ (\text{C}) - 0.28200\ (\text{P}) + 2.38800\ (\text{S}) \\ -2.54200(\text{V})\ (\text{R}^2 = 100\%) \tag{23}$$

$$\text{Porosity (Voids)} = 1.86000 - 0.40000\ (\text{AC}) - 0.85000\ (\text{C}) + 0.95000\ (\text{P}) \\ +0.25000\ (\text{S}) + 0.05000\ (\text{V})\ \left(\text{R}^2 = 99.98\%\right) \tag{24}$$

$$\text{Theoretical density} = 3.05978 - 1.13312\ (\text{AC}) - 1.31884\ (\text{C}) - 0.73978\ (\text{P}) \\ +1.09924\ (\text{S}) + 2.09250\ (\text{V})\ \left(\text{R}^2 = 100\%\right) \tag{25}$$

$$\text{Actual density} = 3.00319 - 1.08789\ (\text{AC}) - 1.26965\ (\text{C}) - 0.74982\ (\text{P}) \\ +1.06949\ (\text{S}) + 2.03787\ (\text{V})\ \left(\text{R}^2 = 100\%\right) \tag{26}$$

In a study by Pratibha et al. [28], biocomposites made from paddy straw crosslinked with corn starch, boric acid, and glycerol exhibited rapid water uptake in the first hour, following Fickian diffusion (**Figure 3**). Similar patterns were observed in pots made from cattle manure and sawdust, crosslinked with cornstarch and sheep's wool (**Figure 4**) [29]. Agricultural residue-based pots, containing hydrophilic cellulose, hemicellulose, and lignin, crosslinked with hydrophobic polymers, typically follow the Fickian model for water absorption.

Porosity in biocomposites (BCs) is primarily influenced by air-filled cavities, with an inverse relationship between porosity and density. BCs with higher porosity tend to exhibit greater water uptake, aligning with their density and moisture absorption capacities (**Figure 5**) [28].

2.2 Isotherm and kinetic analytical models of biosorbent

Biosorption is a critical process for removing pollutants or contaminants from various environments [30], with applications extending to controlled ion release, such

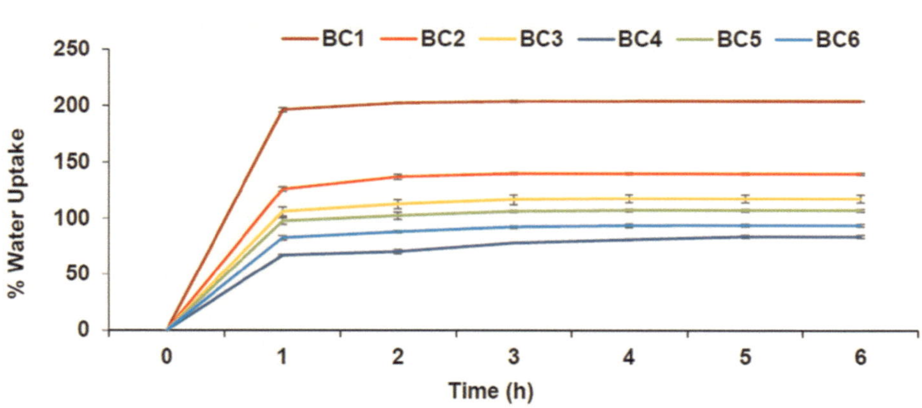

Figure 3.
Fickian's diffusion of the biocomposites made from paddy straw [28].

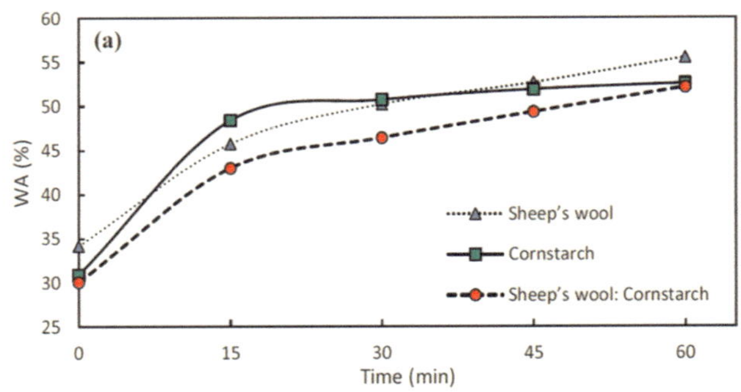

Figure 4.
Water absorption (WA) pattern of the biocomposites made from cattle manure and sawdust [29].

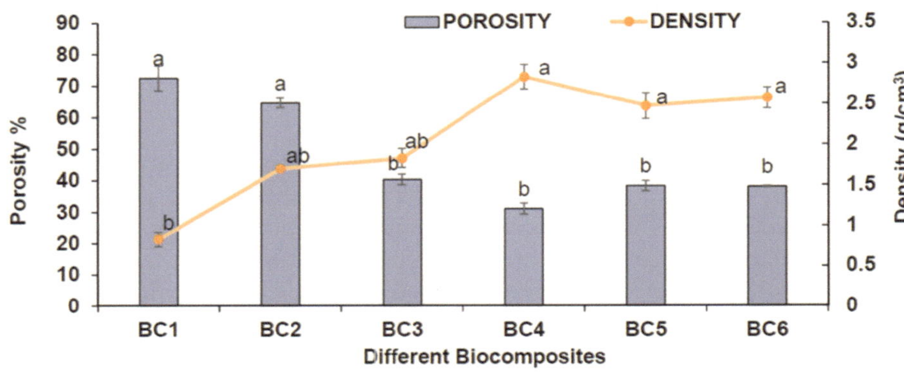

Figure 5.
Porosity and density of the biocomposites made from paddy straw [28].

as the use of keratin nursery pots for controlled copper ion release [31]. The analysis of biosorption behaviors primarily relies on both isotherm and kinetic models to describe the interaction between the biosorbent and the contaminant.

2.2.1 Biosorption isotherms

Biosorption isotherms describe the distribution of biosorption molecules and the interaction between biosorbent and adsorbate in both liquid and solid phases when equilibrium is achieved under constant temperature and pressure. Commonly used models include Langmuir, Freundlich, Temkin, and Dubinin-Radushkevich (D-R) isotherms [30, 32].

2.2.1.1 Langmuir isotherm model

Assumes that the biosorbent surface is homogeneous, with a constant number of active sites. The maximum biosorption corresponds to the total number of active centers on the surface [33]. The Langmuir equation is expressed as follows:

$$Q = \frac{X_L C_e}{1 + k_L C_e} \quad (27)$$

where Q is the amount of biosorbed material (mg g^{-1}); X_L is the maximum biosorption capacity (mg g^{-1}); C_e is the equilibrium concentration; and k_L is the Langmuir isotherm constant (L mg^{-1}).

2.2.1.2 Freundlich isotherm model

Assumes a heterogeneous and multilayered surface, with varying binding energies across different regions of the biosorbent surface. The Freundlich equation is as follows:

$$Q = X_f C_e^{\beta} \tag{28}$$

where X_f is the Freundlich constant, and β is biosorbent surface heterogeneity.

2.2.1.3 Temkin isotherm model

Accounts for the effects of indirect adsorbate-adsorbate interactions, with adsorption energy linearly decreasing as the coverage increases.

$$Q = \left(\frac{RT}{b}\right) ln\, K_t C_e \tag{29}$$

B is the Temkin constant related to adsorption heat (J mol^{-1}), K_t is the Temkin constant (L mg^{-1}), R is the ideal gas constant (8.314 $Jmol^{-1}\,K^{-1}$), and T is the absolute temperature.

2.2.1.4 Dubinin-Radushkevich (D-R) isotherm model

This model is used to explain biosorption mechanisms based on surface porosity and energetics. It provides insight into whether adsorption is physical or chemical in nature based on the mean free energy E_{DR} of biosorption:

$$Q = X_{DR} e^{-\left(K_{DR}\varepsilon^2\right)} \tag{30}$$

where X_{DR} is a measure of biosorption capacity (mg g^{-1}); K_{DR} is the activity coefficient ($-K_{DR} \times 10^9/\text{mol}^2$); and ε is the Polanyi potential (KJ^{-2}). The mean energy E_{DR} and ε are calculated as follows:

$$E_{DR} = (2K_{DR})^{-0.5} \tag{31}$$

$$\varepsilon = RTln\left(1 + \frac{1}{C_e}\right) \tag{32}$$

If $8 < E_{DR} < 16$, the adsorption is chemical, while $E_{DR} < 8$ kJ mol^{-1} indicates physical adsorption [34]. These isotherms are essential for analyzing the performance of biosorbents and determining their adsorption capacity and mechanism for contaminant removal.

2.2.2 Biosorption kinetics

Kinetic models provide a deeper understanding of the adsorption rate, which is key to modeling the adsorption process and determining the interactions between the adsorbate and biosorbent. Several models are commonly employed, including pseudo-first-order (PFO), pseudo-second-order (PSO), intraparticle diffusion (IPD), and Elovich models [30, 35].

2.2.2.1 Pseudo-first-order kinetic (PFO) model

This model applies to reversible adsorption reactions that eventually reach equilibrium between the liquid and solid phases at $H_1 = K_1$. The PFO equation is given by the following:

$$Q_t = Q_e\left(1 - e^{K_1 t}\right) \tag{33}$$

where Q_t is the biosorbed amount at the time t, Q_e is the biosorbed amount at equilibrium, K_1 is the rate constant, and $H_1$is the initial rate constant.

2.2.2.2 Pseudo-second-order kinetic (PSO) model

This model is applied to chemical adsorption processes where covalent forces are generated by the sharing or exchange of electrons between the biosorbent and adsorbate. The PSO equation is at $H_2 = K_2 Q_t^2$:

$$\frac{t}{Q_t} = \frac{1}{K_2 Q_e^2} + \frac{t}{Q_t} \tag{34}$$

where K_2 is the rate constant, and H_2 is the initial rate constant.

2.2.2.3 Intraparticle diffusion (IPD) kinetic model

This model accounts for diffusion within the pores of the biosorbent and is expressed as:

$$Q_t = K_i t^{-0.5} \tag{35}$$

where K_i is the IPD rate constant.

2.2.2.4 Elovich kinetic model

Commonly used to describe adsorption on heterogeneous surfaces, the Elovich equation is as follows:

$$Q_t = \frac{1}{\beta} ln\,(\alpha\beta) + \frac{1}{\beta} ln\,(t) \tag{36}$$

where α (mg g^{-1} min^{-1}) and β (g mg^{-1}) are the Elovich constants.

3. Mechanical analytical models for predicting elastic properties of composites

The mechanical behavior of composite materials, especially in predicting their elastic properties, can be determined using various analytical models. Among the earliest and most fundamental methods is the Rule of Mixtures (ROM), which has evolved to incorporate various other models, including those proposed by Voigt [36], Reuß [37], and modifications introduced by Hill [38]. These models are crucial in determining how the elastic modulus of a composite material is influenced by its constituents, namely the fiber and the matrix.

3.1 Short fiber modeling

The following models are often employed to predict the mechanical properties of short fiber composites: (ROM), inverse rule of mixtures (IROM), Guth Model, Halpin-Tsai equation, modified Halpin-Tsai equation, Hirsch model, Bowyer and Badar's model, Nairn's generalized shear-lag analysis, Mendel's stress transfer model for single fiber and platelet composites. Voigt's parallel model (1889) predicts the composite modulus (E_f by assuming that the strain is constant across the composite). The equation for the modulus is:

$$E_1 = E_f V_f + E_m V_m \tag{37}$$

where E_f and E_m are the moduli of the fiber and matrix, respectively, and V_f and V_m are their corresponding volume fractions [36].

Reuss's serial model (1929) assumes constant stress throughout the fiber and matrix, typically applied when the load is transverse to the fibers:

$$E_2 = \frac{E_f E_m}{E_f V_m + E_m V_f} \tag{38}$$

This model predicts lower bounds for the elastic modulus [38].

For composite laminates made from different fabric structures, such as bidirectional symmetric woven fabrics or random mats, the efficiency of fiber reinforcement can be accounted for using a factor α in the Voigt model:

$$E_1 = \sum_i \alpha_i v_i E_i \tag{39}$$

The value of α depends on the fiber orientation: 1 for unidirectional fibers, 0.5 for bidirectional fabrics, and 0.375 or 0.2 for random in-plane or random in-space layers, respectively [39].

The Halpin-Tsai model is an empirical micromechanics model that accounts for discontinuous, randomly oriented fibers and short-fiber composites. The tensile modulus is given by the following:

$$E_c = E_m \left(\frac{1 + \zeta \eta V_f}{1 - \eta V_f} \right) \tag{40}$$

$$T_c = T_m \left(\frac{1 + \zeta \eta V_f}{1 - \eta V_f} \right) \tag{41}$$

where T_c and T_m are the tensile strength of the composite and matrix, respectively, and η determined by the fiber aspect ratio and orientation, given by the following:

$$\eta = \frac{(E_f/E_m) - 1}{(E_f/E_m) + \zeta} \tag{42}$$

$$\eta = \frac{(T_f/T_m) - 1}{(T_f/T_m) + \zeta} \tag{43}$$

The model is flexible enough to be applied to composites with different fiber shapes, including ribbons and particulates, as long as the appropriate *Empirical factor* shape factor dependent on fiber geometry and packing ζ value is used such that [40],

$$\zeta = 2\left(\frac{L}{T}\right) \text{ (for rectangular shape)}; \zeta = 2\left(\frac{L}{D}\right) \text{ (for circular fibers)}. \tag{44}$$

where L is the fiber length, D is the fiber diameter, and T is the thickness of the rectangular fiber cross section.

The modified Halpin-Tsai equation introduces an additional correction factor (ψ) to improve accuracy for certain composite systems by adjusting the impact of the fiber volume fraction on the overall modulus and strength:

$$E_c = E_m\left(\frac{1 + \zeta\eta V_f}{1 - \eta\psi V_f}\right) \tag{45}$$

$$T_c = T_m\left(\frac{1 + \zeta\eta V_f}{1 - \eta\psi V_f}\right) \tag{46}$$

Hirsch's model combines elements from both the Voigt (parallel) and Reuss (serial) models and incorporates an interfacial adhesion parameter x, which accounts for the fiber-matrix interaction:

$$E_c = x\left(E_f V_f + E_m V_m\right) + (1 - x)\frac{E_f E_m}{E_f V_m + E_m V_f} \tag{47}$$

The value of x reflects the bonding quality between the fiber and matrix, with x = 0.5 for medium adhesion and x = 1 for poor bonding [41, 42].

Short fiber composites, extensively used in automotive and consumer goods due to their low cost and ease of manufacturing, can have their mechanical properties predicted by the same micromechanical models applied to continuous fiber composites, with modifications. For instance, Rao et al. demonstrated that micromechanical models, such as the Halpin-Tsai and shear-lag models, accurately predicted the tensile modulus of short sisal fiber composites, particularly at lower fiber volume fractions $\left(V_f = 0.13\right)$ [42]. However, at higher fiber volume fractions ($V_f = 0.23$), discrepancies between experimental and predicted values were observed, largely due to the impact of interfacial bonding. Cox's shear-lag theory was introduced to account for fiber length and radius, with a length-dependent efficiency factor (ηL) to reflect the influence of fiber length on composite strength. For sisal short fiber composites, the model required the inclusion of shear modulus and interfacial adhesion parameters to improve the prediction accuracy [43, 44].

Cox's shear-lag model:

$$E_c = E_f \left(1 \quad \frac{-\tanh\left(\frac{\eta L}{2}\right)}{\left(\frac{\eta L}{2}\right)} \right) V_f + E_m V_m \tag{48}$$

$$T_c = T_f \left(1 \quad \frac{-\tanh\left(\frac{\eta L}{2}\right)}{\left(\frac{\eta L}{2}\right)} \right) V_f + E_m V_m \tag{49}$$

Where η is the shear-lag parameter, calculated as follows:

$$\eta = \frac{1}{r}\left(\frac{2E_m}{E_f(1+v_m)\ln\left(\frac{P_f}{V_f}\right)} \right)^{\frac{1}{2}} \tag{50}$$

r is the fiber radius, v_m is the Poisson's ratio of the matrix, and P_f is the packing geometry factor, which is π for square packing and 2 π/$\sqrt{3}$ for hexagonal packing.

Nairn's modification of the shear-lag model improves its applicability by accounting for interfacial adhesion and shear modulus variations between fibers and matrix:

Nairn modified shear-lag model:

$$\eta = \left[\frac{2}{r^2 E_m E_f} \left[\frac{E_f V_f + E_m V_m}{\frac{V_m}{4G_f} + \frac{1}{2G_m}\left[\frac{1}{V_m} \ln\left(\frac{1}{V_f+\chi}\right) - 1 - \frac{V_m}{2} \right] + \frac{1}{rD_s}} \right] \right]^{\frac{1}{2}} \tag{51}$$

χ is a correction factor for fiber volume, G_m and G_f are the shear moduli of the matrix and fiber, and D_s is the adhesion parameter.

For predicting stress transfer in composites, Mendels et al. [45] proposed a model for both single-fiber and platelet-reinforced thermoplastics:

For single fiber:

$$\eta = \left[k_C \left(\frac{r^2 E_f (r_1^2 - r^2) E_m}{r E_f (1 + V_m)} \right) \right]^{\frac{1}{2}} \tag{52}$$

For platelet-reinforced thermoplastic:

$$\eta = \left[k_P \left(\frac{T_A E_f + (T_1 - T_A) E_m}{T_A E_f (1 + V_m)} \right) \right]^{\frac{1}{2}} \tag{53}$$

The Bowyer–Bader model is a well-known approach for predicting the tensile strength and modulus of short fiber-reinforced thermoplastic composites. The model is expressed as follows:

$$E_C = a_v E_f V_f + E_m V_m \tag{54}$$

$$T_C = a_v T_f V_f + T_m V_m \tag{55}$$

where a_v is the reinforcement factor, representing the product of fiber orientation and fiber length, with an overall value of 0.5 for natural fiber composites [41, 46].

In modified Bowyer–Bader model, tensile strength is further divided into contributions from subcritical and supercritical fibers [46], given by the following:

$$E_C = K_1K_2E_fV_f + E_mV_m \quad (56)$$

$$T_C = K_1K_2T_fV_f + T_mV_m \quad (57)$$

Where K_1 is the fiber orientation factor, and K_2 is the fiber length factor.
For fibers with $l > l_1$.

$$K_2 = l - l_C/2l \quad (58)$$

For fibers with $l < l_1$.

$$K_2 = l/2l_C \quad (59)$$

where l is the fiber length and l_C is the critical fiber length.

Variability in physical properties such as density and moisture content significantly impacts the mechanical performance of natural fiber composites. To address these challenges, Facca et al. [47] introduced modifications to micromechanics models, accounting for fiber densification and changes in moisture content. In their study of composites made from hemp, hardwood, and rice hull fibers with high-density polyethylene (HDPE), the corrected fiber modulus (E_F) was calculated as follows:

$$E_F = E_{Mc}\frac{\rho_{cw}}{\rho_{MC}} \quad (60)$$

where ρ_{MC} is the initial fiber density, ρ_{cw} is the fiber cell wall density, and E_{Mc} is the fiber modulus at the moisture content of interest [47].

3.2 Particulate composites modeling

In addition to long and continuous fibers, reinforcements of polymeric composites can also take spherical, needle-like, or square shapes. These forms, with the aspect ratios close to one and sub-micrometer sizes, are termed particulate reinforcements, and the composite systems containing such reinforcements are referred to as particulate composites. Initially, the primary goal of adding particulates to polymeric composites was to enhance their brittleness. However, advances in composite science and technology have broadened the role of particulates, which now offer functional improvements. For instance, magnesium hydroxide particles are added to composites to impart fire resistance, and bentonite clay is used to enhance mechanical properties. Other advantages of particulate composites include manufacturing flexibility, high creep resistance, and high toughness. Several models have been proposed to predict the mechanical properties, such as elastic modulus, strength, hardness, and modulus using nanoindentation, for particulate composites. Below are key models and equations used to predict the elastic modulus. Numerous theoretical models exist for processes such as compression molding and injection molding of polymer composites. These models predict the modulus of particulate-reinforced composites. Efficient predictions depend on obtaining accurate input parameters, such as maximum packing fraction (Φ_m), particulate aspect ratio (α), and Einstein's coefficient

(KE). Common semi-empirical equations used for modulus prediction include the Guth-Smallwood model, the Budiansky model, and the Ponte-Castaneda model [48].

Einstein's model for Young's modulus: In 1905, Einstein developed a model to predict the Young's modulus of particulate composites, assuming the particulates are rigid [49]. The equation is as follows:

$$E_C = E_m(1 + 2.5V_P) \tag{61}$$

where E_C is the Young's modulus of the composite, E_m is the Young's modulus of the matrix, and V_P is the volume fraction of particulates. The model assumes uniform particulate distribution and perfect adhesion between particulates and the matrix, providing a linear relationship between modulus and particulate volume fraction. However, it is valid only for low particulate loading.

Ponte-Castaneda model: Another model by Ponte-Castaneda builds on Einstein's work by predicting a slightly higher modulus for composites with rigid particles in a neo-Hookean matrix:

$$E_C = E_m(1 + 3V_P) \tag{62}$$

The Guth-Smallwood equation modifies Einstein's model by including particulate interactions (second power term), providing the equation [50]:

$$E_C = E_m\left(1 + 2.5V_P + 14.1V_P^2\right) \tag{63}$$

For tensile strength:

$$T_C = T_m\left(1 - V_P^{2/3}\right) \tag{64}$$

This model is widely applicable for elastic modulus predictions in filled composites, considering matrix-filler adhesion [48]. It yielded the highest elastic modulus values for sugarcane bagasse, compost, peat moss, activated carbon, and vermiculite-based pots (2.67, 5.94, 4.08, 4.94, and 2.31 MPa, respectively) [27].

Cohan [51] introduced a particle shape factor (S) for non-spherical particles by modified Guth equation, given by the following:

$$E_C = E_m\left(1 + 0.675SV_P + 1.62S^2V_P^2\right) \tag{65}$$

Modified Kerner equation [52] predicts the modulus of composites with spherical particulates:

$$E_C = E_m\left[1 + \frac{V_P}{V_m}\frac{15(1 - \sigma_m)}{(8 - 10\sigma_m)}\right] \tag{66}$$

where V_m is the matrix volume fraction, and σ_m is the Poisson's ratio of the matrix.

Halpin and Tsai Halpin and Tsai developed a model to predict the modulus of particulate composites:

$$E_C = E_m\left(\frac{1 + ABV_P}{1 - BV_P}\right) \tag{67}$$

Nielsen [53] generalized the Kerner and Halpin-Tsai models, incorporating a maximum packing fraction and filler geometry factors:

$$E_C = E_m \left(\frac{1 + ABf}{1 - \Psi Bf} \right) \tag{68}$$

Here, A accounts for filler geometry and matrix Poisson's ratio determined from the Einstein coefficient k as the following equations [24], while B relates to the moduli of the filler and matrix.

$$A = K - 1 \tag{69}$$

where K is denoted as Einstein's coefficient and determined from the l /d ratio of the filler.

$$K = l + (2l/d) \tag{70}$$

$$B = \frac{(E_P/E_m) - 1}{(E_P/E_m) + A} \tag{71}$$

$$\Psi = 1 + \left(\frac{1 - \Phi_m}{\Phi_m^2} \right) f \tag{72}$$

Φm is the maximum packing equals 0.785, 0.907, and 0.82 for square arrangement, hexagonal array, and random packing of fibers, respectively. The modulus of composites can be predicted using models that assume isostress or isostrain conditions in a system of series and parallel-oriented fibers within the matrix. Two key equations often applied to estimate the lower and upper bounds of composite modulus are based on Voigt's model, which are as follows [54]:

For the upper bound:

$$E_c^{upper} = E_\rho V_\rho + E_m (1 - V_P) \tag{73}$$

For the lower bound:

$$E_c^{lower} = \frac{E_p E_m}{\left[E_P (1 - V_P) + E_m V_p \right]} \tag{74}$$

These models are effective in estimating the modulus of particulate composites but tend to overestimate E_C, presenting a linear relationship between E_C and V_p. As noted by Fu et al., the actual modulus of particulate composites lies somewhere between the upper and lower bounds of these equations [55].

Verbeek [56] developed an alternative model, assuming perfect adhesion between the particulates and polymer matrix, with stress transfer occurring *via* shear mechanisms, expressed as follows:

$$E_\varepsilon = f E_m MRF + f_P E_p \tag{75}$$

where modulus reduction factor MRF $= 1 - \frac{tanh \Psi}{\Psi}$, $\Psi = \alpha \sqrt{\frac{(1-\chi)^3 G_p}{E_m} \frac{f}{1-f}}$, $\chi = \frac{\Phi}{f_P (1-\Phi) + \Phi}$, $\Phi = \frac{(1-f_P)^2 \Phi_m}{1 - f_P \Phi_m}$ Here, f_P is the volume fraction of matrix, α is the particle aspect ratio, G_p is the shear modulus of the matrix, E_p and E_m are the moduli of the particles and matrix, respectively, and f is the reinforcement particle volume fraction. The terms Φ and χ denote voidage and modified voidage, respectively, where voidage accounts for particle packing efficiency [56].

Another approach, based on the Mooney equation [57], derives from Einstein's equation and was later modified by Brodnyan to include the particulate aspect ratio, as follows:

$$E_C = E_m \exp[kf/(1 - f/\Phi_m)] \tag{76}$$

According to the Brodnyan model [57], α, the particle aspect ratio, must range between 1 and 15 for non-spherical particles:

$$E_C = E_m \exp[(2.5f + 0.407(\alpha - 1)1.508f)/(1 - f/\Phi_m)] \tag{77}$$

The strength of composites is governed by the weakest fracture path through the material, which defines its load-bearing capacity. The addition of particulates to polymers affects composite strength in two opposing ways: reinforcement and weakening. Particulates reinforce the material by impeding crack propagation, while they may also weaken it by creating stress concentration points. If reinforcement dominates, the composite strength will exceed than that of the neat polymer; conversely, weakening effects will result in lower strength [57]. Unlike modulus prediction, strength prediction is more complex, as it depends on the fracture behaviors of the composite. Factors such as interfacial adhesion, stress concentration, and defect distribution are significant in determining fracture behavior. Although no universal model exists for predicting strength, various semi-empirical equations have been proposed. One such model by Danusso and Tieghi [58] and Levita et al. [59] assumes no stress transfer between the polymer and particulates. The effective load-bearing cross-sectional area of the matrix is $\sigma_c = \sigma_m(1 - V_p)$ where σ_c and σ_m are the composite and matrix strength. This model suggests a linear decrease in composite strength as particulate loading increases [58, 59]. Das et al. [60] found that the tensile strengths of biochar-reinforced polypropylene (PP) composites remained consistent despite increased biochar content (0–35 wt%), which contradicts the negative predictions of the previous equation. To address this limitation, Nicolais and Nicodemo introduced a power-law function for the particulate volume fraction:

$$\sigma_c = \sigma_m\left(1 - aV_P^b\right) \tag{78}$$

where a and b are constants related to particulate shape. However, this still predicts a linear reduction in strength with increasing particulate loading [61].

Nielsen proposed another model for poorly bonded particulate-polymer systems:

$$\sigma_C = \sigma_m\left(1 - aV_P^{2/3}\right)Q \tag{79}$$

where Q accounts for structural weaknesses, ineffective stress transfer, and stress concentrations at the interface [62].

Nicolais and Nicodemo further refined their model to assume no bonding between particulates and matrix, with the matrix solely bearing the load:

$$\sigma_c = \sigma_m\left(1 - 1.21V_P^{2/3}\right) \tag{80}$$

This equation provides a lower bound for composite strength, while the upper bound assumes perfect adhesion between particulates and polymer, equating the composite strength to than that of the neat polymer [63].

Bigg [64] proposed a model incorporating the effects of adhesion and stress transfer:

$$\sigma_c = \sigma_m \left(1 - aV_P^b + cV_P^d\right) \tag{81}$$

where c and d are constants related to adhesion quality [64].

Pukanszky et al. [65] developed an empirical model based on strong interfacial adhesion:

$$\sigma_C = \left(\frac{1 - V_P}{1 + 2.5V_P}\sigma_m\right) exp\,(BV_P) \tag{82}$$

where B is a constant linked to particulate surface area, density, and interfacial bonding energy. For poor bonding, B = 0. These models demonstrate that composite strength depends on particulate size, loading, and interfacial bonding quality.

3.3 Impact of filler type on the mechanical properties

The mechanical properties of cultivating pots depend on different fillers, including tensile strength and elastic modulus. The results reveal that the tensile strength for sugarcane bagasse, compost, peat moss, activated carbon, and vermiculite were 6.84, 11.02, 4.00, 10.97, and 13.24 MPa, respectively. The elastic modulus for these materials was measured at 4.85, 10.66, 0.11, 8.37, and 15.00 MPa for sugarcane bagasse, compost, peat moss, activated carbon, and vermiculite. The selection of filler exerted a significant influence on the mechanical attributes, with vermiculite exhibiting the highest values most substantial mechanical properties, while peat moss demonstrated the lowest values and least [27].

3.4 Theoretical mechanical analysis of filler types

Figure 6 shows the theoretical tensile strength for different fillers made from different (sugarcane bagasse, compost, peat moss, activated carbon, and vermiculite) by using different models (parallel, series, Guth, Halpin-Tsai, modified Halpin-Tsai, Hirsch, and Bowyer-Bader). The results indicated that the theoretical tensile strength for sugarcane bagasse by using parallel, series, Guth, Halpin-Tsai, modified Halpin-Tsai, Hirsch, and Bowyer-Bader models were 19.93, 16.14, 1.99, 2.57, 2.40, 16.51, and 0.29 MPa, respectively. The theoretical tensile strength for compost by using parallel, series, Guth, Halpin-Tsai, modified Halpin-Tsai, Hirsch, and Bowyer-Bader models were 0.60, 3.24, 1.94, 1.74, 1.83, 2.98, and 0.39 MPa, respectively. Also, the theoretical tensile strength for peat moss by using parallel, series, Guth, Halpin-Tsai, modified Halpin-Tsai, Hirsch, and Bowyer-Bader models was 0.36, 2.46, 1.97, 1.65, 1.88, 2.25, and 0.35 MPa, respectively. The theoretical tensile strength for activated carbon by using parallel, series, Guth, Halpin-Tsai, modified Halpin-Tsai, Hirsch, and Bowyer-Bader models were 37.12, 15.82, 1.96, 3.86, 2.67, 17.95, and 0.37 MPa, respectively. The theoretical tensile strength for vermiculite by using parallel, series, Guth, Halpin-Tsai, modified Halpin-Tsai, Hirsch, and Bowyer-Bader models were 0.26, 3.87, 2.00, 1.88, 1.96, 3.51, and 0.27 MPa, respectively.

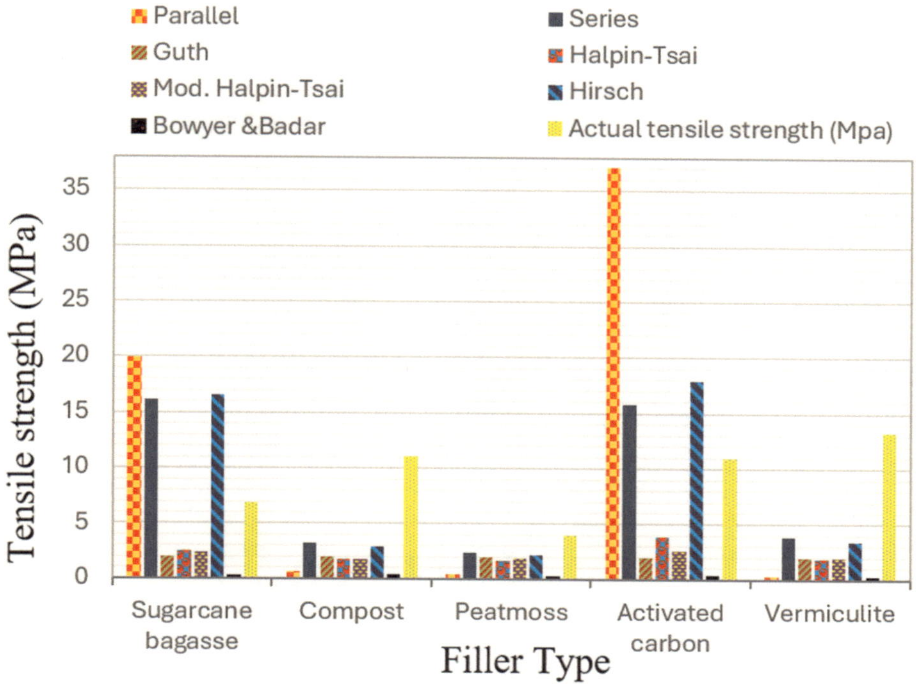

Figure 6.
Theoretical tensile strength of various fillers [27].

Figure 7 shows the theoretical elastic modulus for different fillers made from different (sugarcane bagasse, compost, peat moss, activated carbon, and vermiculite) by using different models (parallel, series, Guth, Halpin-Tsai, modified Halpin-Tsai, Hirsch, and Bowyer-Bader). The results indicated that the theoretical elastic modulus for sugarcane bagasse by using parallel, series, Guth, Halpin-Tsai, modified Halpin-Tsai, Hirsch, and Bowyer-Bader models were 1.53, 1.21, 2.67, 0.19, 0.18, 1.24, and 0.06 MPa, respectively. The theoretical elastic modulus for compost by using parallel, series, Guth, Halpin-Tsai, modified Halpin-Tsai, Hirsch, and Bowyer-Bader models were 2.09, 1.16, 5.94, 0.32, 0.27, 1.25, and 0.15 MPa, respectively. Also, the theoretical elastic modulus for peat moss by using parallel, series, Guth, Halpin-Tsai, modified Halpin-Tsai, Hirsch, and Bowyer-Bader models were 2.19, 1.19, 4.08, 0.26, 0.22, 1.29, and 0.10 MPa, respectively. The theoretical elastic modulus for activated carbon by using parallel, series, Guth, Halpin-Tsai, modified Halpin-Tsai, Hirsch, and Bowyer-Bader models were 1.35, 1.15, 4.94, 0.28, 0.24, 1.17, and 0.12 MPa, respectively. The theoretical elastic modulus for vermiculite by using parallel, series, Guth, Halpin-Tsai, modified Halpin-Tsai, Hirsch, and Bowyer-Bader models were 0.71, 1.21, 2.31, 0.17, 0.17, 1.16, and 0.25 MPa, respectively.

The results indicate that the highest values of theoretical elastic modulus for sugarcane bagasse, compost, peat moss, activated carbon, and vermiculite (2.67, 5.94, 4.08, 4.94, and 2.31 MPa) were found with Guth model, while the lowest values of theoretical elastic modulus for sugarcane bagasse, compost, peat moss, activated carbon, and vermiculite (0.06, 0.15, 0.10, 0.12, and 0.25 MPa) were found with Bowyer-Bader model.

The IROM model exhibited superior predictive accuracy in estimating tensile strength across a variety of fillers, except for sugarcane bagasse, for which the Halpin-Tsai model

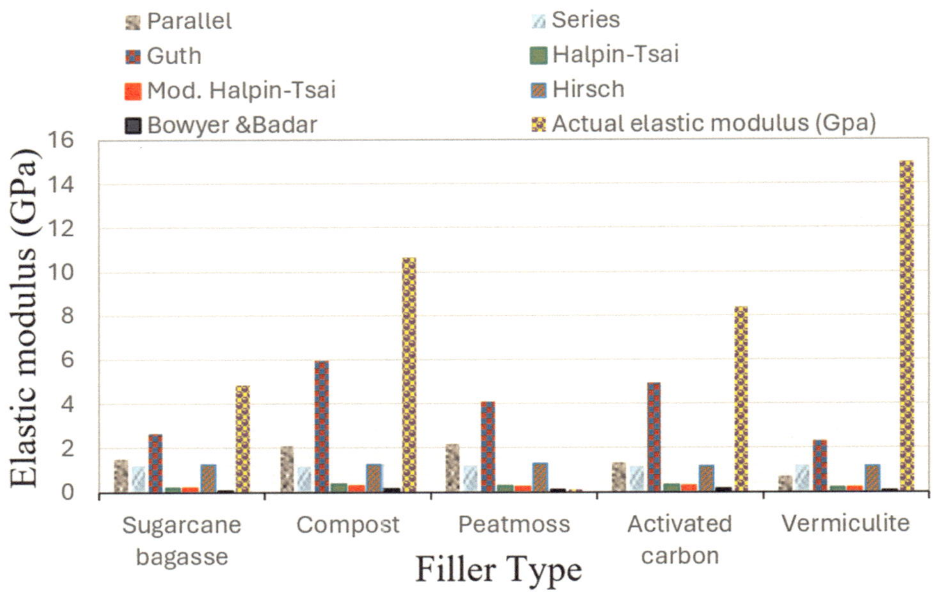

Figure 7.
Theoretical elastic modulus of various fillers [27].

demonstrated a better fit. This discrepancy in model performance may stem from the series model's compatibility with fine particulate structures, whereas sugarcane bagasse comprises long fibers. Conversely, regarding elastic modulus, the Guth model emerged as the most suitable for the diverse range of fillers, except for peat moss, where Bowyer and Badar's model proved to be the optimal choice [27].

3.5 Finite element analysis modeling

The finite element analysis (FEA) of composite materials, including natural fiber-reinforced polymer composites (NFRPCs), follows a structured approach that involves several key steps to ensure accuracy and efficiency in solving complex physical problems. Finite element analysis remains a powerful tool for simulating the behavior of NFRPCs. Through careful consideration of factors such as material properties, meshing, boundary conditions, and loading, FEA provides insights that guide the redesign and improvement of composite products. The use of multiscale models and advanced material simulations enables accurate prediction of real-world performance and failure mechanisms. In finite element computations, key parameters such as load, stiffness matrix, nodal displacements, element stresses, and element strains are organized in matrix form. After meshing in the preprocessor stage, the software forms the element stiffness matrix, global stiffness matrix, and global force vectors. Commercial FEA software, such as ANSYS or LS-DYNA, employs methods such as Gaussian elimination to solve these equations. Proper meshing, loading, and boundary conditions are crucial, as any error can require a complete remeshing. Computation time is influenced by the number of elements, nodes, and element types (higher-order elements take longer to compute compared to linear ones). Accurate solving of the model allows for determining key outcomes such as displacements, temperatures, pressures, and velocities. Inaccuracies prompt remeshing and further refinement [66].

FEA of PMCs can be carried out using specialized software or custom coding (e.g., MATLAB, C). Feeding the finite element model with experimentally determined material properties is essential for realistic simulations. American Society for Testing and Materials (ASTM) standards guide the experimental determination of these properties (e.g., ASTM D3039 for tensile strength, modulus, and Poisson's ratio, ASTM D3518 for in-plane shear strength, and ASTM D790 for flexural strength). These experimental results are critical inputs for the finite element model, ensuring the virtual simulation reflects the actual behavior of the composite. Selecting the appropriate analysis type (e.g., structural, thermal, fluid, electromagnetic) is the first step in any FEA. The choice of analysis activates relevant elements in the FEA software and dictates the appropriate coupled-field analysis for multi-physics problems. Once the analysis type is set, part modeling begins. For simplicity and to reduce computation time, 3D structures can be simplified into 2D models using techniques such as plane stress, plane strain, and axisymmetry. For NFRPCs, which are orthotropic, material modeling requires defining separate elastic moduli along the X, Y, and Z axes, unlike isotropic materials like steel. Experimentally obtained unidirectional material properties such as tensile strength, Poisson's ratio, and shear modulus are necessary inputs for finite element models. Bidirectional properties can be predicted but require advanced computational methods. Mesh size and element choice impact model accuracy and computation time. Coarse mesh is typically refined based on experimental observations. Accurate application of structural or thermal boundary conditions is crucial, with conditions applied at the nodal level. After preprocessing, the FEA model is solved using the software's solver. Postprocessing involves extracting results such as displacements, stresses, and strains, which allow for virtual simulations of real-world behaviors. These outcomes provide valuable insights for redesign or further experiments [66]. Typical failure mechanisms in fiber-reinforced composites include fiber fracture, interface failure, fiber/matrix debonding, fiber pullout, matrix cracking, and delamination. These failure modes can be modeled through FEA to predict the behavior under various loads [66]. Due to the complex structures of natural fibers, which vary across length scales, FEA models must account for both macro- and microstructural behaviors. Multiscale models, such as the representative volume element (RVE), are widely used. RVEs can range from macromechanical to micromechanical analyses and provide an efficient way to homogenize properties across different scales [67]. For instance, Silva et al. simulated bamboo's tensile, flexural, and torsional behavior using different material models. Fu et al. investigated bamboo's shear capability using brittle fracture mechanics [68, 69]. Chand et al. validated bamboo's tensile and bending behavior experimentally and through simulations [9]. In micromechanical analyses, researchers have focused on the effects of microstructures on mechanical properties. For example, Palombini et al. used 3D finite element models of bamboo meshed *via* X-ray microtomography to evaluate axial compressive strength [70]. FEA allows mechanical analysis of NFRPCs, which is essential for evaluating properties such as stress-strain behavior under various loading conditions. Researchers have developed various approaches, including direct 3D RVE models and orientation averaging methods, to simulate these behaviors [71, 72]. For instance, the shape optimization of biodegradable cultivating pots made from rice husk fiber/melamine formaldehyde resin has been successfully modeled using FEA. Simulations demonstrated that increasing the curvature radius at the pot's bottom from 0 to 100 mm reduced the maximum equivalent stress from 21.29 MPa to 20.46 MPa by 4.036% (**Figure 8**), improving durability and structural integrity [73].

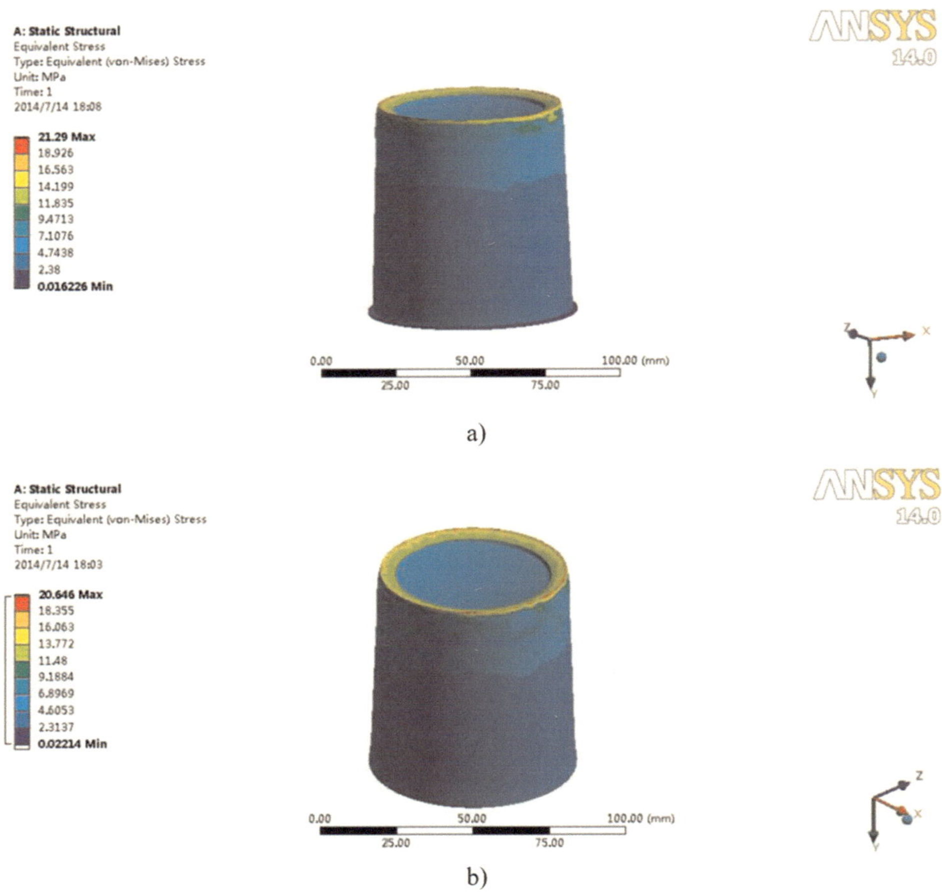

Figure 8.
Equivalent stress (a) 0 mm curvature radius, and (b) 100 mm curvature radius [73].

4. Optimization approaches of composite materials and process

Optimizing the composite materials and process is a complex challenge, requiring the development of reliable models and robust multi-response optimization techniques. Šibalija and Majstorović [74] categorized optimization methods into two major groups:

a. Conventional (statistical or mathematical) methods:

- Iterative search techniques: Based on statistical or mathematical approaches.
- Design of experiment (DOE): Techniques such as response surface methodology (RSM), factorial design, and Taguchi design.

b. Non-conventional (AI-based) methods:

- Fuzzy logic-based methods.
- Artificial neural networks (ANN).

- Metaheuristic algorithms: Including genetic algorithm (GA), simulated annealing (SA), particle swarm optimization (PSO), ant colony optimization (ACP), tabu search (TS), artificial bee colony algorithm (ABC), biogeography-based optimization (BBO), and teaching-learning based optimization (TLBO).
- Expert systems (ES).

Among these, the most prevalent optimization methods in composite manufacturing processes include the following:

1. Conventional optimization techniques, such as the Taguchi method, RSM, and various iterative mathematical search techniques (linear, non-linear, and dynamic programming).

2. Non-conventional techniques, particularly AI-based approaches such as heuristic search techniques, with GA and SA being the most frequently applied [74].

This chapter will emphasize methods based on the Taguchi method, AI-based techniques such as ANN, GA, and SA, as these methods are foundational in multi-response optimization models such as the Intelligent System for Multi-response Robust Process Design (IS-MR-RPD).

4.1 Conventional (statistical or mathematical) methods

4.1.1 Response surface methodology (RSM)

RSM is a powerful statistical tool for multi-response optimization, enabling simultaneous analysis of multiple variables in both manufacturing process factors or in the manufactured product itself. It is especially useful in biocomposite research, where factors such as constituent type, ratio, and experimental conditions can greatly influence material performance. RSM experiments often use central composite design (CCD) or the Box-Behnken design to model relationships between variables. Models can be linear, quadratic, or higher order, depending on system complexity. Regression approaches such as multiple linear regression and polynomial regression are commonly employed. After ensuring residual normality and verifying model fit, optimization techniques such as desirability functions or numerical algorithms can determine optimal parameter values [75]. For instance, Nadhari et al. [76] applied RSM with rotatable central composite design (RCCD-RSM) to optimize binderless particleboard production. The process involved 24 factorial runs to optimize internal bond strength, modulus of rupture, thickness swelling, and water absorption. The primary variables, steaming temperature, steaming time, hot-pressing temperature, and hot-pressing time were adjusted in 30 experimental runs. The empirical mathematical model developed followed a polynomial quadratic fit:

$$Y = \alpha_0 + \sum\nolimits_{i=1}^{n} \alpha_i A_i + \sum\nolimits_{i=1}^{n} \alpha_{ii} A_i^2 + \sum\nolimits_{j=2} \sum\nolimits_{i<1}^{n} \alpha_{ii} A_i A_j + \ldots \quad (83)$$

where Y represents the response, α_0 is the intercept, α_{ii}, α_{ij}, α_{ik}, α_{il}, α_{jk}, α_{jl}, α_{kl} are interaction coefficients, and A_i, A_j, A_k, A_l are coded values of independent variables. Results showed that increasing hot-pressing time reduced thickness swelling and improved dimensional stability, with optimal settings of 120°C (steaming temperature), 46 minutes (steaming time), 215°C (hot-pressing temperature), and 29 minutes (hot-pressing time) [76]. It is concluded that high-performance binderless particleboard can be manufactured with proper manufacturing conditions using steamed oil palm trunk waste as the raw material. The mechanical properties improved significantly with increasing time of steaming of the raw oil palm trunk waste and hot-pressing temperature during board formation but will decrease at long duration of steaming time and high hot-pressing temperature. An increase in hot-pressing time contributes to less thickness swelling and better dimensional stability. The optimized value of manufacturing variables was 120°C, 46 min, 215°C, and 29 min for steaming temperature, steaming time, hot-pressing temperature, and hot-pressing time, respectively. Leading to optimized conditions from actual experiments are the binderless particleboard has internal bond strength, 0.49 MPa; modulus of rupture, 8.18 MPa; thickness swelling, 22%; and water absorption, 51.43%.

Despite its effectiveness, RSM has limitations in multi-response optimization, especially when dealing with highly nonlinear systems with many variables. It may not simultaneously optimize the mean and variance of responses and can struggle with more than three responses, leading to potential trapping in local minima [74]. In the development of biodegradable containers, Sirodariya et al. [77] employed RSM with central composite rotatable design (CCRD) to optimize the mechanical and chemical properties of containers produced using orange peel powder (OPP), wheat flour (FL), and tamarind seed powder (BA). The study optimized compression strength and water activity using design expert software, leading to an ideal formulation of 50.28% OPP, 35.71% FL, and 14.01% BA. The process was further refined by adjusting the baking temperature 131°C for 64 min, yielding a container with 12.63 kgf compression strength and 0.21 water activity. This optimization highlighted the importance of ingredient ratios and process parameters in achieving desired biodegradable container properties.

The effect of each independent variable on the compression strength of the biodegradable containers shows that various factors contribute to improved mechanical properties. Orange peel powder slightly enhances compression strength due to the presence of cellulose, which reinforces the structure. Wheat flour also increases compression strength as the gluten content helps create a more cohesive and durable structure. Tamarind seed powder significantly boosts strength because its stickiness binds the ingredients together without leaving gaps in the material. This powder is rich in xyloglycan starch molecules, which form strong interfacial bonding and contribute to mechanical properties. The process parameters, such as time and temperature, also influence compression strength. High temperatures for shorter periods can reduce strength, likely due to burning of some particles, while optimized temperature and time ranges can yield containers with desirable mechanical properties. Thus, the proportions of wheat flour and binding agents play a major role in determining the compression strength, with proper ratios of gluten, cellulose, and the binding agent leading to improved strength. On the other hand, the effect of independent variables on water activity suggests that containers made from dry ingredients with low inherent water activity and prepared at higher temperatures exhibit reduced water activity. An increase in the binding agent further lowers the water activity, as it minimizes free water in the structure. Though wheat flour and orange peel powder have a minor effect on water activity, the gluten content helps retain water within the structure, which, along with the binding agent, keeps the water activity low. This lower water activity is beneficial for prolonging the shelf life of the containers [77].

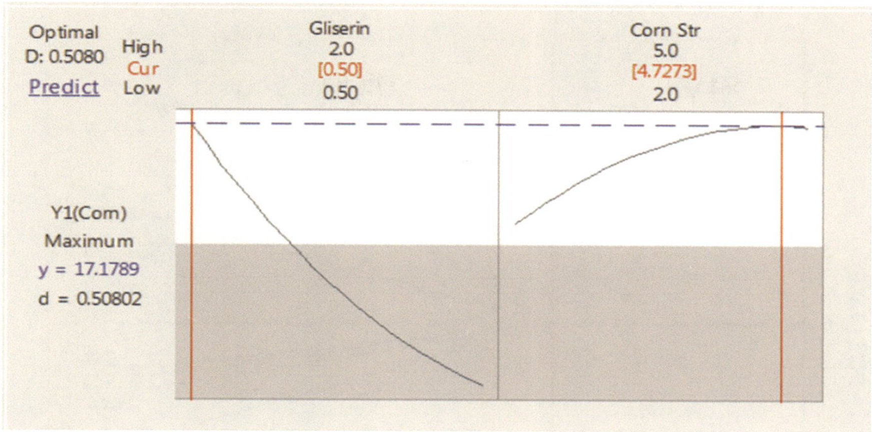

Figure 9.
D-optimal surface response of tensile strength [78].

In terms of tensile strength, glycerin is the common plasticizer used in starch-based biocomposite pots. Optimization of the glycerin-to-starch ratio is essential for achieving optimal tensile strength. For instance, as demonstrated in Ref. [78], a D-optimal response approach was used to achieve a tensile strength of 17.18 MPa under optimal conditions of 0.5 ml glycerin and 4.72 g corn starch (**Figure 9**).

4.1.2 Taguchi's robust parameter design

The Taguchi method is renowned for its focus on minimizing variation and improving quality through signal-to-noise (SN) ratios and the quality loss function (QL). These are employed to assess experimental runs by quantifying the robustness of a system's performance under varying conditions. However, a notable limitation of the Taguchi method is its constraint to single-response optimization problems. To address this, several approaches have been proposed for multi-response optimization within the Taguchi framework, including weighting factors assigned to process responses, regression analysis, desirability function analysis, data envelopment analysis (DEA), principal component analysis (PCA), gray relational analysis (GRA), and technique for order preference by similarity to ideal solution (TOPSIS). These methods have been useful, though many assume that responses are uncorrelated, which limits their applicability in complex systems where interactions among responses may be significant.

For instance, Pratibha et al. [28] applied 2D-PCA to group the biodegradable composite (BC) pots into three distinct clusters based on their physico-chemical attributes, primarily related to the untreated and treated paddy straw (PS). These clusters were as follows: (cluster I: raw PS, cluster II: alkali-treated PS, and cluster III: alkali and autoclaved PS). Within cluster III, BC4 stood apart from BC3, BC5, and BC6, likely due to the cross-linked starch's influence, differentiating its characteristics from other BCs (**Figure 10**).

In another study, Manafi-Dastjerdi et al. [29] compared the average physical properties, including water absorption (WA) and thickness swelling (TS), along with

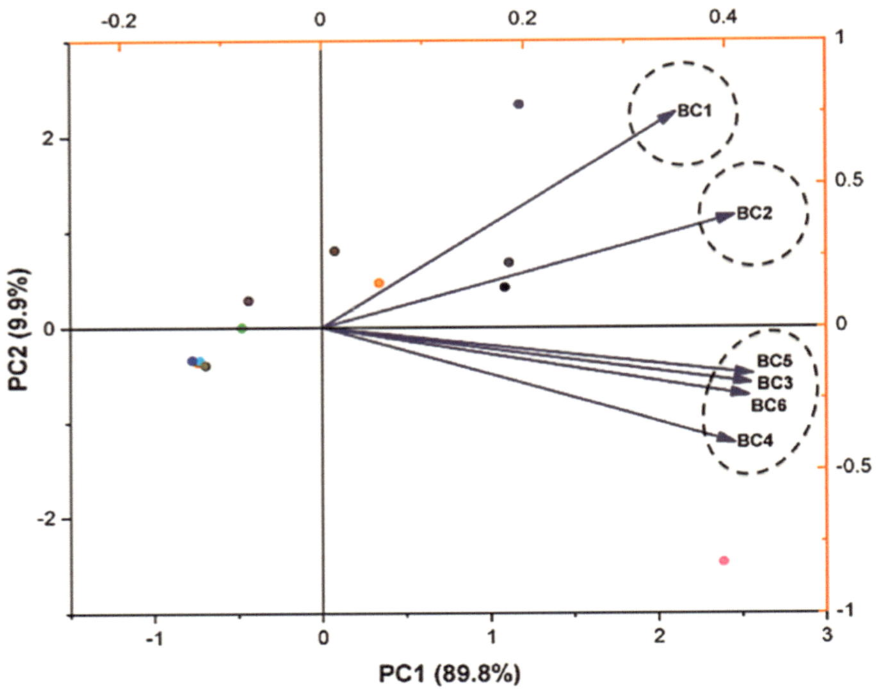

Figure 10.
2D-PCA of the biodegradable composite (BC) pots [28].

Variable	Sheep's wool		Cornstarch		Sheep's wool: cornstarch	
	Score	Rank	Score	Rank	Score	Rank
85:15	0.76	2	0.73	2	0.6	2
80:20	0.99	1	0.82	1	0.83	1
75:25	0.43	5	0.38	4	0.2	4
70:30	0.3	3	0.66	3	0.49	5
65:35	0.47	4	0.26	6	0.42	3
60:40	0.1	6	0.36	5	0.18	6

Table 1.
Ranking pots performance according to percentage of cattle manure to sawdust [29].

mechanical properties such as load failure (P) and internal bonding (IB), across six treatment ratios of cattle manure and sawdust-based biodegradable pots. By utilizing the TOPSIS method, the study established a final ranking to prioritize the optimal percentage of manure and sawdust for constructing biocomposite pots (**Table 1**), emphasizing how varying binders can significantly affect the resulting physical and mechanical properties.

The signal-to-noise (S/N) ratio, a crucial metric in performance optimization, is employed to improve quality characteristics, with the objective of achieving higher performance based on the optimization function category. The S/N ratio can be expressed using the following equations For two distinct categories, depending on the desired optimization target [79]:

Smaller is better:

$$S/N = 10 \log \left(\frac{1}{n} \sum_{i=1}^{n} y_i^2 \right) \tag{84}$$

Larger is better:

$$S/N = 10 \log \left(\frac{1}{n} \sum_{i=1}^{n} \frac{1}{y_i^2} \right) \tag{85}$$

4.2 Multiple-criteria decision-making (MCDM) in composite material selection and optimization

The multiple-criteria decision-making (MCDM) approach plays a crucial role in selecting and optimizing materials for composite systems, addressing the inherent complexity of balancing multiple performance criteria. For instance, Ghaleb et al. [80] presented a comparative assessment of MCDM approaches in manufacturing process selection, including methods such as the Technique for Order Preference by Similarity to Ideal Solution (TOPSIS), Analytic Hierarchy Process (AHP), and VIKOR. MCDM tools like these are essential for systematically evaluating alternatives based on various criteria (e.g., mechanical properties, cost, durability), which are critical in the material selection for composite applications.

4.2.1 TOPSIS method for decision-making

The TOPSIS method, one of the most popular MCDM techniques, is based on the concept of finding an optimal solution that is closest to the positive ideal solution and farthest from the negative ideal solution in Euclidean space. The methodology has been widely applied to optimize various aspects of composite material systems and software management. TOPSIS provides an effective framework for evaluating multiple criteria in complex systems, allowing for a compromise solution that balances trade-offs between competing objectives. In composite material research, TOPSIS is employed to improve software efficiency and to optimize material formulations based on a set of methodological approaches. It identifies the best alternative variant by calculating the Euclidean distance to both ideal and non-ideal solutions. For example, in Ref. [81], a TOPSIS-Taguchi approach was utilized to optimize the proportions of fillers in biocomposite pots, which helps minimize experimental efforts while maximizing composite performance (**Figure 11**).

The combination of the TOPSIS method and Taguchi's robust design provides an effective way to streamline decision-making by optimizing multiple criteria with fewer experimental trials. This hybrid approach uses orthogonal arrays to reduce the number of experimental runs while achieving robust results, enabling the identification of optimal configurations efficiently [81, 82].

4.2.1.1 Water absorption property optimization

Using the Taguchi method, an optimal filler composition for biocomposite seedling pots was determined. Through software-assisted analysis, the ideal mixture was found

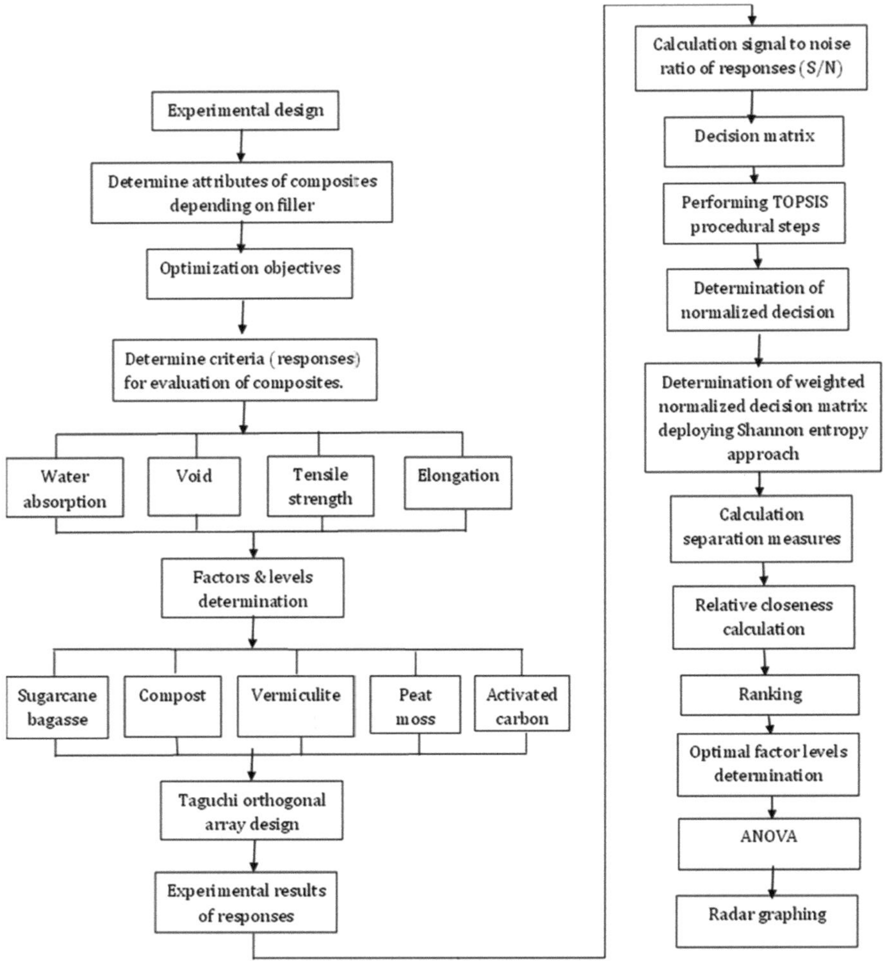

Figure 11.
Flow chart of the TOPSIS-Taguchi [81].

in Experiment 7, consisting of 30% sugarcane bagasse (S2), 30% compost (C2), 20% peat moss (P1), 20% vermiculite (V1), and 30% activated carbon (AC2). This combination minimized water absorption to 15% [26, 83].

4.2.1.2 Density optimization

The same experiment yielded a density of 0.711 g/cm^3. According to ANOVA, vermiculite was the dominant factor (72.54%) in affecting density, followed by activated carbon (14.19%), sugarcane bagasse (10.45%), compost (2.49%), and peat moss (0.31%). These findings are consistent with those from Taguchi's single response analysis in **Table 2**. The influence of vermiculite and activated carbon constituent factors on the pot density was contour-plotted as shown in **Figure 12**. The optimal combination for maximizing density was observed in Run 5, resulting in a value of 0.70 g/cm^3 [81]. Despite the goal for lightweight pots, the findings suggest that higher density results from fiber-matrix reinforcement, requiring consolidation of the fibers.

Level	S	C	P	V	AC
1	−3.872	−3.802	−3.755	−3.331	−3.898
2	−3.583	−3.653	−3.700	−4.125	−3.557
Delta	0.288	0.149	0.056	0.794	0.340
Rank	3	4	5	1	2

Table 2.
Response Table for Signal to Noise Ratios "Larger is better" for density [81].

4.2.1.3 Tensile strength optimization

The highest tensile strength (13.92 MPa) occurred in Experiment 7, while the lowest (12.72 MPa) was found in Experiment 1. Vermiculite and activated carbon were identified as key contributors to tensile strength, with contributions of 71.15 and 19.43%, respectively. The combination of TOPSIS and Taguchi models provided higher predictive accuracy in optimizing tensile strength, with a regression equation yielding an R^2 of 100% as presented in the following equations [81]:

$$\text{Tensile strength} = 9.20200 + 1.74800\,(\text{AC}) + 1.80800\,(\text{C}) - 5.19200\,(\text{P}) - 2.38200\,(\text{S}) + 4.01800\,\left(\text{R}^2 = 100\%\right) \tag{86}$$

$$\text{Tensile elastic modulus (MPa)} = 7.79200 + 0.55800\,(\text{AC}) + 2.85133\,(\text{C}) - 7.67200\,(\text{P}) - 2.96533\,(\text{S}) + 7.22800\,(\text{V})\,\left(\text{R}^2 = 100\%\right) \tag{87}$$

4.2.1.4 Optimization of filler constituents using TOPSIS-Taguchi

The TOPSIS method, when combined with Taguchi's design, identified the most optimal composite configuration for natural fiber materials. The eighth experimental set was deemed the best, based on physical and mechanical attributes (**Figure 13**). Vermiculite and activated carbon significantly influenced the relative closeness coefficient (Ci) with contributions of 71.15 and 19.43%, respectively (**Figures 14** and **15**). The TOPSIS-Taguchi model outperformed the traditional Taguchi approach by fine-tuning the constituent proportions, improving the final composite properties, such as density and tensile strength [11]. These results emphasize the superiority of the TOPSIS-Taguchi model in achieving more accurate and refined optimization outcomes, particularly when compared to traditional Taguchi methods applied to a single response criterion.

4.2.2 Gray relational analysis (GRA) MCDM techniques in composite materials

The field of composite materials benefits significantly from various MCDM methods. Some of the prominent methods include analytic hierarchy process (AHP), gray relational analysis (GRA), TOPSIS, VIKOR, fuzzy approaches (Fuzzy-TOPSIS, Fuzzy-VIKOR), and hybrid approaches such as TOPSIS-Taguchi or GRA-Taguchi. For instance, Sumesh and Kanthavel [84] used ANN modeling alongside GRA to predict and optimize mechanical properties of bio-based composites by adjusting key factors such as fiber type, content, and processing conditions. Similarly, GRA is a popular

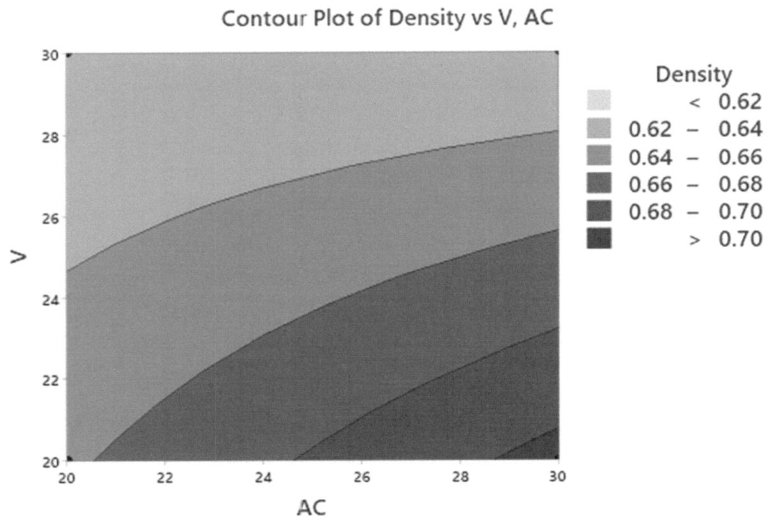

Figure 12.
Influence of vermiculite and activated carbon constituent factors on the pot density [81].

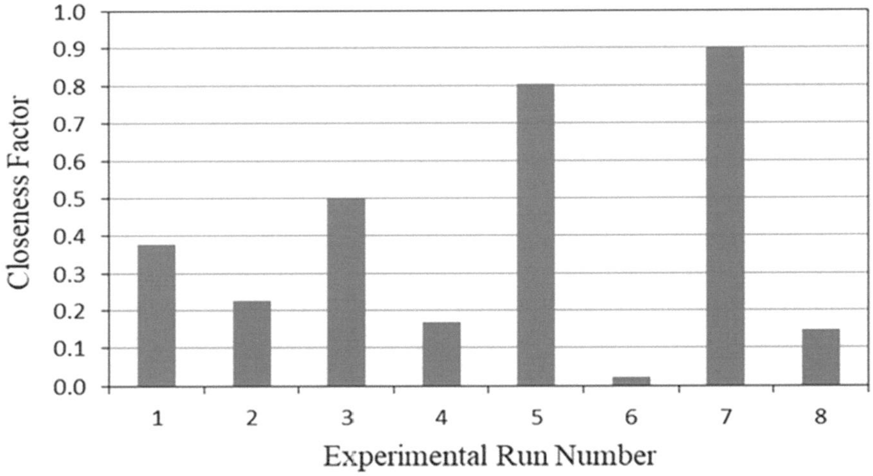

Figure 13.
TOPSIS composites ranking of the pots [81].

method for multi-response optimization, as it converts multiple mechanical property outputs (e.g., tensile, flexural, and impact results) into a single composite response for easier analysis. The gray relational coefficient (GRC) and gray relational grade (GRG) are computed through a sequence of normalization and comparison steps, as shown in the following equations:

Normalization equation

$$y_i k = \frac{x_i^0 k - min\, x_i^0 k}{max\, x_i^0 k - min\, x_i^0 k} \tag{88}$$

Gray relational coefficient (GRC):

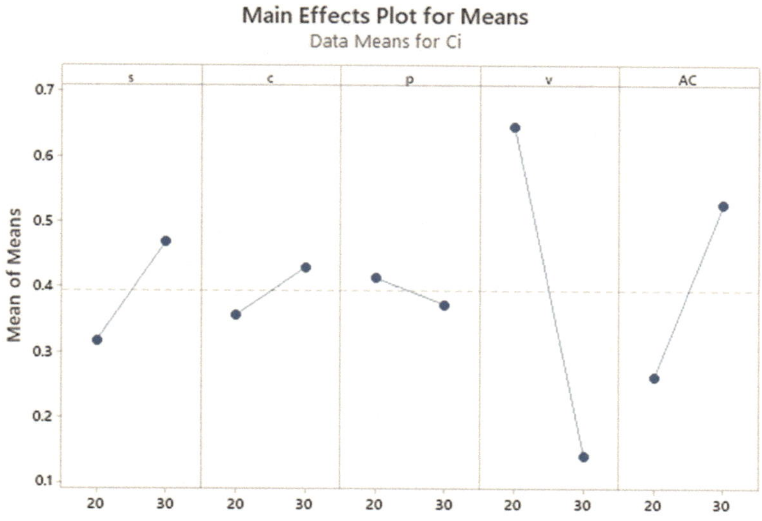

Figure 14.
SN ratios for closeness using the Taguchi for different fillers [81].

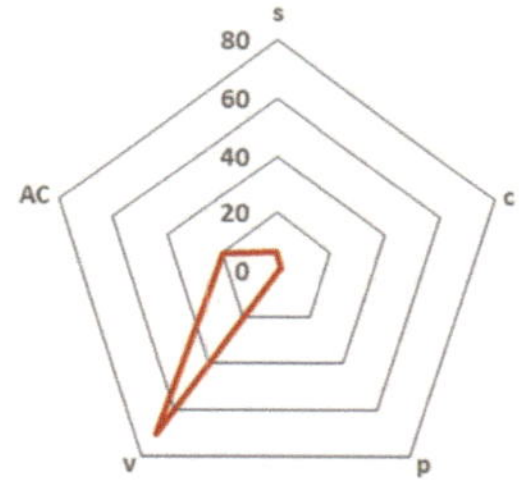

Figure 15.
Percentage contributions of the mixture ingredients to the closeness radar graph [81].

$$\xi_i k = \frac{\Delta_{min} + \zeta \Delta_{max}}{\Delta_{0i} k + \zeta \Delta_{max}} \tag{89}$$

where $\Delta_{0i}k$ is the offset data, ζ is the characteristic coefficient (typically set at 0.5), and $\Delta_{min} \wedge \Delta_{max}$ are the minimum and maximum deviations, respectively. This analysis yields a gray relational grade (GRG), which helps derive multi-response optimized data for bio-composite formulations. Artificial neural networks (ANNs) have been extensively used to validate and predict the mechanical properties of composites, particularly through methods like feedforward backpropagation in MATLAB. For example, in a study using ANN to optimize filler proportions for biocomposites, trials were conducted where the Levenberg-Marquardt backpropagation algorithm trained ANN models with hidden layers ranging from 1 to 21. Training iterations reached up to 10,000 epochs for better accuracy. The model's accuracy was validated using mean square error (MSE) and mean absolute error (MAE). The ANN model optimized by GRA results showed a high correlation between actual and predicted values for mechanical properties, with the best-performing topology being 5-3-1 (five factors, three hidden nodes, and one response). Validation trials further confirmed that the ANN's predictive capability was efficient, with regression analysis yielding 98.42% R^2, demonstrating strong predictive reliability. Thus, the integration of MCDM methods such as TOPSIS and GRA with advanced computational techniques such as ANN provides a robust framework for optimizing and predicting the mechanical properties of composite materials. These methodologies offer systematic approaches to handle multiple criteria and responses, making them invaluable in the design and analysis of high-performance composites [27].

4.3 Non-conventional multiple response process optimization approaches based on artificial intelligence (AI) techniques

Recently, the integration of artificial intelligence (AI) techniques in solving multi-response process optimization has gained prominence due to AI's capability to handle continuous process parameters effectively. AI-based techniques, such as fuzzy logic and expert systems, have been complemented by artificial neural networks (ANNs) and metaheuristic algorithms (e.g., genetic algorithms (GA), simulated annealing (SA), particle swarm optimization (PSO), artificial bee colony (ABC), and ant colony optimization (ACO)). These approaches provide significant advantages in modeling and optimizing multi-response systems without relying on rigid mathematical models. Artificial neural networks (ANNs) have emerged as powerful tools for predicting outcomes in composite materials, especially bio-based composites. ANNs simulate a network of linked nodes that mimic the behavior of the human brain to detect complex relationships between variables. For example, Refs. [85, 86] developed an ANN model to predict the mechanical characteristics of bio-based composites based on fiber type, fiber content, and processing parameters. These ANN models enable a more dynamic and adaptable approach to optimization, especially when combined with evolutionary algorithms such as GA or SA.

In a study of ANN in multiple response optimizations on concrete containing recycled coarse aggregate, Felix et al. [87] proposed a new formulation using ANN to estimate the elastic modulus of concrete. The ANN was trained using 243 datasets with feedforward neural networks and the Levenberg–Marquardt backpropagation algorithm. The optimal ANN architecture (6-4-2-1) achieved a root-mean-square error of 2.4 GPa and a coefficient of determination (R^2) of 0.91, indicating its efficiency in predicting outcomes. The developed model was then applied to 46,656 concrete samples to optimize mix proportions for both natural and recycled aggregates, showing the ANN's potential as a simulation tool for biocomposite industry.

4.3.1 Artificial neural networks (ANNs)

Artificial neural networks (ANNs) are designed to mimic biological neural networks, excelling in tasks such as pattern recognition, learning, and data processing. Structurally, ANNs are composed of neurons or nodes, which process information in a manner similar to the human brain (**Figure 16**). The five fundamental elements of an ANN include the following:

1. Input set x_j data to the network, each carrying carrying a corresponding synaptic, w_{kj}.

2. Summing add $\sum$ to the weighted input signals.

3. Activation function, F (·), to limit the output amplitude.

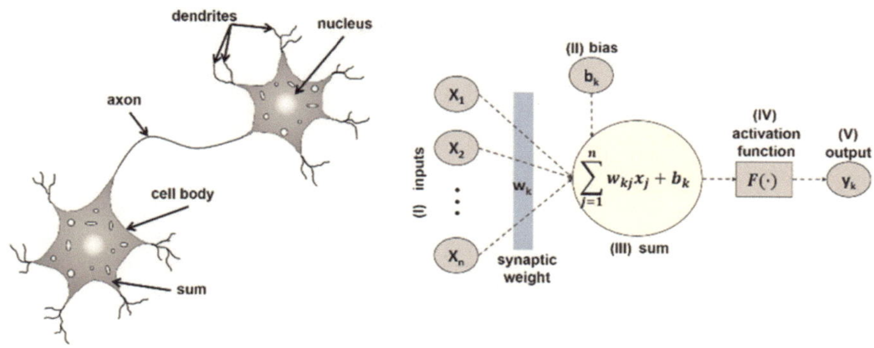

Figure 16.
Artificial neural network (ANN) [87].

4. Adjusting bias b_k that is represent the net input of the activation function to allow for a horizontal translation of the output.

5. Resulting the final output of the network y_k.

The output of a neuron is described by the following equation:

$$y_k = F\left(\sum_{j=} n w_{kj} + b_k\right) \tag{90}$$

where y_k is the output, F is the activation function, and w_{kj} represents the synaptic weights.

4.3.2 Model development using ANN

A comprehensive process for developing an ANN model involves several key stages as illustrated in (**Figure 17**):

1. Data collection: In the case of recycled aggregate concrete, experimental data on the elastic modulus is gathered and divided into training (60%), validation (20%), and testing (20%) datasets.

2. Statistical analysis: The data is analyzed to evaluate distribution and identify key input parameters.

3. Training and validation: Multiple ANN topologies are trained and validated to find the best-performing network.

4. Generation of new data: Using the optimized ANN, new datasets are generated by varying input parameters.

5. Model formulation: Finally, an analytical model is formulated using nonlinear regression, allowing the ANN to predict outcomes across a broader range of input conditions. The resulting model is tested through parametric studies to ensure its validity.

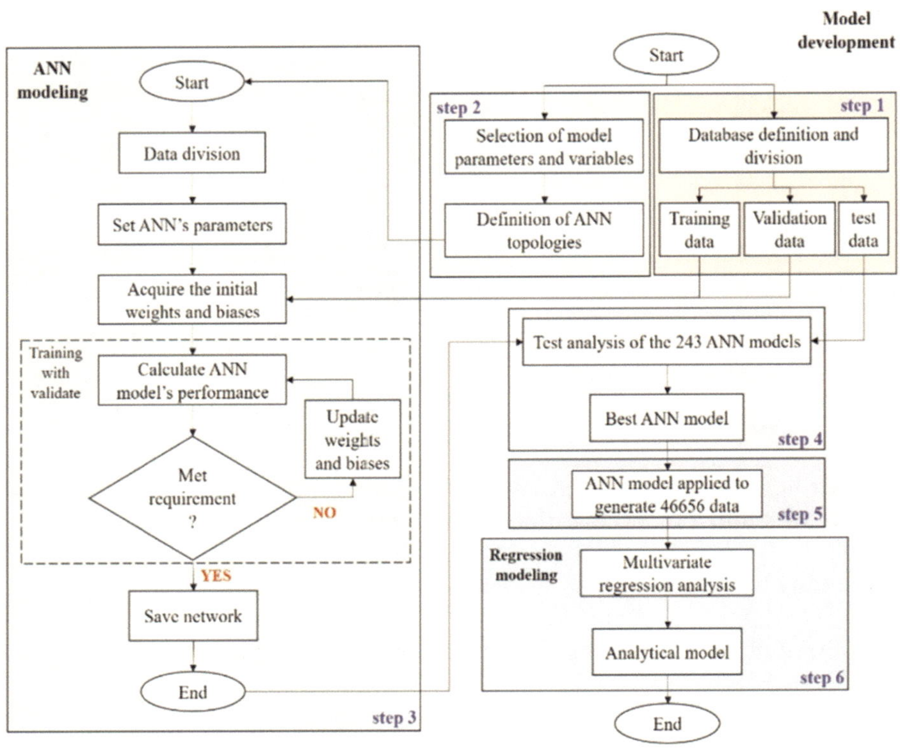

Figure 17.
ANN flowchart methodology [87].

ANNs, with their ability to learn from data and adapt to complex, nonlinear systems, offer a powerful alternative to traditional optimization techniques in multiresponse process optimization. Their flexibility and capacity for handling vast datasets make them particularly useful in fields requiring sophisticated modeling, such as biocomposite in different material researches.

5. Validation of theoretical models with experimental results

Model validation is an essential component in confirming the accuracy of predictions made using theoretical and computational approaches. Common metrics such as the root-mean-square error (RMSE) and coefficient of determination (R^2) are widely used for model validation and verification, as well as coefficient of variance (CV) and chi-square χ^2 test.

The RMSE is calculated as follows:

$$RMSE = \sqrt{\frac{1}{n}\sum_{i=1}^{n}(y_i - t_i)^2} \quad (91)$$

where y_i represents the predicted values from the model, t_i are the known target values, and n is the number of data points. This metric is often used in non-conventional modeling, including AI-based models such as artificial neural networks (ANNs), which have been employed to predict composite material properties such as tensile strength, flexural modulus, and impact resistance.

6. Conclusion

Green composite materials have gained substantial attention across global industries, particularly in agriculture. However, despite their promise, their application in the development of innovative pots for broad-spectrum plants remains underexplored. This is primarily due to the extensive research, prolonged development cycles, and resource-intensive processes required for their widespread adoption. Engineering these materials for agricultural use requires numerous experiments, trials, and optimization efforts, which can be time-consuming and resource heavy. To streamline this process, it is crucial to begin with a well-defined hypothesis, supported by the selection of an appropriate model, its optimization, and subsequent validation through experimental trials.

Bio-based composites, in particular, offer significant potential when integrated with advanced modeling tools. These tools provide valuable insights into material behavior, enabling performance predictions under varying conditions and facilitating the optimization of key attributes. Modeling is essential for optimizing manufacturing processes such as extrusion, compression molding, and injection molding, ultimately enhancing composite properties, reducing waste, and improving production efficiency. Furthermore, modeling plays a critical role in material selection, application design, and the overall optimization of biocomposite design, ensuring these materials meet both performance and sustainability demands effectively.

This chapter provided a comprehensive analysis of the modeling and optimization approaches for the physicomechanical properties of biocomposite cultivating pots. A specific focus was placed on the role of filler materials, optimized through decision-making tools such as Taguchi, TOPSIS, and artificial neural networks (ANN). The combined TOPSIS-Taguchi approach proved particularly effective, reducing the number of experimental trials and improving the efficiency of multi-response optimization. Key findings demonstrated that fillers such as vermiculite and activated carbon were significant contributors to the enhancement of mechanical strength and density in biocomposites. These results reinforce the importance of employing advanced analytical models to fine-tune biocomposite properties.

The chapter highlights the critical role of filler composition in improving both the mechanical performance and sustainability of biocomposite pots. Physical properties, such as moisture diffusion and nutrient retention, play a vital role as they influence both plant growth and the pot's degradation during cultivation. Simultaneously, mechanical properties must ensure that the pots are strong enough to be handled yet soft enough to degrade in synchrony with plant growth, underscoring the need for a balanced approach to biocomposite design.

Future research should focus on the exploration of new materials and the application of modeling design simulations to optimize biocomposite formulations. Utilizing desirability functions, the design and optimization of these materials can be tailored to meet broader agricultural applications, further advancing the development of sustainable biocomposite technologies.

Author details

Manar E. Elashry*, Elsayed G. Khater and Samir A. Ali
Agricultural and Biosystems Engineering Department, Faculty of Agriculture, Benha University, Egypt

*Address all correspondence to: majewreshy@gmail.com

References

[1] Campilho RD. Introduction to natural fiber composites. In: Natural Fiber Composites. Boca Raton, Florida: CRC Press; 2015. pp. 1-33

[2] Kuruppalil Z. Green plastics: An emerging alternative for petroleum-based plastics. International Journal of Engineering Research & Innovation. 2011;**3**(1):59-64

[3] Balaguer MP et al. Compostability assessment of nano-reinforced poly (lactic acid) films. Waste Management. 2016;**48**:143-155

[4] Bhat KM et al. Eco-friendly and biodegradable green composites. In: Kumar B, editor. Biocomposites. London, UK: IntechOpen; 2021

[5] Bledzki AK, Mamun AA, Volk J. Physical, chemical and surface properties of wheat husk, rye husk and soft wood and their polypropylene composites. Composites Part A: Applied Science and Manufacturing. 2010;**41**(4): 480-488

[6] Jahan MS, Mun SP. Studies on the macromolecular components of nonwood available in Bangladesh. Industrial Crops and Products. 2009; **30**(3):344-350

[7] TG YG et al. Biopolymer-based composites: An eco-friendly alternative from agricultural waste biomass. Journal of Composites Science. 2023;7(6):242

[8] Achmad M, Sapsal M. Compost applicators for horticulture. IOP Conference Series: Earth and Environmental Science. 2018;**157**(1): 012007, 1-6

[9] Chand N, Shukla M, Sharma MK. Analysis of mechanical behaviour of bamboo (*Dendrocalamus strictus*) by using FEM. Journal of Natural Fibers. 2008;**5**(2):127-137

[10] Warda MA et al. The applicability of TOPSIS-and fuzzy TOPSIS-based Taguchi optimization approaches in obtaining optimal fiber-reinforced concrete mix proportions. Buildings. 2022;**12**(6):796

[11] Noryani M, Sapuan S, Mastura M. Multi-criteria decision-making tools for material selection of natural fibre composites: A review. Journal of Mechanical Engineering and Sciences. 2018;**12**(1):3330

[12] Célino A et al. Characterization and modeling of the moisture diffusion behavior of natural fibers. Journal of Applied Polymer Science. 2013;**130**(1): 297-306

[13] Ma X et al. An inner boundary condition of moisture diffusion model for simulating transient nonlinear moisture transport in Chinese fir. Heliyon. 2022;**8**(9):1-15

[14] Li R et al. An aging theory-based mathematic model for estimating the wax content of wax deposits using the Fick's second law. AICHE Journal. 2020; **66**(4):e16892

[15] Crank J. The Mathematics of Diffusion. Bristol, England: Oxford University Press; 1979. pp. 254-265

[16] Lubkowski K. Coating fertilizer granules with biodegradable materials for controlled fertilizer release. Environmental Engineering & Management Journal (EEMJ). 2014;**13**(10):1-12

[17] Ritger PL, Peppas NA. A simple equation for description of solute release

I. Fickian and non-fickian release from non-swellable devices in the form of slabs, spheres, cylinders or discs. Journal of Controlled Release. 1987;**5**(1):23-36

[18] Sharma R et al. Assessment of water retention performance of pectin-based nanocarriers for controlled irrigation in agriculture. Agricultural Research. 2017; **6**(2):139-149

[19] Cheng M et al. Environmental durability of carbon/flax fiber hybrid composites. Composite Structures. 2020; **234**:111719

[20] Yu H et al. The two stage moisture diffusion model for non-fickian behaviors of 3D woven composite exposed based on time fractional diffusion equation. Mathematics. 2023; **11**(5):1160

[21] Yu B, Yang J. Hygrothermal Effects in Composites, in Comprehensive Composite Materials II. United Kingdom: Elsevier, Oliver Walter; 2018

[22] Harper B, Staab G, Chen R. A note on the effects of voids upon the hygral and mechanical properties of AS4/3502 graphite/epoxy. Journal of Composite Materials. 1987;**21**(3):280-289

[23] Marom G. The role of water transport in composite materials. In: Polymer Permeability. Netherlands: Springer; 1985. pp. 341-374

[24] Loh W et al. Modelling anomalous moisture uptake, swelling and thermal characteristics of a rubber toughened epoxy adhesive. International Journal of Adhesion and Adhesives. 2005;**25**(1): 1-12

[25] Carter HG, Kibler KG. Langmuir-type model for anomalous moisture diffusion in composite resins. Journal of Composite Materials. 1978;**12**(2):118-131

[26] Werber F. In: Agarwal BD, Broutman LJ, editors. Analysis and Performance of Fiber Composites. New York: Wiley-Interscience, Wiley Online Library; 1980. 355 pp

[27] Elashry M et al. Theoretical and experimental physicomechanical properties of cultivating pots made from different materials. Annals of Agricultural Science, Moshtohor. 2023; **61**(3):599-610

[28] Pratibha SS, Hariprasad P. Paddy straw-based biodegradable horticultural pots: An integrated greener approach to reduce plastic waste, valorize paddy straw and improve plant health. Journal of Cleaner Production. 2022;**337**: 130588

[29] Manafi-Dastjerdi M et al. Production of biodegradable pots from cattle manure and wood waste: Effects of natural binders on mechanical performances and biodegradability. Environmental Science and Pollution Research. 2022;**29**(14):20265-20278

[30] Keskin ZS et al. Prunus mahaleb shell as a sustainable bioresource for carminic acid removal from aqueous solution: Experimental and theoretical studies. Journal of Molecular Structure. 2023;**1275**:134618

[31] Codling E et al. Keratin nursery pots as potential medium for controlled release of copper ions in root growth control in *Theobroma cacao* L. Journal of Horticulture. 2019;**6**(256):2376-0354

[32] Xue H et al. Adsorption of methylene blue from aqueous solution on activated carbons and composite prepared from an agricultural waste biomass: A comparative study by experimental and advanced modeling analysis. Chemical Engineering Journal. 2022;**430**:132801

[33] Langmuir I. The constitution and fundamental properties of solids and liquids. Part I. Solids. Journal of the American Chemical Society. 1916;**38**(11):2221-2295

[34] Gündüz F, Bayrak B. Biosorption of malachite green from an aqueous solution using pomegranate peel: Equilibrium modelling, kinetic and thermodynamic studies. Journal of Molecular Liquids. 2017;**243**:790-798

[35] Low M. Kinetics of chemisorption of gases on solids. Chemical Reviews. 1960;**60**(3):267-312

[36] Voigt W. Ueber die Beziehung zwischen den beiden Elasticitätsconstanten isotroper Körper. Annalen der Physik. 1889;**274**(12):573-587

[37] Reuß A. Berechnung der fließgrenze von mischkristallen auf grund der plastizitätsbedingung für einkristalle. ZAMM-Journal of Applied Mathematics and Mechanics/Zeitschrift für Angewandte Mathematik und Mechanik. 1929;**9**(1):49-58

[38] Hill R. The elastic behaviour of a crystalline aggregate. Proceedings of the Physical Society. Section A. 1952;**65**(5):349

[39] Halpin J. Stiffness and expansion estimates for oriented short fiber composites. Journal of Composite Materials. 1969;**3**(4):732-734

[40] Affdl JH, Kardos J. The Halpin-Tsai equations: A review. Polymer Engineering & Science. 1976;**16**(5):344-352

[41] Kalaprasad G et al. Theoretical modelling of tensile properties of short sisal fibre-reinforced low-density polyethylene composites. Journal of Materials Science. 1997;**32**(16):4261-4267

[42] Rao S, Jayaraman K, Bhattacharyya D. Micro and macro analysis of sisal fibre composites hollow core sandwich panels. Composites Part B: Engineering. 2012;**43**(7):2738-2745

[43] Cox HL. The elasticity and strength of paper and other fibrous materials. British Journal of Applied Physics. 1952;**3**(3):72

[44] Nairn JA. On the use of shear-lag methods for analysis of stress transfer in unidirectional composites. Mechanics of Materials. 1997;**26**(2):63-80

[45] Mendels D, Leterrier Y, Manson J. Stress transfer model for single fibre and platelet composites. Journal of Composite Materials. 1999;**33**(16):1525-1543

[46] Bowyer W, Bader M. On the re-inforcement of thermoplastics by imperfectly aligned discontinuous fibres. Journal of Materials Science. 1972;**7**:1315-1321

[47] Facca AG, Kortschot MT, Yan N. Predicting the elastic modulus of natural fibre reinforced thermoplastics. Composites Part A: Applied Science and Manufacturing. 2006;**37**(10):1660-1671

[48] Jaggi HS et al. Analytical interpretations of structural and mechanical response of high density polyethylene/hydroxyapatite bio-composites. Materials & Design (1980-2015). 2012;**36**:757-766

[49] Einstein A. ber die von der molekularkinetischen Theorie der W rme geforderte Bewegung von in ruhenden Fl ssigkeiten suspendierten Teilchen. Annalen der Physik. 1905;**322**(8):549-560

[50] Guth E. Theory of filler reinforcement. Journal of Applied Physics. 1945;**16**(1):20-25

[51] Cohan L. India Rubber World. 1947; **117**:N-3

[52] Kerner E. The elastic and thermo-elastic properties of composite media. Proceedings of the Physical Society. Section B. 1956;**69**(8):808

[53] Nielsen LE. Generalized equation for the elastic moduli of composite materials. Journal of Applied Physics. 1970;**41**(11):4626-4627

[54] Das O, Kim N, Bhattacharyya D. The mechanics of biocomposites. In: Biomedical Composites. Auckland, New Zealand: Elsevier; 2017. pp. 375-411

[55] Fu S-Y, Hu X, Yue C-Y. A new model for the transverse modulus of unidirectional fiber composites. Journal of Materials Science. 1998;**33**:4953-4960

[56] Verbeek C. The influence of interfacial adhesion, particle size and size distribution on the predicted mechanical properties of particulate thermoplastic composites. Materials Letters. 2003;**57**(13–14):1919-1924

[57] Jandas P, Mohanty S, Nayak S. Mechanical properties of surface-treated banana fiber/polylactic acid biocomposites: A comparative study of theoretical and experimental values. Journal of Applied Polymer Science. 2013;**127**(5):4027-4038

[58] Danusso F, Tieghi G. Strength versus composition of rigid matrix particulate composites. Polymer. 1986;**27**(9):1385-1390

[59] Levita G, Marchetti A, Lazzeri A. Fracture of ultrafine calcium carbonate/ polypropylene composites. Polymer Composites. 1989;**10**(1):39-43

[60] Das O, Sarmah AK, Bhattacharyya D. Nanoindentation assisted analysis of biochar added biocomposites. Composites Part B: Engineering. 2016;**91**: 219-227

[61] Nicolais L, Nicodemo L. The effect of particles shape on tensile properties of glassy thermoplastic composites. International Journal of Polymeric Materials. 1974;**3**(3):229-243

[62] Nielsen LE. Simple theory of stress-strain properties of filled polymers. Journal of Applied Polymer Science. 1966;**10**(1):97-103

[63] Nicolais L, Nicodemo L. Strength of particulate composite. Polymer Engineering & Science. 1973;**13**(6):469-469

[64] Bigg D. Mechanical properties of particulate filled polymers. Polymer Composites. 1987;**8**(2):115-122

[65] Pukánszky B, Turcsanyi B, Tudos F. Effect of interfacial interaction on the tensile yield stress of polymer composites. Interfaces in Polymer, Ceramic and Metal Matrix Composites. 1988:467-477

[66] Naveen J et al. Finite element analysis of natural fiber-reinforced polymer composites. In: Modelling of Damage Processes in Biocomposites, Fibre-Reinforced Composites and Hybrid Composites. Woodhead Publishing; 2019. pp. 153-170

[67] Flores EIS, Haldar S. Micro–macro mechanical relations in palmetto wood by numerical homogenisation. Composite Structures. 1988;**154**:1-10

[68] Silva ECN, Walters MC, Paulino GH. Modeling bamboo as a functionally graded material: Lessons for the analysis

of affordable materials. Journal of Materials Science. 2006;**41**:6991-7004

[69] Fu WS et al. Research on finite element model for parallel to bamboo culms axial shear. Applied Mechanics and Materials. 2014;**477**:986-989

[70] Nilsson T, Gustafsson PJ. Influence of dislocations and plasticity on the tensile behaviour of flax and hemp fibres. Composites Part A: Applied Science and Manufacturing. 2007;**38**(7): 1722-1728

[71] Trivaudey F et al. Nonlinear tensile behaviour of elementary hemp fibres. Part II: Modelling using an anisotropic viscoelastic constitutive law in a material rotating frame. Composites Part A: Applied Science and Manufacturing. 2015;**68**:346-355

[72] Del Masto A et al. Nonlinear tensile behaviour of elementary hemp fibres: A numerical investigation of the relationships between 3D geometry and tensile behaviour. Journal of Materials Science. 2017;**52**(11):6591-6610

[73] Liu X-D et al. Finite element simulation and optimization design of biodegradable seedling pot. In: Energy, Environment and Green Building Materials. London: Taylor & Francis Group; 2015. pp. 223-226

[74] Šibalija TV, Majstorović VD. Advanced Multiresponse Process Optimisation. New York: Springer; 2016. pp. 21-63

[75] Faisal MF et al. Modeling approaches for bio-based composites. In: Comprehensive Materials Processing (Second Edition). Vol. 12. Elsevier; 2024. pp. 672-692

[76] Nadhari WNAW et al. Utilization of oil palm trunk waste for manufacturing of Binderless particleboard: Optimization study. BioResources. 2013; **8**(2):1675-1696

[77] Sirodariya F, Tapre AR, Nandane AS. Formulation optimization and characterization of biodegradable containers incorporated with orange peel powder and tamarind seed powder. Journal of Food Science and Technology. 2024:1-8

[78] Novianti T, Anna I, Cahyadi I. Optimization of bioplastic's tensile strength. Journal of Physics: Conference Series. 2019;**1211**(1):012048, 1-7

[79] Mustapha AN et al. Taguchi and ANOVA analysis for the optimization of the microencapsulation of a volatile phase change material. Journal of Materials Research and Technology. 2021;**11**:667-680

[80] Ghaleb AM et al. Assessment and comparison of various MCDM approaches in the selection of manufacturing process. Advances in Materials Science and Engineering. 2020; **2020**(1):4039253

[81] Elashry M et al. Physical and mechanical properties of composite materials to enhance cultivating pot mix proportions via Topsis-Taguchi approach. MISR Journal of Agricultural Engineering. 2024;**41**(1):125-140

[82] Balasubramaniyan S, Selvaraj T. Application of integrated Taguchi and TOPSIS method for optimization of process parameters for dimensional accuracy in turning of EN25 steel. Journal of the Chinese Institute of Engineers. 2017;**40**(4):267-274

[83] Behera AK et al. Effect of nanoclay on physical, mechanical, and microbial degradation of jute-reinforced, soy milk-based nano-biocomposites. Polymer

Engineering & Science. 2014;**54**(2):345-354

[84] Sumesh K, Kanthavel K. Optimizing various parameters influencing mechanical properties of banana/coir natural fiber composites using grey relational analysis and artificial neural network models. Journal of Industrial Textiles. 2022;**51**(4_Suppl.):6705S-6727S

[85] Abueidda DW et al. Prediction and optimization of mechanical properties of composites using convolutional neural networks. Composite Structures. 2019;**227**:111264

[86] Vineela MG, Dave A, Chaganti PK. Artificial neural network based prediction of tensile strength of hybrid composites. Materials Today: Proceedings. 2018;**5**(9):19908-19915

[87] Felix EF, Possan E, Carrazedo R. A new formulation to estimate the elastic modulus of recycled concrete based on regression and ANN. Sustainability. 2021;**13**(15):8561

Chapter 9

Transformation Journey of Monomers into Copolymers by Free Radical Polymerization (FRP), Atom Transfer Radical Polymerization (ATRP), and Ionic Polymerization (IP)

Bulend Ortaç, Saliha Mutlu, Ahmet Hakan Yilmaz and Sevil Savaskan Yilmaz

Abstract

Within the scope of this chapter, we provide a quick introduction to a variety of polymerization techniques that transform monomers into copolymers. We also emphasize the distinctive mechanisms that these strategies employ in order to exert control over the structure of polymers. Several different polymerization techniques that are utilized to convert monomers into copolymers are discussed in this chapter. Particular attention is paid on the specific processes and structural control capabilities of each of these strategies. Free radical polymerization (FRP), atom transfer radical polymerization (ATRP), and ionic polymerization (IP) are the primary topics of discussion. The argument addresses the basic ideas guiding these three forms of polymerization and evaluates their respective strengths and constraints. FRP is widely used in industrial applications because of its simplicity and efficiency. Conversely, ATRP is quite perfect for the development of novel materials since it gives exact control over polymer architecture. Furthermore, the combination of anionic and cationic polymerization simplifies the synthesis of block copolymers with well-defined properties and unique properties. This chapter also looks at the circumstances of synthesis, molecular weight control, and feasible polymer forms made possible by various polymerization techniques. Comparisons between FRP, ATRP, and IP indicate differences in their structural compositions as well as their levels of efficiency. The establishment of environmentally friendly catalytic systems, the combination of hybrid techniques to expand polymer functionality for industrial and scientific advances, and the enhancement of polymerization efficiency will undoubtedly play a central role in the research that will be conducted in the future.

Keywords: free radical polymerization, atom transfer radical polymerization, ionic polymerization, copolymers, polymerization mechanism

1. Introduction

A polymer is fundamentally a molecule composed of an extended, repeated sequence of smaller monomers, usually comprising four to ten atoms. Polymers can have linear sequences of regularly repeating monomers, chains of varying lengths, or branching structures extending in several directions. The study of polymerization in FRP, an "initiator" material decomposes to produce a free radical, a molecule characterized by an unpaired electron. In this unstable state, the free radical seizes an electron from another molecule. The majority of FRP techniques do not utilize an external catalyst complex; rather, they generate radicals during polymerization by the thermal decomposition of a radical initiator at an appropriate rate. FRP is employed to manufacture polymers such as polystyrene and polyvinyl acetate. Atom transfer radical polymerization (ATRP) is a widely utilized technique for controlled, or "living," radical polymerization that inhibits processes that would otherwise interfere with the formation of polymer chains. As a result, the chain can continuously "grow" without constraint. ATRP synthesizes polymers with defined shapes and characteristics by employing an initiator and a catalyst to commence a reaction. This reaction produces radicals that can be neutralized to form latent species, which can later be reactivated as required. This method facilitates the incremental and uniform development of the polymer chain, which may be modified at various stages to yield a polymer with the precise characteristics sought by the end user. A diverse array of monomers can react through a radical mechanism to generate polymers with desirable features. Included are ethylene, vinyl chloride, vinyl acetate, vinylidene chloride, tetrafluoroethylene, acrylonitrile, methacrylonitrile, methyl acrylate, methyl methacrylate, styrene, butadiene, chloroprene, and numerous others. The mechanism of radical polymerization is a chain-radical process comprising initiation, propagation, termination, and transfer reactions as fundamental phases [1]. Ionic chain polymerization is a chain-growth reaction characterized by a charged active center, which can be either negatively charged, resulting in anionic polymerization, or positively charged, resulting in cationic polymerization. Anionic polymerization is fundamentally restricted to monomers with electron-withdrawing groups, while cationic polymerization is confined to monomers with electron-releasing substituents. Ionic polymerization necessitates rigorous reaction conditions, including elevated monomer purity. The solvent's polarity significantly affects the mechanism and rate of polymerization. Ionic polymerization entails the sequential incorporation of monomer molecules between an ionic chain terminus (negative in anionic polymerization or positive in cationic polymerization) and a counterion of contrasting charge [2]. The discovery of living anionic polymerization made it the most reliable and flexible method for creating model polymers with controlled structures, exact molecular weights, narrow molecular weight distributions, and chemical and compositional uniformity [3]. Only a few types of monomers, such as styrene, dienes, methacrylates, acrylates, ethylene oxide (EO), and vinyl pyridines, are suitable for anionic polymerization. Recent advancements have facilitated the extension of the method's applicability to a wide array of monomers, including methacrylates with bulky or functional ester groups, hexamethylcyclotrisiloxane, D3, isocyanates, and others. Block copolymers comprising methyl vinyl ether (MVE) and isobutyl vinyl ether (IBVE), namely

PMVE-b-PIBVE, PMVE-b-PIBVE-b-PMVE, and PIBVE-b-PMVE-b-PIBVE, were synthesized using live cationic polymerization and sequential monomer addition [4].

Amphiphilic block copolymers comprising poly(2-hydroxyethyl vinyl ether) (POHVE) and poly(vinyl ethers) with fluorinated alkyl groups were synthesized through the successive addition of monomers [5]. POHVE was synthesized through the polymerization of acetoxy vinyl ether following the hydrolysis of the acetoxy masking groups using sodium hydroxide in 1,4-dioxane. The 2-(p-methoxyphenyl)-ethanol-BF_3OEt_2 starting method, $CH_3CH(C_6H_4\text{-p-}OCH_3)OH\text{-}BF_3OEt_2$, was effectively utilized for the polymerization of p-alkoxystyrenes and p-hydroxystyrene (PHOS) without the necessity of protecting the hydroxyl groups [6]. Copolymers consist of at least two monomers, categorized according to their relative arrangement along the chain or monomer distribution. Copolymers are categorized based on the arrangement of monomers along the main chain. They are categorized as linear and branched copolymers [7, 8]. Two unique linear triblock copolymers, ABA and BAB, are made by linking two identical AB diblock copolymers at their B or A termini, respectively. Two unique linear triblock copolymers, ABA and BAB, are made by linking two identical AB diblock copolymers at their B or A termini, respectively. This technique produces three homologous structures: AB diblock, ABA, and BAB triblock copolymers, all with identical chemical compositions but varying in topology [9]. Block copolymers, composed of covalently bonded separate blocks, are soft materials capable of self-assembling into ordered nanoscale structures [10]. The equilibrium structure of block copolymers is affected by the volume fractions of the component blocks, copolymer architecture, and temperature. These dependencies provide "knobs" for regulating the equilibrium morphology self-assembled from block copolymers. Due to their intricate phase features, block copolymers are not only advantageous for diverse applications such as lithography [11], photonics [12], and quantum materials [13], but also offer an exceptional platform for exploring the spontaneous ordering of matter. Statistical copolymers classify linear polymers, which have major chains arranged in a linear configuration. Statistical copolymers classify linear polymers, which have major chains arranged in a linear configuration. Statistical polymers consist of many constitutional units within segments of macromolecules, where adjacent segments/units differ from one another. **Figure 1** shows the experimental setup of the Schlenk system for copolymer synthesis.

Controlled living radical polymerization techniques offer an interesting area of research and adaptable business tools that make it possible to make polymers with precise chemical structures and compositions. In the last 20 years, a lot of research has been done on the reactions and mechanisms of living radical polymerization. This has led to the creation of techniques like reversible addition-fragmentation chain transfer polymerization (RAFT) [14], nitroxide-mediated polymerization (NMP) [15], and atom transfer radical polymerization [16]. A transition metal catalyst typically activates an alkyl halide in a reduced oxidation state through ATRP, a redox catalytic mechanism that generates the corresponding alkyl radical and the transition metal in an elevated oxidation state. The process is controlled or "alive" because the cycles of activation and deactivation happen over and over again. This makes the growth rate consistent for most of the polymer chains [14]. The predominant technique for manufacturing statistical copolymers is through free radical polymerization. The free radical copolymerization of styrene and butadiene yields synthetic rubber (**Figure 2**).

Random copolymers consist of several repeating units arranged in a random sequence. The configuration of the monomeric repeating unit is unaffected by the

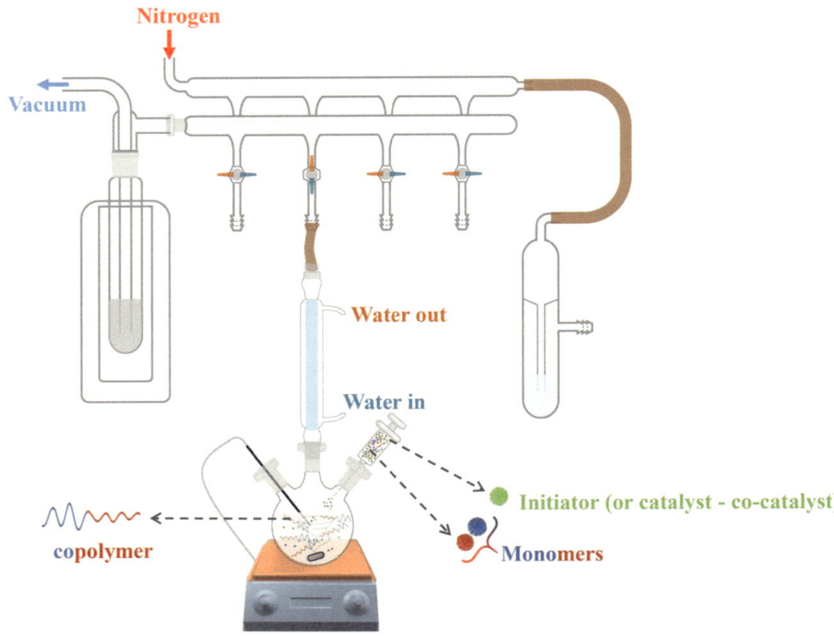

Figure 1.
Experimental setup of schlenk system for copolymer synthesis.

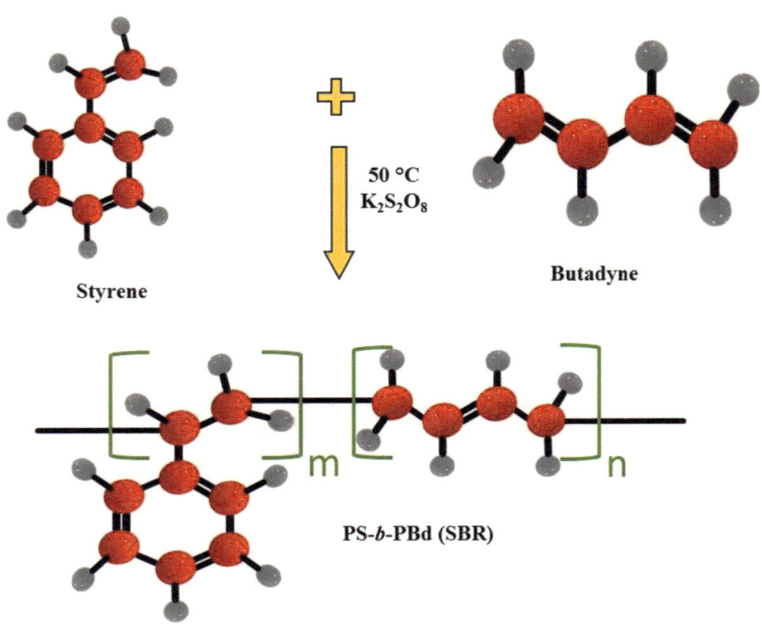

Figure 2.
Synthetic rubber mechanism made of styrene and butadien.

orientation or pattern of the neighboring unit. **Figure 3** contains examples of the polymer chains. A random copolymer is shown in **Figure 3a**, which is made when vinyl chloride and vinyl acetate copolymerize through free radical polymerization. Alternating copolymers consist of multiple equimolar monomer units arranged in

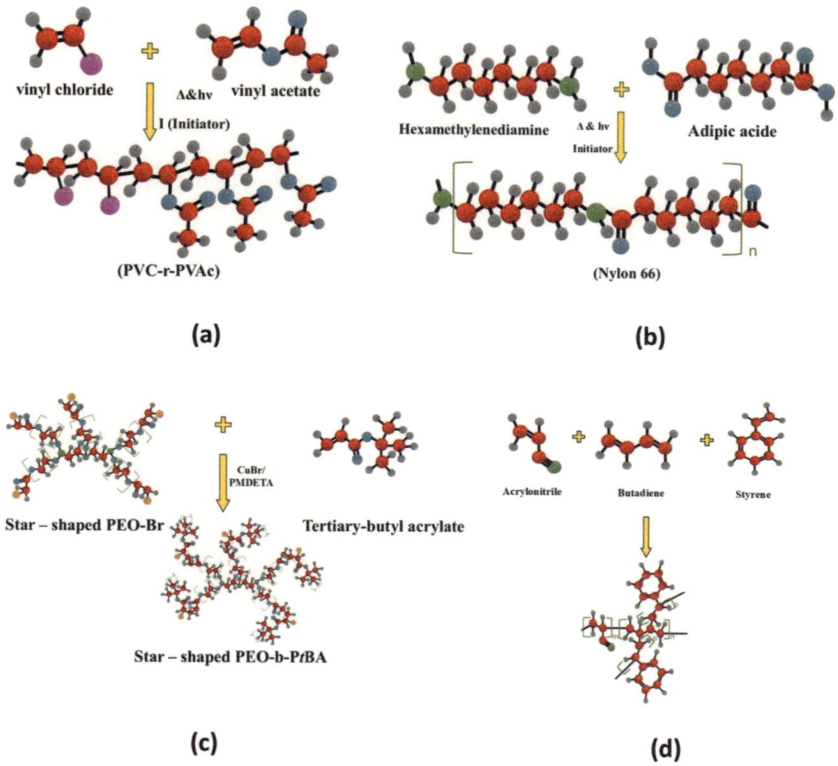

Figure 3.
Examples of copolymer chains.

a distinct, alternating sequence. They comprise a singular principal chain interconnected with alternate monomers. The formula can express them: (-A-B-)n. An instance of an alternating copolymer is nylon 6,6, composed of hexamethylene diamine and adipic acid (**Figure 3b**). Maekawa et al. reported the results of a study that could stably form sub10 nm patterns presented findings from a study that achieved stable formation of sub10 nm patterns through structural modifications of polystyrene-block-poly(methyl methacrylate) (PS-b-PMMA) [17]. Star copolymers consist of a central core polymer connected to several polymeric chains, as illustrated in **Figure 3c**. Block copolymers consist of extended sequences of each monomer type inside a polymer chain. The constituent monomers or polymers are organized in a branching configuration in branched copolymers. The branched copolymer can be classified into the following categories: Graft copolymers consist of the branching of one or more blocks (extended sequences of monomer units) onto the primary polymer chain backbone. An instance of a graft copolymer is poly(acrylonitrile-co-butadiene-co-styrene). **Figure 3d** depicts a representative example of a graft copolymer.

In the last 60 years, block copolymers derived from two or more monomers have become a central focus of modern macromolecular science and engineering. Controlled polymerization, first with live anionic polymerization in the late 1950s and subsequently increasing with ATRP, RAFT, ROMP, and other methodologies during the 1990s, has resulted in substantial advancements in various related domains [18]. In recent decades, the self-assembly of block copolymers (BCPs) has emerged as a viable method for the controlled fabrication of various functional materials, including diverse mesoporous and organized structures, through the soft-templating approach. BCP porous materials have garnered significant attention in energy storage and conversion (ESC) applications due to their capacity to absorb, store, and interact with guest species on their external and interior surfaces, as well as within the pore space. In comparison to alternative synthetic methods, including template-free and hard-templating techniques, BCP soft-templating protocols exhibit significant advantages in the formation of large mesopores with diameters ranging from 10 to 60 nm, making them ideal for applications necessitating the accommodation of larger species or molecules. This technique has unparalleled advantages in the adaptable regulation of pore size, architecture, and wall thickness, which are critical to the ultimate efficacy of mesoporous materials in ESC devices [19].

Table 1 presents the copolymer samples synthesized using FRP, ATRP, and IP methods.

Conducting polymers (CPs) are synthesized by chemical oxidation, electrochemical, and photopolymerization methods. The monomer is oxidized using chemical oxidants, including $FeCl_3$ and $(NH_4)_2S_2O_8$. The optimal approach for large-scale production of CPs, such as polypyrrole (Ppy) and polyaniline (PANI), is through

Copolymer name	Polymerization method	Initiator kind
Polystyrene-block-polyethylene oxide (PS-b-PEO)	Anionic	Benzyl potassium
Methyl vinyl ether (MVE) and isobutyl vinyl ether (IBVE) of the types PMVE-b-PIBVE, PMVE-b-PIBVE-b-PMVE, and PIBVE-b-PMVE-b-PIBVE	Cationic	Acetal/trimethylsilyl iodide/ZnI_2 as activator
Polystyrene-block-poly(glycidyl methacrylate) (PS-*b*-PGM)	Anionic	*sec*-BuLi(*sec*-butyllithium)/ excess LiCl (lithium chloride) and DPE(1,1-diphenylethylene)
Methoxypoly-(ethylene glycol)-b-poly(D,L-latide-co-glycolide) (mPEG-PLGA)	ROP	mPEG
Poly(methyl methacylate)-co-poly(hydroxyethyl methacrylate)-g-poly(ε-caprolactone) (PMMA-co-PHEMA)-g-PCL	Photoinduced metal-free ATRP and ROP	EBPA, P_2-*t*-Bu
PVP-stabilized PsS latex	Living free radical	TEMPO (2,2,6,6-tetramethylpiperidine 1-oxyl, 2,2,6,6-tetramethyl-1-piperidinyloxy)
Poly(oligo(ethylene glycol) methyl ether methacrylate)-*b*-poly(methyl methacrylate) (POEGMA-b-PMMA)	Free radical	AIBN (azobisisobutyronitrile)
Poly(N-isopropyl acrylamide-b-2-methacryloyloxyethylphosphorylcholine-b-N-isopropyl acrylamide) (PNIPAM-*b*-PMPC-*b*-PNIPAM)	ATRP	CuBr/bby, CuBr/Me_4Cyclam

Table 1.
The copolymer examples obtained with FRP, ATRP, and IP methods.

chemical oxidative synthesis. Monomers undergo oxidation in the electrochemical method. The oxidation of the monomer occurs through the light polymerization process, utilizing a photoinitiator and light [20, 21].

The chemical oxidative polymerization process is executed using chain initiation and chain addition mechanisms. The monomer is dissolved by the dopant acid solution in this instance. After dissolving the oxidant in a suitable solvent (often water) and introducing it to the mixture dropwise, continuous agitation is sustained for 4–5 hours. The precipitate undergoes drying, purification, and filtration thereafter. Chemical synthesis has several advantages over electrochemical synthesis. Chemical synthesis significantly enhances the large-scale production of CPs. CPs are produced by potentiodynamic methods, constant current, or constant potential electrochemical polymerization. An electrolyte and a monomer are used in an electrochemical cell to carry out the electrochemical polymerization process. There are reference, counter, and working electrodes in the electrochemical cell. Chemical synthesis methods provide complementary other options for the covalent modification of the CP backbone [21]. The electrochemical method has various drawbacks, such as the challenge of removing the formed layer from the electrode surface and the variability in bulk-conducting polymers. Monomers are subjected to energy sources such as lasers, visible light, and ultraviolet (UV) radiation to promote photopolymerization. Nevertheless, the synthesis of CPs is not a suitable method for photopolymerization. Therefore, it is important to conduct a thorough evaluation of the merits and demerits of chemical and electrochemical techniques in CP synthesis, in accordance with material and application requirements.

Table 1 shows the results of the FRP, ATRP, and IP copolymer samples that were produced.

2. Polymerization methods of copolymer formation

2.1 Free radical polymerization (FRP)

Free radical polymerization is recognized as a conventional chain polymerization that employs extremely reactive radicals as active species. Four fundamental reactions classify the polymerization process: initiation, propagation, termination, and chain transfer. Commonly utilized initiators encompass various peroxides, including alkyl peroxides (e.g., tert-butyl peroxide), hydroxyl peroxides (e.g., tert-butyl hydroperoxide, cumene hydroperoxide), peroxoesters (e.g., tert-butyl peroxobenzoate), and peroxo anhydrides (e.g., benzoyl peroxide). Diazocompounds are extensively utilized as a source of radicals for polymerization. Among these, azobisisobutyronitrile (AIBN) is the most frequently utilized compound. Its disintegration occurs in the elimination of a nitrogen molecule and the production of two carbon-centered radicals (**Figure 4**).

The polymerization temperature and the required rate of radical production dictate the selection of an appropriate initiator. Typically, we use AIBN at 50–70°C, benzoyl peroxide at 80–95°C, and tert-butyl peroxide at 120–140°C for maximum efficacy. The activation energy for initiation is approximately equivalent to the bond dissociation energy of the initiator, often ranging from 105 to 175 kJ/mol. The initiator molecule's decomposition generates a radical, which bonds to the monomer's double bond to form a propagating chain. The concurrent polymerization of multiple monomers within a single vessel yields copolymers. These macromolecules comprise

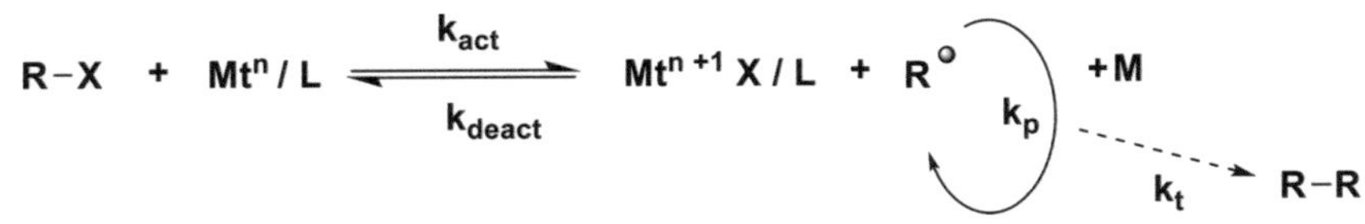

Figure 4.
The mechanism of FRP.

various monomer units linked in a specific sequence. This sequence classifies random, alternating, gradient, grafted, and block copolymers [22].

The synthesis of multiblock copolymers has proven to be an effective method for understanding complex structures, analyzing polymer characteristics, and confirming the end-group integrity of a specific polymerization technique. Even though it is the most commonly used method, reversible addition-fragmentation chain-transfer (RAFT) polymerization often faces difficulties because of the inevitable and excessive use of radical initiators, which can result in increased termination and reduced endgroup integrity. This study utilizes acid in RAFT polymerization to improve the synthesis of multiblock copolymers. The use of a minimal quantity of acid resulted in a fourfold reduction in the overall necessary concentration of radical initiators, facilitating the synthesis of a variety of well-defined multiblock copolymers with differing degrees of polymerization (DP) per block [23].

2.2 Atom transfer radical polymerization (ATRP)

ATRP is an effective controlled radical polymerization (CRP) technique for synthesizing polymers with specified molecular weights, narrow molecular weight distributions (low dispersity [Đ]), diverse chemical compositions, and intricate topologies. This approach employs a catalyst to facilitate the reversible activation and deactivation of polymers having halide end-groups, wherein the halide is eliminated during activation to produce a reactive polymer radical and subsequently reinstated during deactivation to form a "dormant" polymer chain. This reversible deactivation mechanism suppresses bimolecular radical termination events, hence regulating polymer development for the synthesis of well-defined macromolecules with diverse functions and topologies [24–26]. Whittaker was the first person to show how to make multiblock copolymers with very short blocks (DP = 2 per block) when zerovalent copper was present in ATRP [27]. Nonetheless, significant termination was distinctly observable using size exclusion chromatography (SEC), resulting in the progressive accumulation of Cu(II) and the unavoidable stoppage of polymerization, so substantially restricting the number of blocks. **Figure 5** shows the general mechanism of ATRP.

The initial synthesis of a series of amphiphilic triblock copolymers (PAA-b-PS-b-PAA) was reported by Fujitaa et al. using atom transfer radical polymerization (ATRP). The synthesis of macroinitiators (PS) led to the fabrication of the corresponding amphiphilic triblock copolymers, each with varying quantities of hydrophobic and hydrophilic units [28]. Polymerizations in water-based dispersion media are helpful because they allow heat to move quickly, have low viscosity, contain few volatile chemicals, and have been used in industry before. The potential for the development of diverse macromolecular structures and the enhancement of life has been demonstrated by the application of ATRP in disparate media over the past 20 years. The catalyst dosage has been considerably reduced, and the reaction setup has been

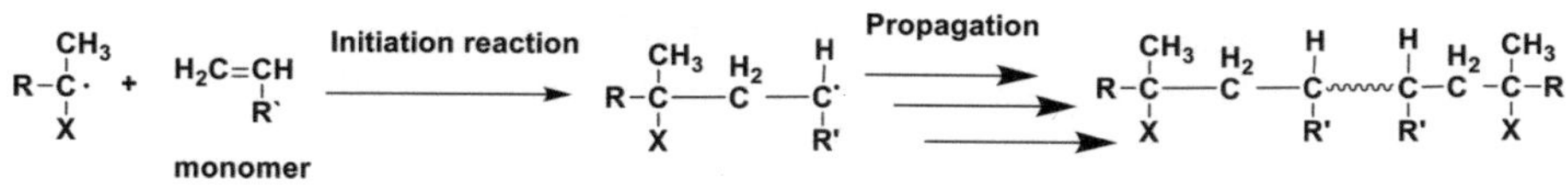

Figure 5.
The general mechanism of ATRP.

streamlined as a result of the introduction of multiple methods for activator regeneration in ATRP. ATRP techniques with low-ppm catalyst loadings have made it easier to create a wider range of polymer architectures and functional polymer particles. These are used for molecular recognition, drug delivery, bioquantification, and making advanced coatings [29]. Flejszar et al. have developed a comprehensive procedure for the synthesis of polymer brushes with complex structures that is based on green chemistry principles. This method is known as the surface-initiated atom transfer radical polymerization (SI-ATRP). A coating made of ampholytic polymers on silicon substrates was made easier by using macromolecular chains that are negatively charged (poly(acrylic acid, PAA)) and positively charged (poly(2-(dimethylamino) ethyl methacrylate, PDMAEMA)). This method, called Si-gPHEMA-g-(PDMAEMA-b-PAA), is very cheap and good for the environment. It makes Si wafers that are grafted with brushes of poly(2-hydroxyethyl methacrylate) (PHEMA) core and PDMAEMA and PAA as block polymer branches. This happens when the synthesis protocol is scaled up [30]. Significant methods for creating precisely defined functional polymer coatings include surface-initiated atom transfer radical polymerization (SI-ATRP) [31] and surface-initiated photoinduced electron transfer-reversible addition-fragmentation chain transfer polymerization (SI-PET-RAFT) [32]. Polymers have a narrow molecular weight distribution (MWD), well-defined molecular weights (MW), and a set composition that comes from the orderly joining of monomers into chain segments [33]. Furthermore, to address constraints in the changed region and minimize the volume of the mixture utilized, microliter-scale diffusion-controlled procedures (μL-scale SI-ATRP) have been developed [34–37]. The features of ATRP, particularly its controlled nature, facilitate the direct synthesis of macromolecules with intricate topology and molecular architecture, including bottlebrushes, also referred to as molecular brushes [25]. The polymerization mechanism of ATRP is dependent on the equilibrium between propagating radicals, which are generated through the homolytic cleavage of the C-X bond (where X represents a halogen atom) by a transition metal complex in a lower oxidation state (CuI/L), and dormant alkyl halides. Additionally, the X atom is transferred from the oxidized form of the complex (X-CuII/L) to the radical [38]. However, "traditional" ATRP requires substantial quantities of a transition metal, typically copper. Consequently, the incorporation of an extra redox cycle enables the reduction of copper concentration to several tens of ppm through the continual renewal of the catalytic complex [39]. Commonly employed internal reducing agents include zerovalent metals (FeO [40], CuO [41]) which function as both a supplemental activator and a reducing agent in SARA ATRP, as well as compounds such as ascorbic acid (AsAc) [42] or glucose [43] utilized in the ARGET ATRP technique. Consequently, external regulation of deactivator reduction may be accomplished through the application of an electric current in electrochemically mediated ATRP (eATRP) [44] or light in photomediated ATRP (photo-ATRP) [45, 46]. Significantly, SI-ATRP enables the grafting of multifunctional and stimuli-responsive polymer brushes, including polyelectrolytes, resulting in anionic, cationic, and zwitterionic surfaces [47, 48]. Similar to inorganic substances, we can classify polyelectrolytes as weak or strong based on their degree of ionization in solutions [49]. Strong polyelectrolytes are predominantly totally dissociated across most pH levels, while weak polyelectrolytes can exist in both ionized and unionized forms, contingent upon the characteristics of the surrounding environment [50]. Consequently, weak polyelectrolytes like PAA or PDMAEMA can be defined by their capacity to respond to variations in pH and ionic strength by altering their conformation, attributable to changes in hydrogen bonding, electrostatic interactions, or hydrophobic interactions. An increase in the ionization degree of the polyelectrolyte, contingent upon the pKa value, leads to brush expansion, whereas a low ionization degree induces the collapse of the polymer chains [51, 52]. In recent years, there has been a scarcity of studies regarding complicated architectural polyelectrolyte brushes grafted from surfaces [53]. A new study describes how to make cationic bottle brush-like polymers on silicon wafer surfaces using SI-ATRP and ring-opening metathesis polymerization, which could be used in solid-state batteries [54]. Initially based on peptide synthesis, solid-phase polymer synthesis has developed into a highly effective method for producing sequence-controlled macromolecules. Szczepaniak and others look into solid-phase polymer synthesis by attaching growing polymer chains to the Chem Matrix (CM) resin, which is made of poly(ethylene glycol) (PEG). CM resin is different from other hydrophobic supports because it can expand in both polar and non-polar solvents. This makes the filtering, rinsing, and drying processes easier. The integration of ATRP with solid-phase methodologies facilitated the high-yield grafting of precisely specified block copolymers. This method is great for sequence-controlled polymer synthesis because it makes di-, tri-, tetra-, and penta-block copolymers quickly and easily while giving you better control over molecular weight and dispersity [55].

In 2000, Guan and Smart published the initial demonstration of the importance of light in ATRP without the use of traditional photoinitiators or dye sensitizers, revealing that visible light enhances the polymerization rate of methyl methacrylate (MMA) in conventional ATRP utilizing CuCl as the catalyst [56]. In 2011, Yagci and colleagues published the inaugural study on UV-mediated ATRP of MMA in bulk, wherein $CuBr_2$ was reduced in situ to CuBr, which subsequently facilitated the polymerization [57]. In 2012, the Mosnacek group revealed that photo-ATRP could be conducted with ppm quantities of $CuBr_2$, facilitating the regulated polymerization of MMA across several wavelengths [58]. In that year, Matyjaszewski and colleagues documented the regulated polymerization of acrylates and methacrylates utilizing ppm levels of $CuBr_2$ under visible and sunlight exposure [59]. In 2014, 2 years later, Haddleton and colleagues employed a $CuBr_2$-Me_6Tren (Tren = tris(2-aminoethyl)amine) complex to facilitate the polymerization of various acrylic monomers under UV irradiation, marking the first report of quantitative monomer conversions for varying degrees of polymerization, exceptionally rapid rates (completion within 2 hours in dimethyl sulfoxide [DMSO]), and the formation of in situ block copolymers [60]. In addition to copper-mediated polymerizations, a notable advancement in photo-ATRP was reported in 2012 by Fors and Hawker, who utilized an iridium (Ir)-based photoredox catalyst to enable the controlled polymerization of methacrylates [61]. In 2014, Boyer and colleagues pioneered the initial instance of PET-RAFT polymerization [32]. The polymerization of both conjugated and unconjugated monomers was achievable by employing minimal quantities of the photoredox catalyst fac-[$Ir(ppy)_3$] (1 ppm to monomers) and utilizing a low-energy visible LED light source (λ = 435 nm) [62].

The researchers led by Yilmaz et al. [63] made new bifunctional macroinitiators for ATRP using poly(methyl methacrylate) (PMMA), poly(styrene) (PS), and poly(ε-caprolactone) (PCL). These were made from a p-xylene-bis(1-hydroxy-3-thia-propanoloxy) core and were used to make ABA-type block copolymers. A new ATRP initiator called 1,4-phenylenebis(methylene-thioethane-2,1-diyl) bis(2-bromo-2-methylpropanoate) (PXTBR) was used. It was made by reacting p-xylene-bis(1-hydroxy-3-thiapropane) (PXTOH) with α-bromoisobutyryl bromide. The PMMA and PS macroinitiators were synthesized using ATRP of MMA and S, utilizing PXTBR as the initiator and copper(I) bromide/N,N,N′,N″,N″-pentamethyldiethylenetriamine (CuBr/PMDETA) as the catalytic system. Yilmaz et al. [64] conducted a study on poly-MMA macroperoxy initiators synthesized via the ATRP of MMA using bromomethyl benzyl t-butyl peroxy ester (t-BuBP) as the initiator and a catalyst system comprising CuX (X: Br or Cl)/2,2′-bipyridine (bpy) at temperatures of 0, 20, 30, and 0°C. The ATRP reaction does not change the peroxygen groups because the low temperatures are not high enough to facilitate their breakdown. The peroxygen groups in PMMA macroperoxy initiators can facilitate the correct reaction with a monomer to produce block or graft copolymers. Poly(MMA-b-S) block copolymers were made with S and graft copolymerized with PBd and RSS-3 natural rubber, making crosslinked poly(MMA-g-PBd) and poly(MMA-g-RSS-3) graft copolymers. In addition, Yilmaz et al. [65] looked at how to make block and graft copolymers using poly(methyl methacrylate) (poly-MMA) as a macroinitiator and peroxygen. This macroinitiator was prepared through the ATRP of MMA in the presence of bis(4,4′-bromomethyl benzoyl peroxide) (BBP). According to Kul et al. [66], the atom transfer radical polymerization of tert-butyl acrylate (tBA) was started by a five-arm star-shaped poly(ethylene oxide) (PEO) with terminal bromide groups. This led to the formation of five-arm star-shaped poly(ethylene oxide)-block-poly(tert-butyl acrylate) block copolymers. A copper(I) bromide-/pentamethyl diethylenetriamine catalytic system was used in acetonitrile as the solvent, and the polymerization happened in a controlled way.

Macrocyclic polymers are new building blocks for complex polymer structures, like cyclic combs and sunflower and jellyfish-shaped topologies, that have better physical properties [67, 68]. The diverse chemical compositions of cyclic polymers were possibly utilized in lithography, emulsion stabilization, self-assembly, dye extraction, drug administration, antimicrobial applications, and topological gels [69–71]. Aswale et al. disclose the synthesis of a triblock copolymer with a spirocyclic mid-block. Using the ATRP to make a spirocyclic polystyrene (cPS) mid-block was then subjected to end-group azidation and a copper (I)-catalyzed azide-alkyne cycloaddition process with a tetra-functional initiator [72]. Double hydrophilic block copolymers (DHBCs) are a type of water-soluble macromolecules that are getting more and more attention because they can be used in many situations and have unique properties when they dissolve in water [73]. They often consist of a single functional block that reacts to external stimuli, such as pH, temperature, or ion complexation, facilitating self-assembly into distinct nanostructures [74].

Two distinct synthesis routes provide the amphiphilic block copolymer poly(ethylene oxide)-block-poly(tert-butoxycarbonylaminomethylacrylate). In the first method, click chemistry connects two end-functionalized segments, while the second method uses ATRP to extend the chain of a poly(ethylene oxide) (PEO)-based macroinitiator. In both instances, the connecting unit comprises an amide group, which is essential for the successful deprotection of the appropriate polymer precursor without cleavage of either segment. Researchers have synthesized a unique double hydrophilic block copolymer, poly(ethylene oxide)-block-poly(dehydroalanine). Various fields, such as medication delivery systems and biomineralization, utilize the relevant polymers [75]. Since it was first introduced in 2014, photoinduced organocatalyzed atom transfer radical polymerization (O-ATRP) has gained a lot of attention. Using organic photoredox catalysts and taking advantage of the many benefits of standard ATRP. O-ATRP makes it easier to make well-defined polymers in moderate reaction conditions [76]. Wang and Zhang made a group of three double-hydrophobic copolymers: PDEAEMA-b-PEG-b-PDEAMA; these were made using the ATRP method with 2-(N,N′-diethylamino)ethyl methacrylate (DEAEMA) and 2-bromoisobutyrate-terminated PEG (Br-PEG-Br), which is made when PEG is esterified with BIBB [77].

2.3 Ionic polymerization (IP)

Ionic polymerization, similar to the established radical polymerization, is a chain polyaddition process. The process starts with a monomer reacting with a species that can give the monomer a highly charged or electrically charged active group. When this group interacts with another monomer molecule, a covalent bond forms between them. At the same time, the active group on the newly added molecule is also renewed. The repeating covalent attachment of monomer molecules to the active end-group, along with the concurrent regeneration of the end-group on the additional unit, results in the elongation of the polymer chain. We classify ionic polymerization as traditional cationic when the active terminal group has a positive charge, or as pseudocationic (or cationic coordination) when this group forms the positive end of an active dipole. In the same way, we call ionic polymerization "traditional anionic" when the active group has a negative charge and "pseudoanionic" (or "anionic coordination") when the active group is the negative pole of an active dipole. The polymerization system must contain the appropriate counterions to maintain electric neutrality when generating electrically charged end-groups. **Figure 6** shows general mechanism of IP.

Monomers that can generate stable carbanionic species are susceptible to anionic polymerization. Most of the time, these monomers are styrenic or diene monomers. They can also be cyclic monomers that can react with nucleophiles to help open rings. For the anions to be safe from possible nucleophilic attacks from other species, the double bond needs to have substituents that can stabilize the negative charge by taking it away. As a result, the monomer unit must be excluded or protected by a suitable derivative in order to prevent interference during anionic polymerization by strong electrophilic groups or relatively acidic proton-donating groups, such as amino-, carboxyl-, hydroxyl-, or acetylene functional groups. Conversely, aromatic rings, double bonds, carbonyl, ester, cyano, and sulfone groups stabilize the negative charge and facilitate anionic polymerization. When polar substituents like carbonyl, cyano, or nitro groups are present, things get complicated because they can react with either the initiator or the spreading anionic species. Living polymerization of monomers with these polar substituents is only possible under very specific conditions, like

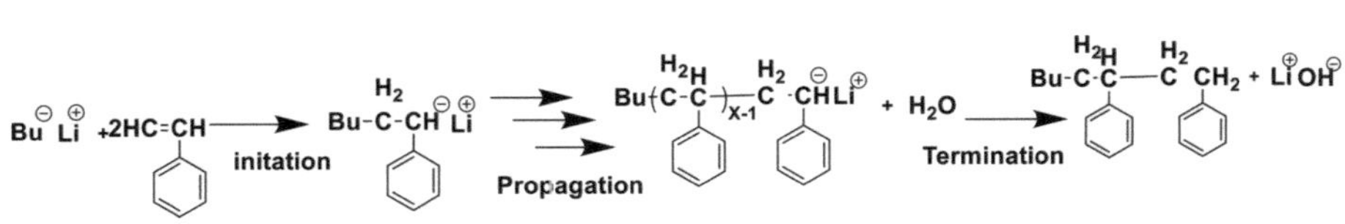

Figure 6.
General mechanism of IP.

using heavy initiators, a specific counterion, and a low temperature for polymerization. Termination reactions frequently transpire, resulting in inadequate control over the molecular weight and molecular weight distribution. Consequently, the anionic polymerization of nitroalkenes, vinyl silanes, halogenated monomers, and acrylonitrile does not occur in a living state. The following monomers have been successfully polymerized under anionic polymerization conditions: styrene and styrene derivatives (a-methyl styrene, p-methyl styrene, t-butyl styrene, etc.), dienes (e.g., isoprene, butadiene, etc.) [78, 79], (meth)acrylates (at low temperatures, using bulky initiators to avoid side reactions with the carbonyl group), vinyl pyridines [80], cyclic siloxanes [81], epoxides [82], and lactones [83]. The alkyllithium compounds are the most versatile and practical anionic initiators. They are commonly available commercially, can be easily prepared by reacting the corresponding alkyl chloride with Li metal, are soluble in a wide range of solvents, including hydrocarbons, and are, most importantly, effective initiators for anionic polymerization. The organolithium compounds are distinguished from other organic compounds containing alkali metals by the fact that the C–Li bond possesses the characteristics of both covalent and ionic bonds. This is due to the fact that lithium has the smallest radius, the highest electronegativity, and the highest ionization potential with respect to the other alkali metals [84]. Furthermore, there are unoccupied p orbitals with relatively low energy that are accessible for bonding. In comparison to the solubility of anions with other alkali metals as a counter ion, the organolithium compounds exhibit a higher solubility in hydrocarbon solvents due to the intense aggregation of the ionic pairs [85], and the covalent character of the C–Li bond. However, researchers discovered that these compounds form aggregates in both the solid state and solution. The structure of the organic moiety significantly influences aggregation. In hydrocarbon solvents, unhindered alkyllithium compounds form hexameric aggregates. However, the aggregates are tetrameric when the alkyl group has a branching point at the a- or b-carbon. The nature of the solvent, the concentration of the solution, and the temperature also affect the degree of association. In general, the degree of association is reduced by decreasing concentration, using a powerfully solvating solvent (aromatic solvents are stronger solvating agents than aliphatic solvents), increasing the temperature, and using an organic group capable of delocalizing electrons [86]. It was demonstrated that the degree of association of the alkyllithium initiators is directly correlated with their reactivity: the initiator's reactivity increases as the degree of association decreases, which compares the relative reactivities of various alkyllithium initiators to the degree of association for styrene and diene polymerizations [87].

For an extended period, cationic polymerization was regarded as the less suitable polymerization method for the synthesis of polymers with controlled molecular weights and narrow molecular weight distributions. This behavior was linked to the carbocations' inherent instability, which is susceptible to termination reactions, isomerization, and chain transfer [88]. The most frequently performed procedure is the elimination of the b-proton of the cation, which is acidic as a result of the vicinal positive charge. Nevertheless, in the past two decades, innovative initiation systems have been created to facilitate the living cationic polymerization of a diverse array of monomers. Monomers that contain electron-releasing groups are susceptible to cationic polymerization [89]. These monomers can be categorized into two primary groups: (1) vinyl monomers, which have a carbocation as their reactive end-group, and (2) heterocyclic monomers, which have one or two heteroatoms within the ring structure and have an onium ion as their reactive end-group (e.g., oxonium, sulfonium, phosphonium). The carbocationic polymerization will be the sole subject of our discussion. The following are representative monomers that can undergo carbocationic polymerization: N-vinylcarbazole, vinyl ethers, 1,1-dialkyl substituted ethylenes, such as isobutylene, styrene, and styrenic derivatives. The stability of the carbocation that is formed, the nucleophilicity of the monomer, and the steric hindrance effects are the most critical parameters that influence the polymerizability of a specific monomer. Two primary classes of compounds are recognized as initiators of carbocationic polymerization. Strong Bronsted acids and Lewis acids are these. Bronsted acids can initiate the cationic polymerization of vinyl monomers. Their initiating capacity is contingent upon the monomer basicity and the acid strength [90]. The most reactive initiator is Bronsted acid, which has the maximum acidity (low pKa). The carbocation undergoes a reaction with strong nucleophilic counteranions, resulting in the formation of nonpropagating covalent compounds. This explains why strong hydrogen halides with highly nucleophilic halide ions, with the exception of N-vinylcarbazole, are unable to promote cationic polymerization. This is due to the formation of extremely stable carbocationic species. Lewis acids are halides and alkyl halides of Group III metals and transition metals, including BF_3, $AlCl_3$, $SnCl_4$, PCl_5, $SbCl_5$, $TiCl_4$, $RAlCl_2$, and R_2AlCl. In only a few instances has the direct initiation of cationic polymerization been observed by Lewis acids. An autoionization procedure has been suggested for AlX_3 (X 1/4 Cld, Brd), which is subsequently added to the monomer [91]. The Lewis acid is most effective when used in conjunction with a second compound, which is referred to as a "coinitiator." The coinitiator is comprised of either a proton donor (protogen), such as water, hydrogen halide, alcohol, and carboxylic acid, or a carbocation donor (cationogen), such as triphenylmethyl chloride and t-butyl chloride. Initiation can also be accomplished through alternative methods, including high-energy electrons, g-irradiation, and light (photoinitiation) [92]. More recently, organometallic complexes, particularly metallocenes and half-metallocenes, have been effectively utilized for the cationic polymerization of isobutylene and other monomers after being activated with appropriate compounds (methylaluminoxane or fluoroarylborates) [93].

We classify ionic polymerization as cationic when the active terminal group has a positive charge or as pseudocationic when this group forms the positive end of an active dipole. We classify ionic polymerization as traditional anionic when the charge of the active group is negative or as pseudoanionic when the active group forms the negative end of an active dipole. When electrically charged end-groups are generated, appropriate counterions must be present in the polymerization system to maintain electric neutrality. The counterions facilitate the formation of ion pairs with the increasing active species during the initiation and propagation reactions. These ion pairs often occur in three thermodynamically different forms: tight ion pairs, loose ion pairs, and free ion pairs, which are crucial for determining polymerization characteristics. When cationic polymerization happens, a counterion that is very nucleophilic will react with the carbenium ion to make a covalent link. This link does not normally start cationic monomers. Ionic liquids have an ionic environment that must change the balance of ion pairs because they are very polar and have a lot of charges. This changes the polymerization process. Cationic polymerization is a significant method for synthesizing (co)polymers with controllable molecular weight and monomer arrangement. Particularly, living/controlled cationic polymerization is an appealing method for the creation of precisely specified polymers, including telechelic, star, and graft copolymers. The synthesis of poly(isobutylene-co-p-methylstyrene) random copolymer, butyl rubber, and poly(styrene-b-isobutylene-b-styrene) thermoplastic elastomer copolymers via conventional cationic polymerization typically employs halogenated alkanes (e.g., methyl chloride) as solvents, which contribute to environmental pollution due to their toxicity, volatility, and corrosiveness. Green solvents should replace them. Furthermore, several Lewis acids (e.g., BF_3, $SnCl_4$, $TiCl_4$, $AlCl_3OBu_2$, and $AlCl_3$) were employed as coinitiators in cationic polymerization. The separation of these Lewis acid catalysts from the reaction products is challenging, and their reuse or disposal poses significant difficulties for the industry [94]. Controlled cationic polymerization has been a significant domain of polymer science for four decades. Higashimura's group [95] in Japan and Kennedy's group [96] in the United States identified living/controlled cationic polymerization between 1984 and 1986.

Yan Zhu et al. have integrated green chemistry with cationic polymerization to investigate environmentally friendly and energy-efficient techniques for synthesizing polyolefin products. Because of this, many eco-friendly starting materials have been made, such as $FeCl_3$, $AgClO_4$, and Maghnite-H^+. These have been combined with eco-friendly reaction media like water, ionic liquids, supercritical CO_2, and non-polar solvents [97]. The variety of monomers is likewise broadening, encompassing isobutylene, vinyl ether, styrene, and its derivatives, as well as N-vinylcarbazole.

Yilmaz et al. studied cationic polymerization of tetrahydrofuran (THF) and epichlorohydrin (ECH) was performed with peroxy initiators synthesized from bis(4,4′-bromomethyl benzoyl) peroxide (BBP) or bromomethyl benzoyl *t*-butyl peroxy ester (*t*-BuBP) and $AgSbF_6$ or $ZnCl_2$ system at 0°C to obtain the poly(THF-*b*-ECH) macromonomeric peroxy initiators. Poly(THF-*b*-ECH-*b*-MMA) and poly(THF-*b*-ECH-*b*-S) block copolymers were synthesized by bulk polymerization of methyl methacrylate (MMA) and styrene (S) with poly(THF-*b*-ECH) initiators [98]. Also, Kul et al., synthesized novel macromonomeric peroxy initiators of S with the cationic copolymerization. Poly(styrene-block-methyl methacrylate) block copolymers [poly(S-b-MMA)] were synthesized by bulk polymerization of methyl methacrylate (MMA) with poly-S initiators [99].

Living polymerization techniques have made it possible to make copolymers with very specific topologies, such as block, star, macrocyclic, and telechelic copolymers. These are examples of big steps forward in molecular design and control in polymer science [97]. The initial step of an optimal live anionic polymerization is rapid relative to the propagation reaction, and neither transfer nor termination events take place. Active chain ends persist in a "living" state and are terminated solely upon the introduction of a terminating agent. The molecular weight of the resultant polymer is determined by the monomer/initiator ratio, yielding narrow molecular weight ranges (M_w/M_n = 1.1) [100]. Since Szwarc's study on the anionic polymerization of S, many more studies have been done on how live polymerization works. Currently, cationic, radical, metal-catalyzed, and group-transfer initiators have all been used to achieve living polymerization [101].

The PMVE-b-PIBVE, PMVE-b-PIBVE-b-PMVE, and PIBVE-b-PMVE-b-PIBVE block copolymers were synthesized using "live" cationic polymerization through the sequential addition of methyl vinyl ether (MVE) and isobutyl vinyl ether (IBVE) monomers by Beatrice Verdonck et al. The polymerizations were conducted using the acetal/trimethylsilyl iodide combination as the initiator and ZnI_2 as the activator [4].

3. The comparison of FRP, ATRP, and IP methods

Vinyl, vinylidene, and diene are the only monomers that can be polymerized using radical mechanisms. Conversely, ionic techniques are employed to polymerize a variety of other monomers, including aldehydes, ketones, and other heterocycles, which are not polymerizable by radical methods. The increased adaptability of ionic polymerization is a result of the ionic or polar end-groups of these polymers' ability to polarize incoming monomers and facilitate their addition, or the ability of counter-ions to coordinate with the monomer prior to its eventual addition, thereby facilitating propagation. Radicals exhibit only moderate interactions with such interactions, which are particularly potent when the propagation involves a reaction between the active end-group and highly polarizable heterobonds of a monomer (e.g., C=O, C=N, etc.). The potential energy barrier that impedes the addition is reduced by the polarization of the reacting monomer. The propagation rate constant of any monomer is substantially higher during polymerization that is induced by free ions than during polymerization that is induced by a free radical process. Particular consideration should be given to the polymerization of isobutene. It is widely acknowledged that this monomer is not polymerized through either an anionic or radical mechanism. The sole process that generates high molecular weight polyisobutene is the cationic polymerization method. The inertness of this monomer in free radical additions is not the reason for the failure of radical polymerization to produce such polymers. The incorporation of methyl radicals into isobutene occurs at a faster rate than into ethylene, which leads to the production of high molecular weight polymers through a radical process. The absence of radical polymerization in isobutene is a result of a relatively rapid chain transfer to the monomer in comparison to its propagation velocity. The activation energy of the initial reaction is higher than that of the subsequent one. Consequently, a radical mechanism could be employed to produce a high molecular weight polyisobutene at extremely low temperatures, provided that the activation energy required for its propagation is reduced. The low activation energy of cationically polymerized iso-butene enables the synthesis of its high molecular weight polymer at low temperatures, such as –70°C, despite the fact that radical propagation occurs too slowly to be detected. The culmination of chain reactions is anticipated by specific events. By coupling or disproportionation, the annihilation of two free radicals is the consequence of their collision. The bimolecular termination of freely mobile radical polymers results in a very brief life, typically only a few seconds, due to the fact that these reactions occur so rapidly and are primarily regulated by diffusion. In contrast, the interactions between the growing polymers in the majority of ionic polymerizations do not impact their activity, thereby preventing their development potential from being impeded by their collisions.

A recognized correlation between the molecular weight of the resultant polymers and the rate of polymerization is established by the inescapable bimolecular termination that is inherent in radical polymerization. Specifically, a decreased molecular weight is correlated with an increased rate. While this relationship is apparent in specific ionic polymerizations, such as the cationic polymerization that is initiated in bulk monomers by γ radiation, it does not apply to the majority of ionic polymerizations. The maximal concentration of developing polymers is also limited by the bimolecular character of the inevitable termination in conventional radical polymerization. When the process produces macromolecules with a high molecular mass, their stationary concentration rarely exceeds 10^{-7} M. In ionic polymerizations, there are no such restrictions on the concentration of developing polymers. It is not uncommon to observe ionic situations in which the concentration of developing polymers exceeds 10^{-2} M. The inevitability of bimolecular termination in radical polymerizations requires a consistent supply of small radicals that produce new developing polymers to replace those that have been terminated. This is necessary for the continuation of polymerization. As a result, the majority of macromolecules that are generated during radical polymerization are inert and incapable of undergoing additional growth, with only a small proportion of them currently participating in the propagation process. This elucidates the nearly uniform degree of polymerization of the resulting macromolecules that are produced rapidly following the initiation of radical polymerization, regardless of the degree of monomer conversion. It is conceivable to evade termination and chain transfer in numerous ionic polymerizations. The resultant polymers are classified as active under these conditions, indicating that they can incorporate monomers and maintain growth for an extended period. Their molecular mass subsequently increases as a result of the monomer conversion. The extent of conversion is directly proportional to the increase in size, which is indicative of a rapid start. The resulting polymers are nearly uniform in size, provided that termination and chain transfer are rigorously prohibited. Sluggish initiation compromises the homogeneity of polymer sizes by generating new polymers while existing ones continue to develop. Numerous similarities exist between anionic and cationic polymerizations. The reason for certain discrepancies is that polymers that are developed cationically are significantly more reactive than those that are developed anionically. The expiration life of cationic polymers is significantly reduced due to the increased likelihood of contamination with hazardous substances compared to their anionic counterparts. Over the past four decades, there has been a substantial increase in the interest in ionic polymerization. The discovery of live polymerization and stereospecific polymerizations of isoprene catalyzed by lithium metal, which both occur through an anionic process, has resulted in a significant increase in research on this subject. The development of ionic polymerization techniques has resulted in the emergence of a diverse array of new, commercially significant products as a result of the polymerization of several inexpensive monomers that are unable to be polymerized using radical methods. The synthesis of numerous specialty macromolecules, such as diversified block polymers and star-shaped polymers, has been facilitated by the use of living polymers, all of which are in accordance with the requirements of modern technology and research. The utilization of these procedures has increased as a result of the increase in their industrial value. The field's rapid expansion is evident in the abundance of publications that appear annually in scientific and professional journals [102].

Cationic and radical polymerizations are typically performed under conditions that are significantly different, utilizing monomers, solvents, and reaction parameters that are significantly different. Controlled "living" carbocationic and radical polymerizations were recently developed using the same fundamental mechanism. They rely on the spontaneous, catalytic, or degenerative equilibration of quiescent covalent species and active species. Traditional carbocationic and radical polymerizations have been difficult to manage to a comparable extent, in contrast to the anionic polymerization of styrene and dienes, which produces polymers with regulated end functionalities, low polydispersities, and specified molecular weights. This behavior is the result of a variety of factors that are inherent in the fundamental properties of carbocations and radicals. Nevertheless, the most critical factor is likely the absence of an effective initiation system that can generate chains that continue to grow. In addition, carbocations demonstrate rapid reactivity with alkenes and participate in transfer reactions by eliminating protons, particularly when basic impurities are present. In contrast, radicals recombine and disproportionate at rates that approach the diffusion-controlled limit. In order to prevent cationic polymerization and/or radical recombination, the quantity of active species (carbocations or radicals) must be negligible (10^{-6} M) during typical carbocationic and radical polymerizations. Nevertheless, the molecular weight of the polymers that form would be exceedingly high (Mn > 108) if the total number of chains is extremely low. This would likely be constrained by transfer mechanisms. Polydispersities, functionalities, and molecular weights can only be regulated for chains that are comparatively short (Mn < 108) in both cationic and radical polymerizations [103].

One of the most frequently employed methods for the production of polymer pearls is suspension polymerization. Advancements in the production of polymer crystals are essential for the development of advanced materials. In order to obtain a granular suspension, suspension polymerization entails the aggressive agitation of a water-insoluble monomer and a monomer-soluble initiator in water, followed by filtration to isolate the resultant polymer post-reaction [104, 105]. The polymerization approach offers the advantages of a straightforward separation process and the ease of temperature regulation for the production of high molecular weight polymers. Droplet coalescence may occur during polymerization, resulting in the formation of larger aggregates, despite the fact that polymers produced during suspension polymerization are fundamentally particulate. In order to produce polymer crystals with precise particle sizes, it is necessary to regulate the size and stability of the molecules. This necessitates the modification of the churning conditions and the addition of dispersion stabilizers [106–110]. As a result, the use of inorganic microparticles or low molecular weight surfactants as stabilizers in suspension polymerization is considered a viable approach [111]. According to reports, this method prevents coalescence by encircling the original particles with stabilizers. Nevertheless, it is crucial to eliminate the inorganic fine particles that envelop the particulate polymers by employing hydrochloric acid following polymerization. This is a critical issue that requires resolution. It has been 100 years since Hermann Staudinger's seminal article in 1920, which established the field of polymer chemistry. Reversible deactivation radical polymerization (RDRP), also known as controlled radical polymerization (CRP), is a significant advancement in the field, as it enables scientists to precisely regulate the molecular weight, dispersity (å), composition, architecture, and end-group fidelity of vinyl polymers. Currently, there are numerous methods for regulating radical polymerization (CRP), such as ATRP [112], NMP [113], and RAFT polymerization [114]. Polymers produced through controlled radical polymerization (CRP) are employed as emulsifiers, dispersants, electrolytes, rheological agents, and surface modifiers in commercial products in the home care, cosmetics, health, paint, energy, and electronics sectors [115]. PET-RAFT polymerization has been utilized in numerous advancements, such as surface PET-RAFT polymerization from DNA [116] and living cells [117] as well as 3D-4D printing methods [118]. Despite being discovered only six years ago, the primary benefit of PET-RAFT polymerization is the use of low-intensity light sources that do not degrade the RAFT agents, which is in contrast to other photo- RAFT polymerization methods. This enables the polymerization of additional monomers without the necessity of photocatalysts, such as vinyl ketones. An additional substantial advantage is the absence of a conventional free radical initiator (e.g., AIBN), which leads to polymers with improved life characteristics. PET-RAFT polymerization is perceived as a more environmentally favorable alternative to conventional RAFT methods that employ elevated temperatures, as it employs lower temperatures, which leads to the complete or partial elimination of by-products and side reactions. PET-RAFT maintains impeccable temporal control and exhibits substantial tolerance to oxygen. Despite these remarkable advancements, there are still issues that necessitate resolution. In situations where orthogonality is advantageous, the identification of novel, efficient, and selective photocatalysts would be beneficial. In addition, it is imperative to develop a universal photocatalyst that is compatible with a variety of monomers, RAFT agents, and wavelengths. In order to optimize the benefits of this adaptable polymerization technique, it is necessary to expand the selection of solvents that are appropriate, with DMSO being the most frequently used. In conclusion, although rudimentary computational investigations have been conducted to identify suitable photocatalysts [119], it is imperative to conduct additional correlations between experimental and theoretical or predictive analyses. The copolymers can be designed to respond to changes in pressure, acidity, light exposure, and other environmental factors. For example, ATRP has the ability to produce a polymer that, upon application, produces a hermetic seal that combines flexibility, adhesion, and resistance to heat and oil. The primary difference between FRP and ATRP is the difficulty of modulating chain end functioning in the former. For example, it may be impossible to reactivate the polymer. As a result, FRP is unable to generate precisely defined segmented copolymers (such as thermoplastic elastomers) that are applicable in a variety of domains. Conversely, in ATRP, all polymer chains experience concurrent development. The chain end functionality, monomer distribution, and chain length are all regulated. ATRP offers additional benefits in comparison to FRP, such as the ability to synthesize more complex polymers and reduced energy requirements. ATRP's reliance on transition metal catalysts, such as copper halides, which may be present in the ultimate product, is a concern. However, CuI is perpetually transformed into CuII throughout the process as a result of inexorable termination reactions, despite the fact that ATRP utilizes a CuI-based catalyst, which is a highly reactive form of copper. It is imperative to incorporate an adequate quantity of CuI to guarantee that the polymerization process is completed, as the reaction may be impeded and ultimately halted by the insufficient reactivity of CuII. A copper accumulation that necessitates removal subsequent to the operation is the result. According to Spanswick, the process of removing copper from a polymer solution can be both labor-intensive and expensive. This involves passing the solution through an ion exchange resin or adsorbent bed, followed by the recycling or regeneration of the media [120]. Despite the ATRP process's adaptability, it is necessary to improve several critical attributes, such as the reaction's sensitivity to oxygen, the recovery and recycling of the catalyst, and the transition metal content in the final polymer, which contributes to its hue [121]. The proposed solutions to these challenges included the use of a variety of techniques for catalyst recovery, the development of methods to reduce the number of copper halides required for the reaction by regenerating catalyst active species through external agents or electrochemical approaches, the utilization of supported catalysts [122], and innovating metal-free ATRP reactions [123]. CuO has been the subject of research by a variety of groups for a variety of applications, such as its use as a sacrificial agent for the reactivation of the catalyst CuI [124], as a catalyst for the development of ultrafast reactions [125] for ultrahigh molecular weight polymers at ambient temperature, the synthesis of well-defined structures [126], and its application in aqueous controlled polymerization to achieve high end-chain fidelity and controlled structures [127].

ATRP necessitates both CuI and CuII. Spanswick elucidates that CuI is essential for the activation of the chain end, whereas CuII is responsible for its deactivation. A balance is maintained during the reaction by the introduction of a reducing agent, which establishes an equilibrium between CuI and CuII. In ARGET, the copper catalyst undergoes a transition from CuI to CuII and is subsequently reduced back to CuI by the addition of reducing compounds, such as glucose or ascorbic acid.

This method reduces the copper requirement for catalysis by as much as 1000 times, from 10,000 to 10 ppm or less. Ascorbic acid functions as an antioxidant, which refers to its capacity to interact with a substance in an oxidized state and convert it to a reduced form. Sodium sulfite, sodium hydrogen sulfite, inorganic salts containing a metal ion, hydrazine hydrate, mercaptoethanol, tetrahydrofuran, dihydroanthracene, 2,3-dimethylbutadiene, silane compounds, borane compounds, aldehydes, and their derivatives are among the numerous agents that can be used to reduce Cu(II). The polymer manufacturers can implement the modification without substantial alterations to their facilities by requiring modest quantities of the reducing agent (typically 50 ppm). ARGET provides the manufacturing process with additional benefits in addition to reducing the quantity of catalyst required. The procedure in ATRP must be performed in deoxygenated environments to prevent the radical from reacting with oxygen, thereby preserving the catalyst. Moreover, in an environment that is not wholly deoxygenated, it can eradicate dissolved oxygen by perpetually reducing the CuII generated when oxygen interacts with CuI. Spanswick suggests that numerous companies are investigating ATRP for the purpose of developing medication delivery systems, coatings for cardiac catheters, protein separation, cosmetics, fabric coatings, paints, adhesives, and pigment dispersants. "Pigment dispersants may not appear to be a concern for the environment or human health." Sequence-controlled polymers can be synthesized through reversible deactivation radical polymerization (RDRP) [26, 128–130]. A vast multitude of polymer topologies can be synthesized with great uniformity and the preservation of end-group functionality using ATRP, a widely used RDRP technology. The creation of a $[X\text{-CuII/L}]^+$ deactivator and a propagating radical ($^{\bullet}R$) is the typical outcome of ATRP, which is typically catalyzed by a transition metal complex, commonly referred to as the activator $[\text{CuI/L}]^+$ (where L is a multidentate amine ligand). This complex interacts with an alkyl halide initiator (R-X). The radical chain ends are deactivated by $[X\text{-CuII/L}]^+$, which results in the formation of X-capped quiescent species and the regeneration of $[\text{CuI/L}]^+$. Radical propagation continues until this occurs [131–136]. ATRP can be improved by incorporating a variety of cocatalysts to increase oxygen tolerance [55, 137] and temporal control [136–140]. The utilization of a small-molecule catalyst is very beneficial for transport inside a restricted system [138, 139]. Notwithstanding the benefits of photo-ATRP, metal contamination in the resultant polymers may restrict various potential applications. Despite the advantages of photo-ATRP, the prospective applications of the resulting polymers may be limited by metal contamination. It was observed that the weight average molecular weight experienced a modest decrease during the polymerization kinetics, suggesting a deviation from controlled polymerization [140].

It is enlightening to compare the radical and ionic mechanisms of polymerization. The process of radical polymerization begins with the introduction of a small radical to a monomer molecule, which is generated by an appropriate initiator. The propagation rate constant, selectivity, or stereochemistry of the subsequent propagation are not influenced by the characteristics of the initiator or the introduced radical, which subsequently forms the tail group of the final polymer. Radical propagation characteristics are determined by the polymerized monomer's properties and the reaction conditions, which include temperature, pressure, and, to a lesser extent, the solvent type. In contrast, the rate constant and the mechanism of ionic propagation are significantly influenced by the characteristics of the initiator. During the entire propagation process, the initiator or its fragments, such as the counterion, are in some way connected to the expanding center. As a result, the characteristics of ionic homopropagation can be substantially altered by altering the type of initiator. The active end-groups of homopolymers that expand rapidly are uniform in nature and identical. Ionically developing polymers demonstrate unique characteristics in this regard. Even when the initiator employed has a unique structure, their active end-groups frequently consist of a combination of numerous species. The variable nature of their propagating groups is the unique characteristic of ionic polymerization. For example, the active end-groups of specific ionically polymerizing substances may appear as ion pairs. Ion pairs dissociate into free ions, establishing a balance in which both undissociated pairs and free ions contribute to the process in their own distinctive manners. The observed rate and mode of polymerization are determined by the mole fractions of the species in the reacting mixture, which are influenced by temperature, solvent characteristics, and other factors. Therefore, ionic polymerization is more complex yet more adaptable than radical polymerization. The interaction between radicals and their environment is frequently feeble. As a result, the pace of the majority of radical polymerizations is not significantly affected by the characteristics of the solvent or the penultimate unit, and the resulting effects are often disregarded. This is not the case in ionic polymerization. The process of ionic propagation may be substantially influenced by the solvent, penultimate units, and even more distant units. The mechanism and viability of specific ionic polymerizations are frequently influenced by the solvent's characteristics [141].

4. Conclusions and outlook

The chapter is focused on the synthesis of copolymers derived with FRP, ATRP, and IP methods from the monomers, the comparison of polymerization methods, and the applications of the copolymers. A conventional chain polymerization that utilizes highly reactive radicals as active species is known as FRP. Four fundamental reactions classify the polymerization process: initiation, propagation, termination, and chain transfer. People frequently employ an assortment of peroxides, such as alkyl peroxides (e.g., tert-butyl peroxide) and hydroxyl peroxides, as initiators. ATRP is a good CRP process that can be used to make polymers with specific molecular weights, narrow molecular weight distributions (low dispersity), different chemical compositions, and complex topologies. This method uses a catalyst to enable the reversible activation and deactivation of polymers with halide end-groups. During activation, the halide is removed to create a reactive polymer radical, and during deactivation, it is reintroduced to create a "dormant" polymer chain. IP, akin to conventional radical polymerization, is a chain polyaddition mechanism. The process commences with a monomer interacting with a species that imparts a highly charged or electrically charged active group to the monomer. A covalent bond forms when this group interacts with another monomer molecule. The newly incorporated molecule simultaneously refreshes its active group.

FRP, ATRP, and IP mechanisms offer unique advantages, but further innovation is needed in several areas to overcome their inherent limitations. The precision of ATRP in controlling molecular weight, chain-end functionality, and polymer architecture has enabled the production of complex macromolecules suitable for advanced industrial applications. However, issues such as metal contamination, high-energy requirements for catalyst removal, and sensitivity to oxygen continue to hinder wider adoption. Future research could focus on developing alternative, environmentally friendly catalysts, improving reaction kinetics, and improving scalability. With lower energy demands and reduced metal requirements, ARGET ATRP and PET-RAFT polymerization point to promising avenues for industrial application. Expanding the range of solvents and developing photo-ATRP techniques for controlled polymerization with minimal contamination will further increase the sustainability and versatility of these methods. IP, particularly cationic and anionic types, has proven effective in synthesizing polymers from monomers that are not suitable for radical polymerization. The development of living polymerization has paved the way for specialized polymer architectures such as block copolymers and star-shaped polymers. However, the sensitivity of ionic polymerizations to impurities, complex reaction conditions, and limited shelf life for cationic systems necessitate improved robustness. Addressing these challenges through the design of new initiators, stabilizers, and reaction media could significantly increase the utility and commercial importance of IP. While FRP remains a cost-effective and simple approach, the lack of control over polymer properties limits its usefulness for complex applications. In contrast, ATRP and IP offer improved control but are more complex and sensitive to reaction parameters. Bridging the gap between these mechanisms is a promising research direction, particularly through hybrid techniques that combine the simplicity of FRP with the sensitivity of ATRP or IP. For example, integrating photoinitiated processes with IP can increase reactivity and broaden monomer compatibility. The rapid development of controlled polymerization techniques has opened doors to applications in nanotechnology, medicine, energy storage, and advanced materials. Future research should focus on expanding the library of polymerizable monomers and designing smart polymers that respond to external stimuli such as pH, temperature, and light. Moreover, integrating computational modeling with experimental techniques will accelerate the discovery of efficient photocatalysts and reaction conditions and bridge the gap between theoretical predictions and practical applications.

As a result, the field of polymer chemistry is on a path of rapid growth and innovation. Continuous efforts to improve the sensitivity, sustainability, and scalability of ATRP, IP, and hybrid polymerization methods will not only address current challenges but also unlock new possibilities for creating advanced materials designed to meet the demands of modern technology and industry.

Acknowledgements

The authors thank Ibrahim ERDEMOGLU for their support. The authors would like to thank Karadeniz Technical University for their support. In addition, the authors also would like to express gratitude to Bilkent University UNAM for their kind hospitality.

Author details

Bulend Ortaç[1], Saliha Mutlu[1,2], Ahmet Hakan Yilmaz[3] and Sevil Savaskan Yilmaz[1,2*]

1 UNAM-National Nanotechnology Research Center and Institute of Materials Science and Nanotechnology, Bilkent University, Ankara, Turkey

2 Department of Chemistry, Faculty of Sciences, Karadeniz Technical University, Trabzon, Turkey

3 Department of Physics, Faculty of Sciences, Karadeniz Technical University, Trabzon, Turkey

*Address all correspondence to: sevily@ktu.edu.tr

References

[1] Frazer L. Radical departure: Polymerization does more with less. Environmental Health Perspectives. 2007;**115**:258-261. DOI: 10.1289/ehp.115-a258

[2] Pitsikalis M. Ionic polymerization. In: Reference Module in Chemistry, Molecular Sciences and Chemical Engineering. Amsterdam, Netherlands: Elsevier; 2013

[3] Hong K, Uhrig D, Mays JW. Living anionic polymerization. Current Opinion in Solid State & Materials Science. 1999;**4**:531-538. DOI: 10.1016/S1359-0286(00)00011-5

[4] Verdonck B, Goethals EJ, Du Prez FE. Block copolymers of methyl vinyl ether and isobutyl vinyl ether with thermo-adjustable amphiphilic properties. Macromolecular Chemistry and Physics. 2003;**204**:2090-2098. DOI: 10.1002/macp.200350069

[5] Matsumoto K, Mazaki H, Matsuoka H. Fluorinated amphiphilic vinyl ether block copolymers: Synthesis and characteristics of their micelles in water. Macromolecules. 2004;**37**:2256-2267. DOI: 10.1021/ma021166i

[6] Satoh K, Kamigaito M, Sawamoto M. Direct synthesis of amphiphilic random and block copolymers of p-hydroxystyrene and p-methoxystyrene via living cationic polymerization with BF_3OEt_2/ROH systems. Macromolecules. 2000;**33**:5830-5835. DOI: 10.1021/ma000070q

[7] Müller M. Process-directed self-assembly of copolymers: Results of and challenges for simulation studies. Progress in Polymer Science. 2020;**101**:101198. DOI: 10.1016/j.progpolymsci.2019.101198

[8] Saldívar-Guerra E, Vivaldo-Lima E, editors. Handbook of Polymer Synthesis, Characterization, and Processing. United States of America: Wiley; 2013

[9] Xie J, Shi A-C. Phase behavior of triblock copolymer and homopolymer blends: Effect of copolymer topology. Physical Review Materials. 2024;**8**:015601. DOI: 10.1103/PhysRevMaterials.8.015601

[10] Mai Y, Eisenberg A. Self-assembly of block copolymers. Chemical Society Reviews. 2012;**41**:5969. DOI: 10.1039/c2cs35115c

[11] Bates CM, Maher MJ, Janes DW, Ellison CJ, Willson CG. Block copolymer lithography. Macromolecules. 2014;**47**:2-12. DOI: 10.1021/ma401762n

[12] Lequieu J, Quah T, Delaney KT, Fredrickson GH. Complete photonic band gaps with nonfrustrated ABC bottlebrush block polymers. ACS Macro Letters. 2020;**9**:1074-1080. DOI: 10.1021/acsmacrolett.0c00380

[13] Thedford RP, Yu F, Tait WRT, Shastri K, Monticone F, Wiesner U. The promise of soft-matter-enabled quantum materials. Advanced Materials. 2023;**35**:1-28. DOI: 10.1002/adma.202203908

[14] Matyjaszewski K, Spanswick J. Controlled/living radical polymerization. Materials Today. 2005;**8**:26-33. DOI: 10.1016/S1369-7021(05)00745-5

[15] Keddie DJ. A guide to the synthesis of block copolymers using reversible-addition fragmentation chain transfer (RAFT) polymerization. Chemical Society Reviews. 2014;**43**:496-505. DOI: 10.1039/C3CS60290G

[16] Yang Q, Lalevée J, Poly J. Development of a robust photocatalyzed ATRP mechanism exhibiting good tolerance to oxygen and inhibitors. Macromolecules. 2016;**49**:7653-7666. DOI: 10.1021/acs.macromol.6b01808

[17] Maekawa S, Seshimo T, Dazai T, Sato K, Hatakeyama-Sato K, Nabae Y, Hayakawa T. Chemically tailored block copolymers for highly reliable sub-10-nm patterns by directed self-assembly. Nature Communications. 2024;**15**:5671. DOI: 10.1038/s41467-024-49839-0

[18] Lodge TP. Block copolymers: Long-term growth with added value. Macromolecules. 2020;**53**:2-4. DOI: 10.1021/acs.macromol.9b02069

[19] Li C, Li Q, Kaneti YV, Hou D, Yamauchi Y, Mai Y. Self-assembly of block copolymers towards mesoporous materials for energy storage and conversion systems. Chemical Society Reviews. 2020;**49**:4681-4736. DOI: 10.1039/D0CS00021C

[20] Nguyen D, Yoon H. Recent advances in nanostructured conducting polymers: From synthesis to practical applications. Polymers (Basel). 2016;**8**:118. DOI: 10.3390/polym8040118

[21] Zhang L, Du W, Nautiyal A, Liu Z, Zhang X. Recent progress on nanostructured conducting polymers and composites: Synthesis, application and future aspects. Science China Materials. 2018;**61**:303-352. DOI: 10.1007/s40843-017-9206-4

[22] Das R, editor. Polymeric Materials for Clean Water. Cham: Springer International Publishing; 2019

[23] Antonopoulou M-N, Truong NP, Anastasaki A. Enhanced synthesis of multiblock copolymers via acid-triggered RAFT polymerization. Chemical Science. 2024;**15**:5019-5026. DOI: 10.1039/D4SC00399C

[24] Krys P, Matyjaszewski K. Kinetics of atom transfer radical polymerization. European Polymer Journal. 2017;**89**:482-523. DOI: 10.1016/j.eurpolymj.2017.02.034

[25] Matyjaszewski K. Advanced materials by atom transfer radical polymerization. Advanced Materials. 2018;**30**:1-22. DOI: 10.1002/adma.201706441

[26] Corrigan N, Jung K, Moad G, Hawker CJ, Matyjaszewski K, Boyer C. Reversible-deactivation radical polymerization (controlled/living radical polymerization): From discovery to materials design and applications. Progress in Polymer Science. 2020;**111**:101311. DOI: 10.1016/j.progpolymsci.2020.101311

[27] Soeriyadi AH, Boyer C, Nyström F, Zetterlund PB, Whittaker MR. High-order multiblock copolymers via iterative Cu(0)-mediated radical polymerizations (SET-LRP): Toward biological precision. Journal of the American Chemical Society. 2011;**133**:11128-11131. DOI: 10.1021/ja205080u

[28] Fujita E, Kaneko K, Oshima T, Fujioka D, Kaneko K, Fuchigami K, et al. Synthesis of amphiphilic triblock copolymers by ATRP and coalescence suppression effect in suspension polymerization. Journal of Chemical Engineering of Japan. 2023;**56**:1-7. DOI: 10.1080/00219592.2023.2276424

[29] Wang Y, Lorandi F, Fantin M, Matyjaszewski K. Atom transfer radical polymerization in dispersed media with low-ppm catalyst loading. Polymer (Guildf). 2023;**275**:125913. DOI: 10.1016/j.polymer.2023.125913

[30] Flejszar M, Ślusarczyk K, Hochół A, Chmielarz P, Wytrwal M, Wolski K, et al. Sequential SI-ATRP in µL-scale for surface nanoengineering: A new concept for designing polyelectrolyte nanolayers formed by complex architecture polymers. European Polymer Journal. 2023;**194**:112142. DOI: 10.1016/j.eurpolymj.2023.112142

[31] Kalelkar PP, Geng Z, Cox B, Finn MG, Collard DM. Surface-initiated atom-transfer radical polymerization (SI-ATRP) of bactericidal polymer brushes on poly(lactic acid) surfaces. Colloids and Surfaces. B, Biointerfaces. 2022;**211**:112242. DOI: 10.1016/j.colsurfb.2021.112242

[32] Li M, Fromel M, Ranaweera D, Rocha S, Boyer C, Pester CW. SI-PET-RAFT: Surface-initiated photoinduced electron transfer-reversible addition–fragmentation chain transfer polymerization. ACS Macro Letters. 2019;**8**:374-380. DOI: 10.1021/acsmacrolett.9b00089

[33] Hui CM, Pietrasik J, Schmitt M, Mahoney C, Choi J, Bockstaller MR, et al. Surface-initiated polymerization as an enabling tool for multifunctional (nano-)engineered hybrid materials. Chemistry of Materials. 2014;**26**:745-762. DOI: 10.1021/cm4023634

[34] Narupai B, Page ZA, Treat NJ, McGrath AJ, Pester CW, Discekici EH, et al. Simultaneous preparation of multiple polymer brushes under ambient conditions using microliter volumes. Angewandte Chemie, International Edition. 2018;**57**:13433-13438. DOI: 10.1002/anie.201805534

[35] Ślusarczyk K, Flejszar M, Chmielarz P. Less is more: A review of µL-scale of SI-ATRP in polymer brushes synthesis. Polymer (Guildf). 2021;**233**:124212. DOI: 10.1016/j.polymer.2021.124212

[36] Flejszar M, Ślusarczyk K, Chmielarz P, Smenda J, Wolski K, Wytrwal-Sarna M, et al. SI-ATRP on the lab bench: A facile recipe for oxygen-tolerant PDMAEMA brushes synthesis using microliter volumes of reagents. Polymer (Guildf). 2022;**257**:125268. DOI: 10.1016/j.polymer.2022.125268

[37] Toozandehjani M. Conventional and advanced composites in aerospace industry: Technologies revisited. American Journal of Aerospace Engineering. 2018;**5**:9. DOI: 10.11648/j.ajae.20180501.12

[38] Flejszar M, Chmielarz P. Surface-initiated atom transfer radical polymerization for the preparation of well-defined organic–inorganic hybrid nanomaterials. Materials (Basel). 2019;**12**:3030. DOI: 10.3390/ma12183030

[39] Pan X, Fantin M, Yuan F, Matyjaszewski K. Externally controlled atom transfer radical polymerization. Chemical Society Reviews. 2018;**47**:5457-5490. DOI: 10.1039/C8CS00259B

[40] Cheng J, Liu K, Li X, Huang L, Liang J, Zheng G, et al. Nickel-metal-organic framework nanobelt based composite membranes for efficient Sr^{2+} removal from aqueous solution. Environmental Science and Ecotechnology. 2020;**3**:100035. DOI: 10.1016/j.ese.2020.100035

[41] Zhang K, Yan W, Simic R, Benetti EM, Spencer ND. Versatile surface modification of hydrogels by surface-initiated, Cu^0-mediated controlled radical polymerization. ACS Applied Materials & Interfaces. 2020;**12**:6761-6767. DOI: 10.1021/acsami.9b21399

[42] Ramu A, Rajendrakumar K. Natural catalyst mediated ARGET and SARA ATRP of N-isopropylacrylamide and methyl acrylate. Polymer Chemistry. 2020;**11**:687-694. DOI: 10.1039/C9PY01770D

[43] Zaborniak I, Chmielarz P. How we can improve ARGET ATRP in an aqueous system: Honey as an unusual solution for polymerization of (meth) acrylates. European Polymer Journal. 2023;**183**:111735. DOI: 10.1016/j.eurpolymj.2022.111735

[44] Wang Y, Lorandi F, Fantin M, Chmielarz P, Isse AA, Gennaro A, et al. Miniemulsion ARGET ATRP via interfacial and ion-pair catalysis: From ppm to ppb of residual copper. Macromolecules. 2017;**50**:8417-8425. DOI: 10.1021/acs.macromol.7b01730

[45] Nardi M, Blasco E, Barner-Kowollik C. Wavelength-resolved photo ATRP. Journal of the American Chemical Society. 2022;**144**:1094-1098. DOI: 10.1021/jacs.1c11259

[46] Wang G, Shao H, Ma J, Liao S. Chalcogenide-doped anthracenes as organophotocatalysts for metal-free atom transfer radical polymerization. Macromolecular Chemistry and Physics. 2023;**224**:1-7. DOI: 10.1002/macp.202200382

[47] Yamazoe K, Higaki Y, Inutsuka Y, Miyawaki J, Takahara A, Harada Y. Critical in-plane density of polyelectrolyte brush for the ordered hydrogen-bonded structure of incorporated water. Langmuir. 2022;**38**:3076-3081. DOI: 10.1021/acs.langmuir.1c02895

[48] Aktas Eken G, Ober CK. Strong polyelectrolyte brushes via alternating copolymers of styrene and maleimides: Synthesis, properties, and stability. Macromolecules. 2022;**55**:5291-5300. DOI: 10.1021/acs.macromol.2c00647

[49] Yu M, Guo X, Zhao W, Zhang K. Single-molecule studies reveal the distinction of strong and weak polyelectrolytes in aqueous solutions. Physical Chemistry Chemical Physics. 2021;**23**:26130-26134. DOI: 10.1039/D1CP03572J

[50] Zhang J, Kou R, Liu G. Effect of salt concentration on the pH responses of strong and weak polyelectrolyte brushes. Langmuir. 2017;**33**:6838-6845. DOI: 10.1021/acs.langmuir.7b01395

[51] Zhang J, Cai H, Tang L, Liu G. Tuning the pH response of weak polyelectrolyte brushes with specific anion effects. Langmuir. 2018;**34**:12419-12427. DOI: 10.1021/acs.langmuir.8b02776

[52] Sanchez-Ballester NM, Sciortino F, Mir SH, Rydzek G. Weak polyelectrolytes as nanoarchitectonic design tools for functional materials: A review of recent achievements. Molecules. 2022;**27**:3263. DOI: 10.3390/molecules27103263

[53] Ng G, Li M, Yeow J, Jung K, Pester CW, Boyer C. Benchtop preparation of polymer brushes by SI-PET-RAFT: The effect of the polymer composition and structure on inhibition of a Pseudomonas biofilm. ACS Applied Materials & Interfaces. 2020;**12**:55243-55254. DOI: 10.1021/acsami.0c15221

[54] Poisson J, Christopherson CJ, Hudson ZM. Cationic bottlebrush brush polymers via sequential SI-ROMP and SI-ARGET-ATRP. Polymer International. 2023;**72**:267-273. DOI: 10.1002/pi.6482

[55] Szczepaniak G, Fu L, Jafari H, Kapil K, Matyjaszewski K. Making ATRP more practical: Oxygen tolerance. Accounts of Chemical Research.

2021;**54**:1779-1790. DOI: 10.1021/acs.accounts.1c00032

[56] Guan Z, Smart B. A remarkable visible light effect on atom-transfer radical polymerization. Macromolecules. 2000;**33**:6904-6906. DOI: 10.1021/ma000646b

[57] Tasdelen MA, Uygun M, Yagci Y. Photoinduced controlled radical polymerization. Macromolecular Rapid Communications. 2011;**32**:58-62. DOI: 10.1002/marc.201000351

[58] Mosnáček J, Ilčíková M. Photochemically mediated atom transfer radical polymerization of methyl methacrylate using ppm amounts of catalyst. Macromolecules. 2012;**45**:5859-5865. DOI: 10.1021/ma300773t

[59] Konkolewicz D, Schröder K, Buback J, Bernhard S, Matyjaszewski K. Visible light and sunlight photoinduced ATRP with ppm of Cu catalyst. ACS Macro Letters. 2012;**1**:1219-1223. DOI: 10.1021/mz300457e

[60] Anastasaki A, Nikolaou V, Zhang Q, Burns J, Samanta SR, Waldron C, et al. Copper(II)/tertiary amine synergy in photoinduced living radical polymerization: Accelerated synthesis of ω-functional and α,ω-heterofunctional poly(acrylates). Journal of the American Chemical Society. 2014;**136**:1141-1149. DOI: 10.1021/ja411780m

[61] Fors BP, Hawker CJ. Control of a living radical polymerization of methacrylates by light. Angewandte Chemie, International Edition. 2012;**51**:8850-8853. DOI: 10.1002/anie.201203639

[62] Xu J, Jung K, Atme A, Shanmugam S, Boyer C. A robust and versatile photoinduced living polymerization of conjugated and unconjugated monomers and its oxygen tolerance. Journal of the American Chemical Society. 2014;**136**:5508-5519. DOI: 10.1021/ja501745g

[63] Mısır M, Savaskan Yılmaz S, Bilgin A. Synthesis and characterization of ABA-type triblock copolymers using novel bifunctional PS, PMMA, and PCL macroinitiators bearing p-xylene-bis(2-mercaptoethyloxy) core. Polymers (Basel). 2023;**15**:3813. DOI: 10.3390/polym15183813

[64] Ozturk T, Yilmaz SS, Hazer B. Synthesis of a new macroperoxy initiator with methyl methacrylate and T-butyl peroxy ester by atom transfer radical polymerization and copolymerization with conventional vinyl monomers. Journal of Macromolecular Science, Part A. 2008;**45**:811-820. DOI: 10.1080/10601320802300495

[65] Ozturk T, Yilmaz SS, Hazer B, Menceloglu YZ. ATRP of methyl methacrylate initiated with a bifunctional initiator bearing bromomethyl functional groups: Synthesis of the block and graft copolymers. Journal of Polymer Science, Part A: Polymer Chemistry. 2010;**48**:1364-1373. DOI: 10.1002/pola.23898

[66] Kul D, Van Renterghem LM, Meier MAR, Strandman S, Tenhu H, YIlmaz SS, et al. Encapsulation and release by star-shaped block copolymers as unimolecular nanocontainers. Journal of Polymer Science, Part A: Polymer Chemistry. 2008;**46**:650-660. DOI: 10.1002/pola.22413

[67] Zhang S, Tezuka Y, Zhang Z, Li N, Zhang W, Zhu X. Recent advances in the construction of cyclic grafted polymers and their potential applications. Polymer Chemistry. 2018;**9**:677-686. DOI: 10.1039/C7PY01544E

[68] Chen C, Weil T. Cyclic polymers: Synthesis, characteristics, and emerging applications. Nanoscale Horizons. 2022;**7**:1121-1135. DOI: 10.1039/D2NH00242F

[69] Golba B, Benetti EM, De Geest BG. Biomaterials applications of cyclic polymers. Biomaterials. 2021;**267**:120468. DOI: 10.1016/j.biomaterials.2020.120468

[70] Xu J, Pu L, Ma J, Kumar SK, Duan H. Antibacterial properties of synthesized cyclic and linear cationic copolymers. Polymer Chemistry. 2020;**11**:6632-6639. DOI: 10.1039/D0PY00755B

[71] Haque FM, Grayson SM. The synthesis, properties and potential applications of cyclic polymers. Nature Chemistry. 2020;**12**:433-444. DOI: 10.1038/s41557-020-0440-5

[72] Aswale S, Kim M, Kim D, Mohanty AK, Jeon HB, Cho HY, et al. Synthesis and characterization of spirocyclic mid-block containing triblock copolymer. Polymers (Basel). 2023;**15**:1677. DOI: 10.3390/polym15071677

[73] Schmidt BVKJ. Double hydrophilic block copolymer self-assembly in aqueous solution. Macromolecular Chemistry and Physics. 2018;**219**:1-15. DOI: 10.1002/macp.201700494

[74] Volkmann L, Köhler M, Sobotta FH, Enke MT, Brendel JC, Schacher FH. Poly(2-acrylamidoglycolic acid) (PAGA): Controlled polymerization using RAFT and chelation of metal cations. Macromolecules. 2018;**51**:7284-7294. DOI: 10.1021/acs.macromol.8b01260

[75] Max JB, Mons PJ, Tom JC, Schacher FH. Double hydrophilic poly(ethylene oxide)-block-poly(dehydroalanine) block copolymers: Comparison of two different synthetic routes. Macromolecular Chemistry and Physics. 2020;**221**:1-9. DOI: 10.1002/macp.201900383

[76] Corbin DA, Miyake GM. Photoinduced organocatalyzed atom transfer radical polymerization (O-ATRP): Precision polymer synthesis using organic photoredox catalysis. Chemical Reviews. 2022;**122**:1830-1874. DOI: 10.1021/acs.chemrev.1c00603

[77] Wang G, Zhang L. Synthesis, self-assembly and pH sensitivity of PDEAEMA–PEG–PDEAEMA triblock copolymer micelles for drug delivery. Reactive and Functional Polymers. 2016;**107**:1-10. DOI: 10.1016/j.reactfunctpolym.2016.08.001

[78] Young RN, Quirk RP, Fetters LJ. Anionic polymerizations of non-polar monomers involving lithium. In: Anionic Polymerization. Berlin/Heidelberg: Springer-Verlag; 2005. pp. 1-90

[79] Xu Z, Mays J, Chen X, Hadjichristidis N, Schilling FC, Bair HE, et al. Molecular characterization of poly(2-methyl-1,3-pentadiene) and its hydrogenated derivative, atactic polypropylene. Macromolecules. 1985;**18**:2560-2566. DOI: 10.1021/ma00154a034

[80] Ozaki H, Hirao A, Nakahama S. Polymerization of monomers containing functional silyl groups. 11. Anionic living polymerization of 3-(tri-2-propoxysilyl) propyl methacrylate. Macromolecules. 1992;**25**:1391-1395. DOI: 10.1021/ma00031a005

[81] Fish D, Khan IM, Smid J. Anionic ring opening polymerization of cyclotetrasiloxanes with large substituents. Makromolekulare Chemie, Macromolecular Symposia. 1990;**32**:241-253. DOI: 10.1002/masy.19900320119

[82] Herman Francis Mark JIK. Encyclopedia of Polymer Science and Engineering. New York: John Wiley & Sons; 1985

[83] Dubois P, Jérôme R, Teyssié P. Aluminium alkoxides: A family of versatile initiators for the ring-opening polymerization of lactones and lactides. Makromolekulare Chemie, Macromolecular Symposia. 1991;**42-43**:103-116. DOI: 10.1002/masy.19910420108

[84] Lambert C, von Ragué Schleyer P. Are polar organometallic compounds "carbanions"? The gegenion effect on structure and energies of alkali-metal compounds. Angewandte Chemie (International Ed. in English). 1994;**33**:1129-1140. DOI: 10.1002/anie.199411291

[85] Dietrich H. Low temperature study of the structure of crystalline ethyllithium. Journal of Organometallic Chemistry. 1981;**205**:291-299. DOI: 10.1016/S0022-328X(00)82431-6

[86] Fraenkel G, Beckenbaugh WE, Yang PP. Exchange and inversion in 2-methylbutyllithium: Proton nuclear magnetic resonance line shapes at 300 MHz. Journal of the American Chemical Society. 1976;**98**:6878-6885. DOI: 10.1021/ja00438a019

[87] Selman CM, Hsieh HL. Effect of aggregate size on alkyllithium initiated polymerizations. Journal of Polymer Science, Part B: Polymer Letters. 1971;**9**:219-224. DOI: 10.1002/pol.1971.110090311

[88] Kennedy JP. Living cationic polymerization of olefins. How did the discovery come about? Journal of Polymer Science, Part A: Polymer Chemistry. 1999;**37**:2285-2293. DOI: 10.1002/(SICI)1099-0518(19990715)37:14<2285::AID-POLA1>3.0.CO;2-P

[89] Shea JJ. Macromolecular design of polymeric materials [book review]. IEEE Electrical Insulation Magazine. 1999;**15**:50-51. DOI: 10.1109/MEI.1999.793835

[90] Vairon J-P, Rives A, Bunel C. Direct observation and stability of active species in cationic polymerization: A reexamination of the polymerization of styrene initiated by triflic acid. Makromolekulare Chemie, Macromolecular Symposia. 1992;**60**:97-105. DOI: 10.1002/masy.19920600109

[91] Balogh L, Wang L, Faust R. Initiation via haloboration in living cationic polymerization. 1. The polymerization of isobutylene. Macromolecules. 1994;**27**:3453-3458. DOI: 10.1021/ma00091a002

[92] Yagci Y, Onen A, Reetz I. Structure and reactivity of addition fragmentation agents in photochemically and thermally induced cationic polymerization. Macromolecular Symposia. 2001;**174**:255-268. DOI: 10.1002/1521-3900(200109)174:1<255::AID-MASY255>3.0.CO;2-K

[93] Kourti M, Vougioukalakis GC, Hadjichristidis N, Pitsikalis M. Metallocene-mediated cationic ring-opening polymerization of 2-methyl- and 2-phenyl-oxazoline. Journal of Polymer Science, Part A: Polymer Chemistry. 2011;**49**:2520-2527. DOI: 10.1002/pola.24679

[94] Wu Y. Ionic polymerization in ionic liquids. In: Recent Advances in Ionic Liquids. London, UK: InTech; 2018

[95] Miyamoto M, Sawamoto M, Higashimura T. Living polymerization of isobutyl vinyl ether with hydrogen iodide/iodine initiating system.

Macromolecules. 1984;**17**:265-268. DOI: 10.1021/ma00133a001

[96] Faust R, Kennedy JP. Living carbocationic polymerization. Polymer Bulletin. 1986;**15**:317-323. DOI: 10.1007/BF00254850

[97] Zhu Y, Yu Z, Liu R, Jin Y, Shi Q, Wu Y. Recent advances in green cationic polymerization. Journal of Polymer Science. 2024;**62**:2549-2573. DOI: 10.1002/pol.20230971

[98] Misir M, Ozturk T, Emirik M, Yilmaz SS. Synthesis of novel tetrahydrofuran-epichlorohydrin [poly(THF-b-ECH)] macromonomeric peroxy initiators by cationic copolymerization and the quantum chemically investigation of initiation system effects. Journal of Polymer Science Part A: Polymer Chemistry. 2010;**48**:2896-2909. DOI: 10.1002/pola.24069

[99] Kul D, Yılmaz SS, Öztürk T, Usta A, Mısır M. Synthesis of novel macromonomeric peroxy initiators of styrene with the cationic copolymerization and the quantum chemically investigation of the initiation system effects. Journal of Applied Polymer Science. 2006;**102**:348-357. DOI: 10.1002/app.23880

[100] Du Prez F, Goethals EJ, Hoogenboom R. Cationic Polymerizations. United States of America: Department of Organic and Macromolecular Chemistry; 2013

[101] Szwarc M. 'Living' polymers. Nature. 1956;**178**:1168-1169. DOI: 10.1038/1781168a0

[102] Szwarc M, Van Beylen M. Ionic Polymerization and Living Polymers. Netherlands, Dordrecht: Springer; 1993

[103] Puskas JE, Michel A, Barghi S, editors. Ionic Polymerizations and Related Processes. Netherlands, Dordrecht: Springer; 1999

[104] Brooks BW. Basic aspects and recent developments in suspension polymerisation. Makromolekulare Chemie, Macromolecular Symposia. 1990;**35-36**:121-140. DOI: 10.1002/masy.19900350110

[105] Dowding PJ, Vincent B. Suspension polymerisation to form polymer beads. Colloids and Surfaces A: Physicochemical and Engineering Aspects. 2000;**161**:259-269. DOI: 10.1016/S0927-7757(99)00375-1

[106] Primo AR, Paiva VUE, Medeiros AMMS, Machado F. Evaluation of process variables on the suspension polymerization of vinyl pivalate. Macromolecular Symposia. 2018;**380**:1-6. DOI: 10.1002/masy.201800134

[107] Sen N, Shaikh T, Singh KK, Sirsam R, Shenoy KT. Synthesis of polyacrylamide (PAM) beads in microreactors. Chemical Engineering and Processing: Process Intensification. 2020;**157**:108105. DOI: 10.1016/j.cep.2020.108105

[108] Jahanzad F, Sajjadi S. Two-stage stabilizer addition protocol as a means to reduce the size and improve the uniformity of polymer beads in suspension polymerization. Journal of Applied Polymer Science. 2018;**135**:1-7. DOI: 10.1002/app.45671

[109] Alroaithi M, Jahanzad F, Sajjadi S. Suppressing coalescence and improving uniformity of polymer beads in suspension polymerization using a two-stage stirring protocol. Industrial and Engineering Chemistry Research. 2018;**57**:11883-11892. DOI: 10.1021/acs.iecr.8b01599

[110] Alroaithi M, Sajjadi S. Uniform polymer beads by membrane emulsification-assisted suspension polymerisation. RSC Advances. 2016;**6**:79745-79754. DOI: 10.1039/C6RA09807J

[111] Fuchigami K, Taguchi Y, Tanaka M. Synthesis of calcium carbonate vaterite crystals and their effect on stabilization of suspension polymerization of MMA. Advanced Powder Technology. 2009;**20**:74-79. DOI: 10.1016/j.apt.2008.10.003

[112] Matyjaszewski K, Xia J. Atom transfer radical polymerization. Chemical Reviews. 2001;**101**:2921-2990. DOI: 10.1021/cr940534g

[113] Hawker CJ, Bosman AW, Harth E. New polymer synthesis by nitroxide mediated living radical polymerizations. Chemical Reviews. 2001;**101**:3661-3688. DOI: 10.1021/cr990119u

[114] Moad G, Rizzardo E, Thang SH. Living radical polymerization by the RAFT process—A third update. Australian Journal of Chemistry. 2012;**65**:985. DOI: 10.1071/CH12295

[115] Destarac M. Industrial development of reversible-deactivation radical polymerization: Is the induction period over? Polymer Chemistry. 2018;**9**:4947-4967. DOI: 10.1039/C8PY00970H

[116] Lueckerath T, Strauch T, Koynov K, Barner-Kowollik C, Ng DYW, Weil T. DNA–polymer conjugates by photoinduced RAFT polymerization. Biomacromolecules. 2019;**20**:212-221. DOI: 10.1021/acs.biomac.8b01328

[117] Niu J, Lunn DJ, Pusuluri A, Yoo JI, O'Malley MA, Mitragotri S, et al. Engineering live cell surfaces with functional polymers via cytocompatible controlled radical polymerization. Nature Chemistry. 2017;**9**:537-545. DOI: 10.1038/nchem.2713

[118] Zhang Z, Corrigan N, Bagheri A, Jin J, Boyer C. A versatile 3D and 4D printing system through photocontrolled RAFT polymerization. Angewandte Chemie, International Edition. 2019;**58**:17954-17963. DOI: 10.1002/anie.201912608

[119] Wu C, Chen H, Corrigan N, Jung K, Kan X, Li Z, et al. Computer-guided discovery of a pH-responsive organic photocatalyst and application for pH and light dual-gated polymerization. Journal of the American Chemical Society. 2019;**141**:8207-8220. DOI: 10.1021/jacs.9b01096

[120] Matyjaszewski K. Atom transfer radical polymerization (ATRP): Current status and future perspectives. Macromolecules. 2012;**45**:4015-4039. DOI: 10.1021/ma3001719

[121] Boyer C, Corrigan NA, Jung K, Nguyen D, Nguyen T-K, Adnan NNM, et al. Copper-mediated living radical polymerization (atom transfer radical polymerization and copper(0) mediated polymerization): From fundamentals to bioapplications. Chemical Reviews. 2016;**116**:1803-1949. DOI: 10.1021/acs.chemrev.5b00396

[122] Park S, Chmielarz P, Gennaro A, Matyjaszewski K. Simplified electrochemically mediated atom transfer radical polymerization using a sacrificial anode. Angewandte Chemie, International Edition. 2015;**54**:2388-2392. DOI: 10.1002/anie.201410598

[123] Treat NJ, Sprafke H, Kramer JW, Clark PG, Barton BE, Read de Alaniz J, et al. Metal-free atom transfer radical polymerization. Journal of the American Chemical Society. 2014;**136**:16096-16101. DOI: 10.1021/ja510389m

[124] Fantin M, Isse AA, Venzo A, Gennaro A, Matyjaszewski K. Atom transfer radical polymerization of methacrylic acid: A won challenge. Journal of the American Chemical Society. 2016;**138**:7216-7219. DOI: 10.1021/jacs.6b01935

[125] Enayati M, Jezorek RL, Monteiro MJ, Percec V. Ultrafast SET-LRP of hydrophobic acrylates in multiphase alcohol–water mixtures. Polymer Chemistry. 2016;**7**:3608-3621. DOI: 10.1039/C6PY00353B

[126] Whitfield R, Anastasaki A, Truong NP, Wilson P, Kempe K, Burns JA, et al. Well-defined PDMAEA stars via Cu(0)-mediated reversible deactivation radical polymerization. Macromolecules. 2016;**49**:8914-8924. DOI: 10.1021/acs.macromol.6b01511

[127] Zhang Q, Wilson P, Li Z, McHale R, Godfrey J, Anastasaki A, et al. Aqueous copper-mediated living polymerization: Exploiting rapid disproportionation of CuBr with Me_6TREN. Journal of the American Chemical Society. 2013;**135**:7355-7363. DOI: 10.1021/ja4026402

[128] Hakobyan K, Noble BB, Xu J. Solid-phase synthesis of iterative RAFT single unit monomer insertion adducts. Polymer Chemistry. 2023;**14**:4116-4125. DOI: 10.1039/D3PY00828B

[129] Zhang L, Song Y, Cao Y, Wang Z, Huang Z, Xuan S, et al. A photo–thermal dual-regulated latent monomer strategy for sequence control of polymers. Polymer Chemistry. 2021;**12**:4996-5002. DOI: 10.1039/D1PY00961C

[130] Chen TTD, Zhu Y, Williams CK. Pentablock copolymer from tetracomponent monomer mixture using a switchable dizinc catalyst. Macromolecules. 2018;**51**:5346-5351. DOI: 10.1021/acs.macromol.8b01224

[131] Kapil K, Szczepaniak G, Martinez MR, Murata H, Jazani AM, Jeong J, et al. Visible-light-mediated controlled radical branching polymerization in water. Angewandte Chemie, International Edition. 2023;**62**:1-6. DOI: 10.1002/anie.202217658

[132] Li F, Cao M, Feng Y, Liang R, Fu X, Zhong M. Site-specifically initiated controlled/living branching radical polymerization: A synthetic route toward hierarchically branched architectures. Journal of the American Chemical Society. 2019;**141**:794-799. DOI: 10.1021/jacs.8b12433

[133] Matyjaszewski K, Tsarevsky NV. Nanostructured functional materials prepared by atom transfer radical polymerization. Nature Chemistry. 2009;**1**:276-288. DOI: 10.1038/nchem.257

[134] Matyjaszewski K, Tsarevsky NV. Macromolecular engineering by atom transfer radical polymerization. Journal of the American Chemical Society. 2014;**136**:6513-6533. DOI: 10.1021/ja408069v

[135] Lorandi F, Fantin M, Jafari H, Gorczynski A, Szczepaniak G, Dadashi-Silab S, et al. Reactivity prediction of Cu-catalyzed halogen atom transfer reactions using data-driven techniques. Journal of the American Chemical Society. 2023;**145**:21587-21599. DOI: 10.1021/jacs.3c07711

[136] Pintauer T, Matyjaszewski K. Atom transfer radical addition and polymerization reactions catalyzed by ppm amounts of copper complexes. Chemical Society Reviews. 2008;**37**:1087. DOI: 10.1039/b714578k

[137] Kapil K, Jazani AM, Szczepaniak G, Murata H, Olszewski M, Matyjaszewski K. Fully oxygen-tolerant visible-light-induced ATRP of acrylates in water: Toward synthesis of protein-polymer hybrids. Macromolecules. 2023;**56**:2017-2026. DOI: 10.1021/acs.macromol.2c02537

[138] Cuthbert J, Wanasinghe SV, Matyjaszewski K, Konkolewicz D. Are RAFT and ATRP universally interchangeable polymerization methods in network formation? Macromolecules. 2021;**54**:8331-8340. DOI: 10.1021/acs.macromol.1c01587

[139] Szczepaniak G, Łagodzińska M, Dadashi-Silab S, Gorczyński A, Matyjaszewski K. Fully oxygen-tolerant atom transfer radical polymerization triggered by sodium pyruvate. Chemical Science. 2020;**11**:8809-8816. DOI: 10.1039/D0SC03179H

[140] Parkatzidis K, Wang HS, Truong NP, Anastasaki A. Recent developments and future challenges in controlled radical polymerization: A 2020 update. Chem. 2020;**6**:1575-1588. DOI: 10.1016/j.chempr.2020.06.014

[141] Rahman MM, editor. Recent Advances in Ionic Liquids. London, UK: InTech; 2018